NUTRITION, WELL-BEING AND HEALTH

Edited by **Jaouad Bouayed** and **Torsten Bohn**

Nutrition, Well-Being and Health
Edited by Jaouad Bouayed and Torsten Bohn

Published by InTech
Janeza Trdine 9, 51000 Rijeka, Croatia

Publishing Process Manager Jelena Marusic

Technical Editor Teodora Smiljanic

Cover Designer InTech Design Team

Additional hard copies can be obtained from orders@intechopen.com

Nutrition, Well-Being and Health, Edited by Jaouad Bouayed and Torsten Bohn
p. cm.
ISBN 978-953-51-0125-3

For EU product safety concerns:
IN TECH d.o.o., Prolaz Marije Krucifikse Kozulić 3, 51000 Rijeka, Croatia,
info@intechopen.com or visit our website at intechopen.com.

Meet the editors

Associate professor Jaouad Bouayed was born in 1977 in Oujda, in Morocco. He is exerting teaching and research activities in Lorraine University in Metz. He has defended his Ph.D. speciality "Phytochemistry and Pharmacology" the 29th October 2007 at Paul Verlaine - Metz University, France. Afterwards, he was a postdoctoral researcher, in which the last position was realized in Luxembourg under the supervision of Dr. Torsten BOHN to investigate the effects of simulated gastric and intestinal digestion on the release of antioxidants from apple food matrix. During this period, Bouayed & Bohn have decided to edit the book "Nutrition, Well-Being and Health", compelling the most recent evidence and state of the art in the relation of bioactive food ingredients and their potential to reduce the burden of chronic diseases, especially with respect to prevention. Jaouad Bouayed has published more than 30 peer reviewed journals including reviews and chapters in books. He is also an ad hoc reviewer in several international journals. Dr Bouayed's current fields of interests are the relationship between behaviour and cognitive function, and oxidative stress, as well as the impact of antioxidants or dietary contaminants on these conditions using mouse model, at different stages of live including juvenile, adult and old ages.

Dr Torsten Bohn was born in Troisdorf, Germany, in 1972 and is currently the head of the Plant and Nutrition Unit at the Public Research Center - Gabriel Lippmann in Luxembourg. Following his PhD in human nutrition which was awarded by the Swiss Federal Institute of Technology (ETH) in Zurich for studies on "Magnesium Absorption in Humans" in 2002, he made his postdoc at the Ohio State University, at Columbus, OH. During this time he focused on human studies investigating potential health benefits of consuming soy and tomato rich foods, which also sparked the interest for a more general relation between Food and Health, some topics of which appear in this present book "Nutrition, Well-Being and Health", aiming to highlight selected areas of nutrition and health interactions. Torsten Bohn has published ca. 50 peer reviewed articles encompassing reviews and book chapters. He is a member of the Editorial Board of the British Journal of Nutrition and has edited several special issues for various Food and Nutrition oriented Journals. His main research focus is the bioavailability of phytochemcials, especially carotenoids and polyphenols, and their relation to inflammation and oxidative stress.

Contents

Preface

Nutrition is an important lifestyle factor that contributes to our general feeling well. Recently, it has even further suggested, based on a number of epidemiological studies, that our diet is also associated with the risk of developing a number of chronic diseases, such as diabetes type II, cardiovascular diseases, osteoporosis, many types of cancer, just to name few. Thus, a balanced nutrition is firmly interwoven with many aspects of our long-term health, including the prevention of diseases, albeit this is typically rather associated with the medicinal areas. Whereas medicine however usually is brought on stage when a disease has already appeared, the strength of dietary strategies would rather rest in its preventive potential.

Apart from focusing on the macro-molecules in our diet, including carbohydrates, proteins, and lipids, a number of micronutrients such as vitamins and minerals, and also phytochemicals (non-nutrients), or secondary plant metabolites, have moved into the focus of attention. Among these are the most prevalent and large group of polyphenols, the lipid soluble carotenoids, but also less well studied groups such as terpenes. Albeit still lacking hard data in terms of randomized control, double blinded intervention studies on large scales, there exist now a number of prospective cohort studies that suggest that many of these phytochemicals, when consumed within e.g. a fruit and vegetable rich diet, are important contributors to our health, and this has been further supported by a number of studies focusing on the mechanisms of their biological activity. In addition to some of the antioxidant properties, which have been attributed to these compounds, additional mechanisms, such as impacting gene transcription and therefore altering the body's own antioxidant defense system, or inflammatory cascades, may eventually be found to be of superior importance. Much research is currently focusing on these topics, and more studies in this area are warranted to reveal the potential of many of these compounds.

The knowledge that many vitamins, minerals, and phytochemicals with no direct nutritive value are important for a healthy development is not new, but can be found in many dietary approaches, such as in the Chinese Traditional Medicine, which aims at closer linking specific food items for specific health conditions and diseases, i.e. targeting to extend our common view of nutrition as merely supplying sufficient energy and essential nutrients. Functional foods and nutraceuticals also aim toward this direction, and many interesting approaches with potential health benefits are under consideration. Among one of these strategies are also improved technological means, such as by increasing bioaccessibility and bioavailability of certain less stable ingredients, by encapsulation. Only time will reveal the potential of these new strategies to combat chronic diseases or aid in their prevention, but new possibilities in sight of the exploding number of chronic diseases, such as the metabolic syndrome, and the increasing age pyramid, are utterly needed.

This special issue is based on selected chapters that deal with the above mentioned topics. Rather then aiming at giving an exhaustive overview over "Nutrition, Well-Being and Health", which will virtually be impossible even in a large compilation of volumes, we chose to highlight some of the recent developments and investigations in this domain.

We appreciate all the efforts that were bundled to bring this book together, and we would like to express our gratitude especially toward all authors and their valuable contributions.

Dr. Jaouad Bouayed and Dr. Torsten Bohn
Centre de Recherche Public - Gabriel Lippmann
Luxembourg

1

Dietary Derived Antioxidants: Implications on Health

Jaouad Bouayed[1,2,*] and Torsten Bohn[1]
[1]*Centre de Recherche Public - Gabriel Lippmann, Environment and Agro-Biotechnolgies Department, Nutrition and Toxicology Unit, Belvaux,*
[2]*Neurotoxicologie Alimentaire et Bioactivité, UR AFPA, Université de Lorraine-INRA, Metz,*
[1]*Luxembourg*
[2]*France*

1. Introduction

Oxidative stress state is involved in the aging process as well as in a vast array of pathological conditions, including atherosclerosis, cardiovascular complications, diabetes, cancer, and neuropsychiatric and neurodegenerative diseases. Oxidative stress is characterized by an imbalance in the cellular redox state in favour of a high formation of reactive oxygen species (ROS), overcoming the reducing capacity of the human antioxidant defence system, which has the role to eliminate excess ROS production, avoiding the oxidative action of such species on cellular components (nucleic acids, lipids, proteins or carbohydrates) and thereby their resulting adverse effects.

In general, oxidative stress can result from a high production of ROS or a poor antioxidant defence system, which is in part depending on exogenous molecules which could act as antioxidants, such as vitamin C, vitamin E, carotenoids and polyphenols. ROS at low or moderate concentrations in human tissues are required for optimum cellular functioning, owing to their crucial role in many physiological functions, such as stimulating cellular signaling, gene expression, the regulation of immune responses and fostering antioxidant defense mechanisms (Valko et al., 2007; Bouayed & Bohn, 2010). While the double-edged effects of ROS are well known, with toxic and deleterious effects at high concentrations, the biphasic effects of antioxidants have been postulated recently (reviewed by Bouayed & Bohn, 2010). Interaction of antioxidants with ROS present at physiological concentrations required for optimal cell functioning could disrupt the balance between oxidant production and antioxidant protection, being believed to be critical in maintaining healthy biological systems. This has been earlier stressed in transgenic animals overexpressing antioxidant enzyme systems (e.g., superoxide dismutase (SOD) and glutathione peroxidase (GPx)) (Mirochnitchenko et al., 1995; Kondo et al., 1997; Bouayed & Bohn, 2010). Exogenous antioxidants at high concentrations could also behave as prooxidants or by activating other cellular responses that could result in detrimental effects such as inflammatory reactions

* Corresponding Author

(Bouayed & Bohn, 2010). However, exogenous antioxidants at nutritional doses, as occurring in their natural matrices such as in fruits and vegetables, are necessary to complete the scavenging action of the endogenous antioxidant defense system. Indeed, several laboratory, epidemiological, and intervention studies have suggested that antioxidants from fruits and vegetables can act as chemopreventive agents against several diseases related to oxidative stress, and are of interest especially in the prevention of chronic diseases (reviewed by Bouayed & Bohn, 2010). The use of antioxidants (e.g. inhibitors of xanthine oxidase) as therapeutic agents is also emerging, e.g. to treat hypertension (reviewed by Fang et al., 2009). However, the health-promoting effects of vitamin C, vitamin E, carotenoids and polyphenols may also occur independently from their antioxidant properties, such as by interacting with cellular signalling pathways, interacting with e.g. inflammatory processes or cell differentiation (reviewed by Bouayed & Bohn, 2010).

In this chapter, important classes of antioxidants occurring in our diet are presented. The necessity of exogenous antioxidants to maintain optimal health and prevent chronic diseases is discussed. The health-promoting effects of antioxidants within fruits and vegetables are also emphasized. Changes occurring during and following ingestion and digestion of bioactive compounds prior to reaching target organs and exerting their activity are also briefly reviewed.

2. ROS and antioxidant defence system

2.1 ROS

ROS are constantly generated during normal and aberrant cell metabolism, which relies on the use of molecular oxygen. Mitochondria constitute the principal cellular site producing ROS, as the majority of intracellular oxygen (*ca.* 85%) is consumed in these organelles. During mitochondrial respiration, high amounts of energy required for our organism are constantly extracted from organic molecules resulting finally in complete reduction of oxygen by 4 electrons leading to water and carbon dioxide formation. However, during oxidative phosphorylation, 1-3% of electrons leak prematurely from the respiratory complexes I and III of the mitochondrial electron transport chain, forming the superoxide free radical anion ($O_2^{\bullet-}$), resulting from monoelectronic reduction of oxygen (Delattre et al., 2005; Valko et al., 2007). The superoxide anion ($O_2^{\bullet-}$) is also generated, e.g. enzymatically by xanthine oxidase, known for its physiologic role in purine metabolism, NAD(P)H oxidase, especially during the oxidative burst stimulated by phagocytosis in immune cells and cytochromes P450 involved in metabolism I phase.

$O_2^{\bullet-}$ is considered to be the main precursor of ROS such as hydrogen peroxide (H_2O_2), hydroxyl radicals ($OH^\bullet$), alkoxyl radicals ($RO^\bullet$), and peroxyradicals ($ROO^\bullet$), among others (reviewed by Bouayed, 2010). Their order of reactivity has been determined as follows: $O_2^{\bullet-} < ROO^\bullet < OH^\bullet$. Thus, the anion superoxyde radical ($O_2^{\bullet-}$) has a low reactivity, contrary to the hydroxyl radical ($OH^\bullet$), which has a high reactivity, making it a very dangerous radical with a very short half-life *in vivo* (Delattre et al., 2005). Indeed, when $OH^\bullet$ radicals are produced *in vivo*, they react close to their site of formation, explaining the non-selectivity of $OH^\bullet$ radicals toward cellular components. Other reactive species including ozone (O_3), peroxynitrite anions ($ONOO^-$), nitrogen dioxide radicals ($^\bullet NO_2$) and hypochlorous acid (HOCl) could react with biomolecules without preference or specificity. Peroxyl radicals

($ROO^{\bullet}$) present an intermediate reactive species with respect to $O_2^{\bullet-}$ and $OH^{\bullet}$ radicals; consequently $ROO^{\bullet}$ may be more rapidly eliminated by antioxidants than $O_2^{\bullet-}$. The negative electric charge of $O_2^{\bullet-}$ impedes its diffusion across membranes, also limiting its range of action. However, its protonated form (hydroperoxyl radical, $HO_2^{\bullet}$), although constituting only *ca.* 0.3% of all superoxide radicals present in the cytosol of cells, is more reactive than its precursor ($O_2^{\bullet-}$), and also possesses the ability to cross cellular membranes (Delattre et al., 2005; Valko et al., 2007; Franco et al., 2009).

Despite being less reactive, the toxicity of $O_2^{\bullet-}$ is mainly attributed to its capacity to generate highly reactive species such as $OH^{\bullet}$ via the Haber-Weiss reaction, or also $ONOO^-$, which are non-radical oxidizing molecules able e.g. to cause DNA fragmentation and lipid oxidation (Fig. 1) (Delattre et al., 2005; Valko et al., 2007). $ONOO^-$ is the result from the reaction of $O_2^{\bullet-}$ with nitric oxide ($^{\bullet}NO$), a nitrogen-centered radical and an important cellular messenger molecule, and thus the product is considered both as an oxidant (ROS), and a nitrating agent (reactive nitrogen species, RNS). The protonated form of $ONOO^-$ (peroxynitrous acid, ONOOH), which could be easily formed at physiological pH, can decompose into $OH^{\bullet}$ and $^{\bullet}NO_2$, another RNS. However, it seems that the formation of nitrosoperoxycarbonate ($ONOOCO_2^-$) is more plausible *in-vivo*, following the reaction of $ONOO^-$ with CO_2, due to

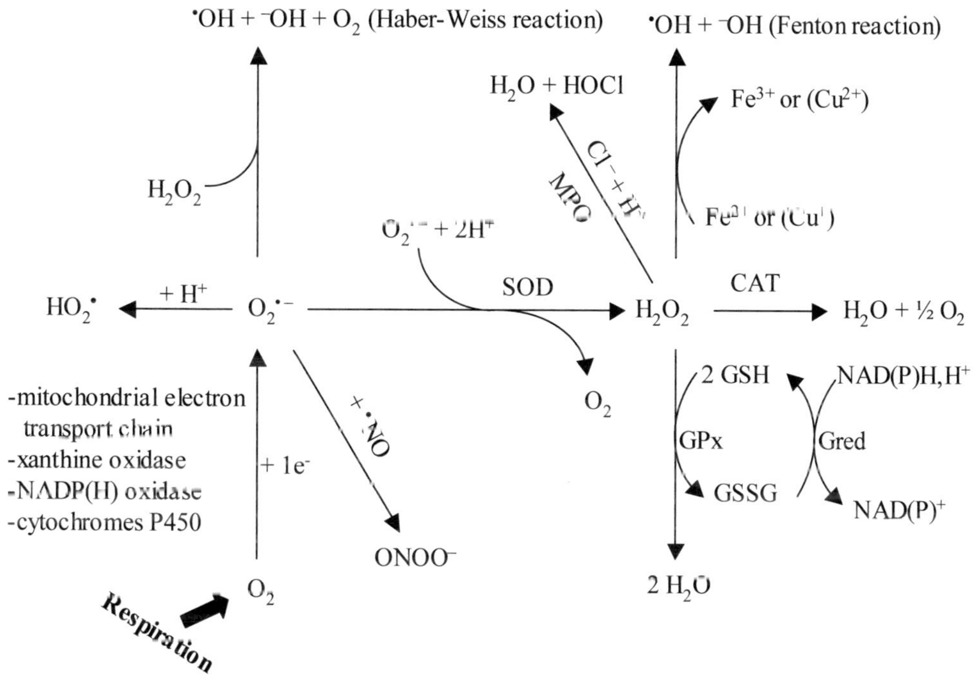

Fig. 1. Principal cellular pathways producing and metabolizing ROS. SOD, CAT, GPx, Gred and MPO mean superoxide dismutase, catalase, glutathione peroxidase, glutathione reductase and myeloperoxidase, respectively. The reaction with MPO is specific for phagocytic cells. Fenton reaction could also involve other transition metals. GSSG and GSH stand for oxidized and reduced glutathione, respectively. $NAD(P)^+$ and NAD(P)H stand for oxidized and reduced nicotinamide adenine dinucleotide phosphate, respectively.

the abundance of CO_2 following respiration. The decomposition of $ONOOCO_2^-$ results in the formation of $^{\bullet}NO_2$ and carbonate radical ($CO_3^{\bullet -}$), a ROS (Halliwell, 2006). For the above reason of multiple possibilities for interaction between nitrogen and oxygen containing reactive species, many authors usually use the collective term reactive oxygen and nitrogen species (RONS) to include both ROS and RNS. The major component of intracellular ROS *in vivo* is considered H_2O_2, formed by its precursor $O_2^{\bullet -}$. Albeit being of non-radical nature, H_2O_2 is more reactive than $O_2^{\bullet -}$, having the ability to freely pass across cell membranes, and to generate more reactive molecules such as $OH^{\bullet}$ via e.g. the Fenton reaction, or HOCl in phagocytic cells involving phagocyte-derived myeloperoxidase (MPO) (Fig. 1) (Splettstoesser & Schuff-Werner, 2002; Halliwell, 2006). Due to its microbicidal activity, HOCl (and OCl^-) contributes to the destruction of internalized bacteria and fungi by phagocytes (Halliwell, 2006).

2.2 Endogenous antioxidants

Cells are equipped with systems allowing for scavenging these oxidative species. This detoxifying system or antioxidant defense system is encompassing enzymatic and non-enzymatic antioxidants, with the latter based on endogenous (e.g. glutathione and coenzyme Q) as well as exogenous reducers (e.g. vitamin C and polyphenols) that are predominantly derived by dietary intake (table 1). In terms of enzymatic antioxidant

Antioxidant defense system	
Endogenous antioxidants	**Exogenous antioxidants**
Enzymatic antioxidants - Superoxide dismutase (SOD): enzyme detoxifying superoxide radical ($O_2^{\bullet -}$) - Catalase (CAT) and glutathione peroxidase (GPx): enzymes involved in the detoxification of peroxides (CAT against H_2O_2, and GPx against both H_2O_2 and ROOH) - Glutathione reductase: enzyme involved in the regeneration of glutathione - Thioredoxin reductase: enzyme involved in the protection against protein oxidation - Glucose-6-phosphate dehydrogenase: enzyme involved in the regeneration of NADPH **Non-enzymatic antioxidants** (principal intracellular reducing agents) Glutathione (GSH), uric acid, lipoic acid, NADPH, coenzyme Q, albumin, bilirubin	**Prinicipal dietary antioxidants from fruits, vegetables and grains** - **Vitamins**: vitamin C, vitamin E - **Trace elements** : zinc, selenium - **Carotenoids**: ß-carotene, lycopene, lutein, zeaxanthin - **Phenolic acids**: chlorogenic acids, gallic acid, cafeic acid, etc - **Flavonols**: quercetin*, kaempferol*, myricetin* - **Flavanols:** proanthocyanidins and catechins - **Anthocyanidins**: cyanidin* and pelagonidin* - **Isoflavones**: genistein*, daidzein* and glycitein* - **Flavanones**: naringenin*, eriodictyol* and hesperetin* - **Flavones**: luteolin* and apigenin*

* and their glucosides

Table 1. Human antioxidant defense systems include endogenous (enzymatic and non-enzymatic) and exogenous antioxidants, with the diet being the main exogenous source (Bouayed & Bohn, 2010)

defense systems, superoxide dismutase (GPx), glutathione peroxidase (GPx) and catalase (CAT) are the most known and prevalent antioxidant enzymes aiming to sequentially reduce $O_2^{\bullet-}$ and H_2O_2, avoiding the formation of oxidative species such as hydroxyl radicals ($^{\bullet}OH$). For example, transgenic mice studies have shown the importance of antioxidative enzymes for optimal health, even for survival. Homozygous mutant mice lacking manganese superoxide dismutase (Mn-SOD) died within the first 10 days of life (Storey, 2004). Mice lacking the enzyme glutathione peroxidase-1 (GPx-1), which among other functions protects the lens of the eye against H_2O_2-mediated oxidative damage, have developed central cataracts (Wang et al., 2009). Homozygous catalase (CAT) knockout mice, although apparently developing normally, have shown differential sensitivity to oxidant tissue injury in comparison to wild-type mice (Ho et al., 2004).

SOD catalyses the dismutation of superoxide radicals ($O_2^{\bullet-}$) into O_2 and hydrogen peroxide (H_2O_2), which is itself detoxified by either CAT or GPx to water. GPx detoxifying activity, which extends also to peroxides (ROOH), requires glutathione (GSH) as the electron donor, depending on glutathione reductase (Gred) that assures the regeneration of GSH from the oxidized form (GSSG) (see Fig. 1). Glutathione transferase (GST) plays a role in the detoxification of prooxidant xenobiotics, as well as peroxidised lipids by catalysing their conjugation with GSH, facilitating their excretion from the cell. GSH is the most prevalent endogenous, non-enzymatic antioxidant. GSH is considered as a major antioxidant in aerobic cells, functioning as an important cellular redox buffer antioxidant. GSH depletion has shown to cause systemic oxidative stress and other detrimental effects such as anxious behaviour in mice (Bouayed et al., 2009; Bouayed, 2011). Besides its role as substrate for GPx, GST and glyoxalase I, GSH can also directly scavenge free radicals by hydrogen donation, resulting in glutathiyl radicals ($GS^{\bullet}$), following the reaction (1). However, other reactive thiyl radicals could be generated from $GS^{\bullet}$ such as $GSO^{\bullet}$ and $GSOO^{\bullet}$ (reactions (2-3)). In addition to GSH, cells contain other endogenous antioxidants including uric acid, lipoic acid, NADPH, coenzyme Q, albumin and bilirubin (table. 1). However, our antioxidant defense system requires exogenous antioxidants, e.g. vitamin C, vitamin E, polyphenols and carotenoids, to efficiently scavenge RONS acting interactively (e.g., additively or synergistically) in order to maintain or re-establish redox homeostasis.

$$GSH + R^{\bullet} \longrightarrow GS^{\bullet} + RH \quad (1)$$

$$GS^{\bullet} + O_2 \longrightarrow GSO_2^{\bullet} \quad (2)$$

$$GSO_2^{\bullet} + GSH \longrightarrow GSO^{\bullet} + GSOH \quad (3)$$

2.3 Exogenous antioxidants

A compound can be defined as an antioxidant if it is able to either delay or prevent free radical-mediated oxidation (or autooxidation) of an oxidizable compound (e.g., DNA, proteins, lipids or carbohydrates), at low concentration compared to the substrate, generating a more stable radical (Halliwell, 1990; Rice-Evans et al., 1996). Several dietary compounds satisfy these two basic conditions of an antioxidant, including polyphenols and carotenoids, and some nutrients such as vitamin C or vitamin E, of which especially fruits, vegetables, and other plant foods such as whole grains, but also fish, meat, and dairy products constitute natural sources (table 2).

dietary antioxidants	rich dietary sources	concentration in foods (mg/100g)
vitamin C	bell pepper, citrus fruits	10-170
quercetin	apples, onions	4-46
carotenoids	leafy vegetables, plums, tomatoes, watermelon, carrots	0.2-10
EGCG	green tea	5-450#
selenium*	fish (dairy products, potato, rice)	1-150*
vitamin E	fish, meat, leafy vegetables	0.2-10
isoflavonoids	soy, beans, peanuts	0.1-155

*µg/100g; #mg/cup (ca. 225 mL of tea beverage)

Table 2. Examples of antioxidant concentrations in fruits and vegetables (Bouayed & Bohn, 2010)

2.3.1 Polyphenols

This group constitutes the majority of dietary antioxidants (and also of secondary plant metabolites). Plants typically produce polyphenols as a defence against herbivores and various stresses in general. It is estimated that in Westernized countries, polyphenol intake is approx. 0.4-1g/d and capita (reviewed by Bouayed and Bohn, 2010), with higher intake for persons following a vegetarian diet. Food sources that are especially rich in polyphenols include, among others, potato, plums, leafy vegetables, whole grain products, and coffee (Souci et al., 2000).

Several *in vitro* studies have shown that polyphenols are the major contributors with respect to the total antioxidant activity of the majority of fruits and vegetables. Over 8.000 polyphenolic molecules have been identified and can be classified into flavonoids and non-flavonoid phenolics. Certain authors consider (poly)phenols to be all secondary phytochemicals that have at least two phenol subunits. However, (poly)phenols are also defined as all secondary metabolites possessing an aromatic benzene ring that is substituted by at least two hydroxyl groups, including their functional derivatives (reviewed by Bouayed, 2010). Moreover, in the classification given by Manach et al. (2004), even phenolic acids bearing only one hydroxyl group on the aromatic ring and acrylic acids, such as coumaric acids, are included in the polyphenol definition.

Flavonoids have a common structure consisting of 2 aromatic rings that are linked together by 3 carbon atoms that form an oxygenated heterocycle. This large group, which constitutes the most prevalent in the human diet, is divided into 6 subgroups, including flavonols (e.g. quercetin), flavones (e.g. apigenin), isoflavones (e.g. daidzein), flavanones (e.g. hesperetin), anthocyanidins (e.g. cyanidin), and flavanols (catechins and proanthocyanidins). Non-flavonoid polyphenolics include phenolic acids (e.g. chlorogenic acid), lignans (e.g. secoisolariciresinol) and stilbenes (e.g. resveratrol) (Manach et al., 2004).

Polyphenols scavenge free radicals ($R^{\bullet}$) possessing an unpaired electron either by donation of hydrogens or electrons, resulting in comparatively stable phenoxyl ($PhO^{\bullet}$) radicals (neutral ($PhO^{\bullet}$) or cationic ($PhO^{+\bullet}$) molecules, respectively), which are stabilized by delocalization of unpaired electrons around the aromatic ring (Rice-Evans et al., 1996, Bouayed et al., 2011a and 2011b). However, from an energetic point of view, it has been debated that phenolics favour hydrogen atom transfer mechanisms, in which lower energies are involved (Leopoldini et al., 2011). The radicals derived from oxygen represent the most important class of radical species generated in living systems. However the term antioxidant is often used to describe the scavenging activity of all reactive radicals including e.g. RNS radicals. The potential scavenging abilities of phenolics mainly depend on the number and the position of hydrogen donating hydroxyl groups on the aromatic cycles of the phenolic molecules (Rice-Evans et al., 1996). For example, aglycones or polyphenols with 2 hydroxyl groups on aromatic residues are better free radical scavengers than their glycoside forms or polyphenols with a single hydroxyl group.

Depending on their structures, polyphenols (e.g. tea polyphenols) could also act by chelating prooxidant transition metal ions such as Fe^{2+}, which are involved in reactions eliciting free radical production, including hydroxyl radicals ($OH^{\bullet}$) and alkoxyl radicals ($RO^{\bullet}$) (Dufresne & Farnworth, 2001). Polyphenols that are able to scavenge lipid peroxyl ($LOO^{\bullet}$) and lipid alkoxyl ($LO^{\bullet}$) radicals or act as singlet oxygen quenchers ($^1O_2$) are effective inhibitors of lipid peroxidation processes, owing to the recognized role of these reactive species to initiate or to propagate free radical lipid peroxidation in cell membranes. The polarity of polyphenols is variable, ranging from water-soluble polyphenols (e.g. catechins), to more poorly water-soluble (e.g. flavonoid aglycones such as quercetin), to lipophilic polyphenols (e.g. curcumin).

Many *in vitro* studies have clearly revealed the potent role of flavonoids in inhibiting lipid peroxidation and oxidation of low-density lipoproteins (LDL). It has been proposed that flavonoids near membrane surfaces are ideally located for scavenging free radicals generated in the aqueous phase. For example, it has been debated that catechins might prevent the oxidation of vitamin E (a lipophilic antioxidant) by scavenging hydrophilic radicals near membrane surfaces, whereas vitamin E scavenges lipid peroxyl radicals ($LOO^{\bullet}$) as hydrogen donor to stop free radical chain reactions (chain-breaking antioxidant). Polyphenols could also play a role in the regeneration of vitamin E through reduction of its oxidized form (vitamin $E^{\bullet}$ radical) (Rice-Evans et al., 1996; Rice evens, 2001), acting synergistically. In some cases, polyphenols could exert their antioxidant activity by inhibiting the catalytic activity of many enzymes eliciting ROS formation, including xanthine oxidase, lipoxygenase, cyclooxygenase and NAD(P)H oxidase (Atmani et al., 2009).

In food matrices, bioactivity of polyphenols like all dietary antioxidants in the human body, depends firstly on their bioaccessibility (i.e. the release from the food matrix) and bioavailability (i.e. absorbable fraction that can be used for specific physiological functions in organs). Polyphenols of comparatively high bioavailability include isoflavonids (absorption cover > 50%, Bohn, 2010), while e.g. anthocyanins are of very low bioavailability, usually ca. 1.7% (Sakakibara et al., 2009). The prerequisite for bioavailability of any compound is its bioaccessibility in the gut. Following their ingestion, native polyphenols may undergo several modifications in the gastrointestinal (GI) tract until absorption, changes that may also affect their antioxidant capacity (Bouayed et al., 2011b; 2011c). This process may concern especially polyphenols of high molecular weight such as tannins or polyphenols not absorbed in the small intestine (e.g. polyphenols linked to a rhamnose moiety) – which could be extensively metabolized by the microflora of the colon. In addition, the majority of polyphenols in nature occur as glycosides or esters, which require typically cleavage prior to the absorption, such as by intestinal and microflora enzymes, especially cytosolic ß-glucosidase, brush border inheritent lactase phlorizin hydrolase or esterases.

However, cellular uptake of aglycones has been suggested to occur in their native form by passive diffusion. Absorbed polyphenols can directly undergo phase II metabolism as phenolic structures are generally unfavorable substrates to the cytochrome P450s (phase I metabolism). At nutritional doses, almost all polyphenols are conjugated to form *O*-glucuronides, sulfate esters and *O*-methyl ethers, by glucuronidation, sulfation and methylation, in the gut mucosa and later in the liver or kidney. Glucuronidation and sulfation of polyphenols may facilitate their rapid urinary and biliary excretion by increasing their hydrophilicity, and also may limit their potential toxicity. Bioaccessible unabsorbed polyphenols may play a role in the protection of the GI tract against RONS prior to their fecal excretion. In contrast to native polyphenols, less data exists on the antioxidant activity of bioavailable polyphenol phase II metabolites (conjugated derivatives). Despite the variable and overall relatively poor bioavailability of polyphenols (concentrations range between high nanomolar and low micromolar in human plasma and organs), polyphenols have been reported to be more efficient than vitamin C, vitamin E and carotenoids (concentration ranges between high micromolar and low millimolar in human plasma and organs) against oxidative stress at tissue levels (Scalbert et al., 2002; Manach et al., 2004; Yang et al., 2008; Pandey & Rizvi, 2009; Bouayed & Bohn, 2010; Bouayed et al., 2011b, 2011c).

2.3.2 Carotenoids

These tetraterpenoid (C-40) compounds are also naturally occurring substances with antioxidant potential, and found especially in colored fruits and vegetables but also eggs, algae, some seafood, and are synthesized by plants, bacteria and several fungi. So far, over 700 carotenoids have been identified, of which however only 40-50 species play a role in the human diet (reviewed by Bohn, 2008). It is estimated that approx. 9-16 mg carotenoids per day are consumed in industrialized countries (O'Neill et al. 2001). Food items rich in carotenoids include for example spinach (11 mg/100 g edible portion), tomatoes (5 mg/100 g), and carrots (20 mg/100 g) (Biehler et al. 2011).

Some carotenoids are considered as nutrients, such as alpha-, beta-, and gamma-carotene, and alpha- and beta-cryptoxanthin, exhibiting vitamin A activity following their metabolisation into retinol by humans (Biehler & Bohn, 2010; Biehler et al., 2010). Carotenoids are pigments with several conjugated double bonds (polyene chain), which could be divided into oxygen containing carotenoids (oxocarotenoids or xanthophylls), and non-oxygen containing carotenoids (carotenes). Despite not showing vitamin A activity, lutein and zeaxanthin posses the ability to stabilize membrane integrity, and can efficiently act as secondary antioxidants by absorbing damaging blue light that enters the eye (Johnson, 2002; Maci, 2010), important e.g. for the prevention of age-related macular degeneration (Maci, 2010). These blue-light filtering phytochemicals are found especially in dark green leafy vegetables such as spinach, broccoli and kale. However, the main mechanism of antioxidant activity of carotenoids is radical scavenging and also quenching excited triplet states of oxygen ($^1O_2$). Carotenoids (CAR-H) neutralize reactive radicals via electron transfer, generating carotenoid radical cations (CAR-H$^{\bullet+}$), which are less reactive due to the ability of their conjugated double-bonded structure to delocalize unpaired electrons. However, some carotenoids such as lycopene interacting with $O_2^{\bullet-}$ may yield the carotenoid radical anion (lycopene$^{\bullet-}$ + O_2). Carotenoids (CAR-H) could also scavenge free radicals (e.g. ROO$^{\bullet}$) by hydrogen atom mechanism transfer, resulting in alkyl radicals (CAR$^{\bullet}$). The potential scavenging effect of carotenoids is depending mostly on the length of the electron rich conjugated double bond system. This system can be as short as 3 double bonds in the case of phytoene, and as long as 11 conjugated double bonds in the case of lycopene, both two predominant carotenoids in tomato products. It has been also proposed that interaction of carotenoids (CAR-H) with some radicals (e.g. ROO$^{\bullet}$) could also occur by radical addition reaction, resulting in adduct formation (e.g. ROO-CAR-H$^{\bullet}$). It has been believed that carotenoids may combine with ROO$^{\bullet}$ to form a large resonance stabilized radical (Palace et al., 1999; Mortensen et al., 2001; Krinsky & Yeum, 2003). It has been suggested that carotenoids have the potential to prevent a number of degenerative diseases including cancer, atherosclerosis and age-related macular degeneration via prevention of lipid peroxidation (Mortensen et al., 2001), see also following chapters. Carotenoids are hydrophobic compounds and thereby act as lipophilic antioxidants preventing polyunsaturated fatty acids from oxidative damages. In fact, carotenoids are incorporated into lipid membranes and thus could act as chain-breaking antioxidants by stopping free radical chain reactions (propagation of lipid peroxidation), scavenging lipid peroxyl radicals (LOO$^{\bullet}$), avoiding the abstraction of allylic hydrogens from neighboring lipids. Inactivation of lipoxygenase activity by carotenoids is also proposed as another mechanism of protection against oxidative stress. Although it is still controversial, it has been debated that vitamin E regenerates the radical form of cartenoids and vice versa (Palace et al., 1999; Mortensen et al., 2001; Splettstoesser & Schuff-Werner, 2002; Krinsky & Yeum, 2003).

It is generally admitted that carotenoids are lipid-soluble antioxidants; however we can find exceptions of this rule with e.g. crocin, which is a water-soluble carotenoid. Differences exist also between the xanthophylls and the carotenes – while the latter typically rests rather deep in the apolar cores of lipid membranes, whereas the more polar oxocarotenoids interact more with the surface of lipid bilayers (Borel et al., 1996). The beneficial role of carotenoid consumption including the antioxidant effect of carotenoids has been questioned owing to previous findings of several studies including results from the CARET and the Finnish ATBC study, where comparatively high, isolated doses (supplements) of ß-carotene resulted in increased lung cancer incidence in smokers. In fact, the general low absorption (ca. 5-20% in most cases, Bohn, 2008), has resulted in the idea of administering high doses in rather isolated form, such as in dietary supplements. The negative effects observed following the administration of supplements over prolonged periods of time however have never been related to regular dietary carotenoid consumption in healthy subjects. In contrast, several epidemiological studies have suggested that when consumed within fruits and vegetable, several beneficial effects can be attributed to these compounds, including reduced incidence of cancer, cardiovascular disease, and perhaps even osteoporosis (reviewed by Bouayed and Bohn, 2010; Bub et al., 2000). It can be hypothesized that safety and benefits of antioxidants rely mainly on their concentration, generally physiologic (nutritional) in their natural matrices such as within fruits and vegetables, which may explain the advantageous effects of phytochemicals and nutrients in plant foods, acting additively and synergistically when consumed in a complex mixture (Bouayed and Bohn, 2010). In contrast to polyphenols, carotenoids appear to be less extensively metabolized, and are mainly excreted via bile and pancreas into the feces, or broken down into shorter apo-carotenals (Khachik et al., 2002a; 2002b; 2006) and further hydroxylated, and later possibly also excreted via the urine (Bohn, 2008).

2.3.3 Vitamin C and vitamin E

Vitamin C (ascorbic acid) is a water-soluble antioxidant, constituting one among the most prevalent dietary antioxidants found in fruits, vegetables and beverages. Dietary intake is usually in the area of 100 mg/d, with a DRI-RDA of 75mg/d (men) and 60mg/d (women) (National Academy of Sciences, 2000). Food items rich in vitamin C include bell peppers (ca. 120 mg/100g) and citrus fruits such as oranges (ca. 50 mg/100g) (Souci, 2000). Vitamin C contribution to the total antioxidant activity conferred e.g. by fruits was estimated to be generally less than 15%, except for kiwi fruits and honeydew melons (Wang et al., 1996), as typically, polyphenol content is up to one magnitude higher, and the antioxidant potential as measured by several tests is about comparable (per mass) to polyphenols. Ascorbic acid and its oxidized form, dehydroascorbic acid, both have vitamin C activity. Vitamin C is essential for the prevention of scurvy, due to its importance as a cofactor in the hydroxylation of proline to hydroxyproline, essential for the structure of collagen and other tissues (Shils et al., 2006). Several advantageous effects of vitamin C on human health have been stressed, such as the relationship between high plasma vitamin C concentration and reduced gastric cancer risk found in EPIC study (Jenab et al., 2006). Besides the antioxidant activity of vitamin C, which may explain its protective role against gastric cancer risk, several other activities could play a role, such as its ability to modulate cell growth kinetics and its putative antimicrobial activity, for example against *Helicobacter pylori,* a bacteria responsible for chronic ulcer and even stomach cancer. However, in the above case-control study, it seems that inhibition of carcinogenic *N*-nitroso compound formation within the stomach is more plausible as the chemopreventive mechanism of action of vitamin C.

The relationship between dietary intake and plasma vitamin C is non-linear, with maximum plasma vitamin C saturation (ca. 80 µmol/l) being reached with dietary intakes >1000 mg/day (Levine et al., 1996). In elderly men, a high dietary intake of both vitamin C and vitamin E, and a higher plasma concentration of vitamin C and ß-carotene were associated with a protection against vascular dementia and improved memory performances, respectively (Masaki et al., 2000; Perrig et al., 1997). It has been reported that vitamin C in plasma dose-dependently increases resistance to lipid peroxidation (reviewed by Flora, 2009). However, at elevated oral intake, vitamin C (e.g. 500 mg/day over 6 weeks) has displayed prooxidant effects by increasing oxidative lymphocyte DNA damage of 30 healthy volunteers (Podmore et al., 1998). Furthermore, high doses can negatively impact the intactness of the gastro-intestinal lining. In contrast, some studies on healthy human volunteers consuming fruits and vegetables rich in vitamin C decreased levels of oxidative DNA damage (reviewed by Halliwell, 2002). In humans, vitamin C is predominantly present in form of ascorbate anions, and its oxidation sequentially leads first to monodehydroascorbate (by loss of an electron) and then dehydroascorbate (by loss of hydrogen), which are relatively stable radicals, and the reaction therefore is reversible (reviewed by André et al., 2010; Flora, 2009). The ascorbyl radical is comparatively stable due to the stabilization of the adjacent vinyl group, transmitting electrons between the hydroxyl and the carbonyl (Flora, 2009).

Vitamin C can be transported into the cell either as its reduced form or dehydroascorbate (oxidized form), using active sodium-dependent transporters (SVCT1 and SVCT2) and facilitative glucose transporters (GLUTs) (André et al., 2010). GLUTs also permit the permeation of vitamin C into mitochondria. Besides the protective role of vitamin C against oxidative injury within the cytosol, this water-soluble nutrient can also confer protection to mitochondria that are targets of oxidative attacks resulting from ROS produced as a side product of the respiratory chain that is active within mitochondria. Several beneficial functions have been attributed to vitamin C, e.g. as regulative factors that may influence gene expression, apoptosis and other cellular functions, and playing a role as a cofactor for several enzymatic steps in the synthesis of monoamines, amino acids, peptide hormones, and carnitine (Santos et al., 2009).

Besides its implication in many biological processes, vitamin C is a powerful reducing agent, participating in several antioxidant mechanisms (Santos et al., 2009) by directly scavenging radicals, mediating electron transfer to ascorbate-dependent peroxidases or regenerating membrane bound vitamin E that has been oxidized, e.g. by lipid peroxyl radicals ($LOO^\bullet$), and thus indirectly limiting lipid peroxidation in cell membranes. Radical stabilization of the oxidized form of vitamin E ($VE^\bullet$) is conferred by electron delocalization around the aromatic ring of vitamin E, which is a phenolic antioxidant. The dietary recommended intake of vitamin E (DRI-RDA) is 12 mg/d (National Academy of Sciences, 2000), and main dietary sources include vegetable oils up to 200-300 mg/kg, and to a lesser extend, leafy vegetables and wholegrain foods (Souci, 2000). It has been considered that vitamin E is the major membrane bound antioxidant employed in humans, and thus this lipophilic chain breaking antioxidant has the ability to inhibit lipid peroxidation. Synergistic actions between vitamin C and vitamin E therefore appear important in their preventive activity against lipid peroxidation. Human trials and *in-vitro* studies have shown that oxidative stress causes a rapid depletion of vitamin C and vitamin E (reviewed by Bouayed & Bohn, 2010). As mentioned above, regeneration of vitamin E may also occur by intervention of other antioxidants such as glutathione dependent enzymes (GSH).

Eight different isomeric forms of vitamin E (4 tocopherols and 4 tocotrienols) have been found in nature. Tocotrienols (α, β, γ and δ) are identical in structure to tocopherols except for the degree of saturation in their side chain. However, α-tocopherol is considered to be the most active form of vitamin E in humans. It has been found that the concentrations of vitamin E isomers in human feces are higher than in plasma - possibly due to its limited absorption from the diet, suggesting that it could play a protective role against RONS produced in the GI tract (Halliwell et al., 2005). Although all vitamin E constituents can be absorbed from the GI tract, α-tocopherol represents the predominant form existing in human plasma with approx. 17 µmol/l, followed by γ-tocopherol at 1 µmol/l (Halliwell et al., 2005). It has been estimated that prior to degradation, one molecule of α-tocopherol can deactivate up to 120 $^1O_2$ molecules by resonance energy transfer (reviewed by André et al.,

2010). Thus, the main antioxidant function of vitamin E is the protection of biological membranes against oxidative damage of lipids either by quenching $^1O_2$ or by intercepting directly free radical intermediates (e.g. $OH^•$, lipid radicals ($L^•$), $LO^•$, $LOO^•$) by hydrogen donation generated during lipid oxidation, terminating this chain reaction of peroxidation. Following their lipophilicity, it is more plausible that vitamin E and ß-carotene cooperate to protect membranes and lipoproteins from oxidative damages. Of course, as all antioxidants, vitamin E exerts other biological functions that are independent from its antioxidant properties, including modulation of cellular signaling, gene expression, immune response, and many more. However, as opposed to other vitamins, lack of vitamin E results in rather unspecific symptoms, also highlighting that this vitamin is mainly needed for its antioxidant activity *in vivo*. For example, deficiency of vitamin E has shown to provoke oxidative stress disturbances in transgenic rats. In another study, vitamin E deficiency in the murine brain caused brain oxidative stress and anxious behaviour (reviewed by Bouayed et al., 2009; Bouayed & Bohn, 2010, Bouayed, 2011).

3. Antioxidants and disease prevention: Potential mechanisms of action

3.1 RONS and chronic diseases

When oxidative injury of cellular bio-components is not repaired by the cellular repair mechanisms, constituting another defense system against oxidative damage, oxidative stress can lead to several dysfunctions that would result in development and progression of several human diseases including cardiovascular diseases (CVD), neurological diseases and cancer, and also to the acceleration of the aging process.

DNA, proteins and lipids represent the major targets of the oxidative action of RONS. For example, the causative link between peroxidation of lipids and lipoproteins, and several multifactorial diseases including CVD and cancer has been stressed in several reports (Halliwell, 2000; Atmani et al., 2009; Gupta et al., 2009; Rice-Evans, 2001). Lipid peroxidation and protein oxidation play also a role in the progression of Alzheimer's disease (Sultana et al., 2006). For many diseases, endothelial changes of the blood vessels do play a role. Peroxidation of circulating low-density lipoproteins (LDL) and especially within blood vessel walls is thought to be a possible initiator of the pathogenesis of atherosclerosis (Halliwell, 2000). RONS promote adhesion of platelets and monocytes to the endothelium, resulting in endothelial lesions, and eventually thrombosis and atherogenesis (Guo & Bruno, 2011).

In addition, lipid peroxidation in liver cells could, over prolonged periods of time, lead to liver diseases and diabetes (Atmani et al., 2009). Peroxides formed during lipid peroxidation processes can decompose into a vast array of toxic carbonyl products such as malondialdehyde (MDA), playing among others a role in the carcinogenesis process by interacting with cellular DNA, yielding e.g. DNA-MDA adducts that appear to be promutagenic (Gupta et al., 2009). The adverse effects of RONS are also involved in different stages of carcinogenesis by causing structural DNA damage, interacting with oncogenes, tumor suppressor genes or immunological mechanisms (Gupta et al., 2009). In this respect, the mitochondrial DNA (mtDNA) is also a target of oxidative injury, which could lead to lethal cell injury following mitochondrial genomic instability (Franco et al., 2009).

Cumulative DNA damage over time, especially in mtDNA, plays an important role in the aging process as well as in the pathogenesis of several acute and chronic neurodegenerative diseases including ischemic brain injury (Fiskum, 2000; Silva et al., 2008). In addition, at the level of mitochondria, lipid peroxidation induces the opening of the mitochondrial permeability transition pores (MPTP), resulting in the loss of the mitochondrial membrane potential (MMP), and subsequent impairments such as ATP synthesis, activating the intrinsic pathway of apoptosis, and playing a significant role in neurodegeneration following neurotrauma such as in ischemic and chronic disease-related neurodegeneration (Fiskum, 2000, Franco et al., 2009).

Oxidative stress could also play a key role in the pathogenesis of hypertension, counteracting the vasodilator effect of $^{\bullet}NO$ in the vascular endothelium by favouring the formation of peroxynitrite ($ONOO^-$). It has been hypothesized that a high production of $O_2^{\bullet-}$, by e.g. xanthine oxidase in the vascular endothelium leads to the rapid formation of $ONOO^-$, as the reaction between $O_2^{\bullet-}$ and $^{\bullet}NO$ is six-fold faster than the dismutation of $O_2^{\bullet-}$ by SOD. In addition, the non availability of $^{\bullet}NO$ would stimulate the release of the vasoconstrictor endothelin-1, and in turn, increases $O_2^{\bullet-}$ production by activating NADPH oxidase. As a result, the lack of $^{\bullet}NO$ in the vascular system could result in hypertension. Endothelial vasodilator dysfunction is related to several diseases including atherosclerosis, coronary artery disease, ischemia, stroke, etc. (reviews: Fang et al., 2009; Guo & Bruno, 2011). On the other hand, due to oxygen-free radicals produced during normal cell respiration, molecular and cellular oxidative damages accumulate over time, and were hypothesized to result in aging, and ultimately death (free radical theory of aging, Harman, 1956).

3.2 Focus cancer

The advantageous effects of antioxidant properties of food ingredients (phytochemicals and nutrients) on human health occur owing to their ability to inhibit (or retard) lipid peroxidation (see chapter 2.2) and oxidation of other sensitive cellular bio-components, including DNA and proteins. Several *in vitro* studies have shown antimutagenicity and anticarcinogenity effects of antioxidants (e.g. tea polyphenols including catechins) at different levels of cancer development, namely initiation, promotion and progression, albeit these properties could also be independent from the antioxidant mechanisms. Besides the direct protective effects of antioxidants against lipid peroxidation and DNA oxidation, resulting in the prevention against mutations and DNA strand breakage, antioxidants (e.g. epigallocatechin gallate (EGCG)) have the ability to inhibit the activation of procarcinogens by phase I enzymes and also to induce detoxification of active carcinogens by phase II enzymes, facilitating their excretion following their conjugation (Dufresne & Farnworth, 2001). Interestingly, rosmarinic acid, a dietary polyphenol, has exhibited another mechanism of protection against oxidative DNA damages *in vitro*, by enhancing DNA repair resulting from strand break formation (Silva et al., 2008).

Despite the promising *in vitro* chemopreventive effects of antioxidants, human supplementation trials with individual antioxidants have generally failed to prevent CVD and cancer formation or progression, even leading to controversial results, except in the

Linxian trial, in which the combined effect of ß-carotene, vitamin E and various minerals (Zn, Se) in a poorly nourished population in China has yielded beneficial effects on the incidence on cancer in general (reviewed by Bouayed & Bohn, 2010). In contrast, earlier retrospective epidemiological studies have suggested preventive effects of colored fruits and vegetables rich in carotenoids, high ß-carotene plasma concentrations against cancer risk, especially lung cancer. Epidemiologists have also presented fruits and vegetables rich in polyphenols including e.g. apples, onion and white grapefruit as protective strategies against lung cancer (reviewed by Bouayed & Bohn, 2010). Although it is thought that anticarcinogenity of these food plant items was related to the entire effects of all ingredients of fruits and vegetables (e.g. vitamins, polyphenols, carotenoids, dietary fiber and many more) acting additively and synergistically, and unfolding several protective mechanisms; many researchers have aimed to attribute the advantageous effects of these plant foods to few or even individual components, such as ß-carotene and quercetin, owing to their antioxidant properties including the chemopreventive activity of quercetin against carcinogens *in vitro* (reviewed by Bouayed & Bohn, 2010). Nevertheless, due to the many confounding factors, it is generally extremely difficult to attribute the observed beneficial health effects to specific ingredients. Moreover, it cannot be excluded that the so far targeted phytochemicals, including carotenoids and polyphenols, are merely indicators for a fruit and vegetable rich diet, or that they are indicative of additional, yet unidentified beneficial health compounds.

Thus, although a high intake of fruits and vegetables is believed to be a good way to prevent against cancer and other chronic diseases, recent prospective epidemiological studies such as the EPIC study have shed some doubt at the strength of this relationship, at least for certain types of cancers such as lung, breast and prostate cancers (reviewed by Key, 2011). However, several case–control studies and few prospective cohort studies have shown an inverse relationship between high fruits and vegetables intake and the risk of several types of cancers including cancers of the oral cavity, pharynx, larynx, oesophagus and stomach (reviewed by Key, 2011). In addition, based on an observational study on Korean dietary habits, showing that "westernization" of diet (i.e. high intake of calories and fats and limited intake of plant foods) has led to a rapid increase of mortality due to several types of cancers including lung (by 53%), breast (by 37%), pancreas (by 63%), prostate (by 200%) and colon (by 75%) cancers within one decade (1990-1999). These cancers were less often fatal when plant-based diets were adopted, thus highlighting the preventive effect of plant foods against cancer development (Lee et al., 2004, review).

3.3 Focus CVD

In addition to cancer, the advantageous effects of antioxidants are also recognized in the prevention of several other human chronic diseases, such as coronary heart disease (CHD) and stroke. Prospective cohort studies showed that the inverse relationship between the potential of prevention of fruits and vegetables against the above diseases depended on the amount of edible portions consumed per day. The protection against coronary heart disease was generally established when fruits and vegetables were consumed at >4 servings/d for several years often >8 years. In a meta-analysis of eight independent cohort prospective studies, it has been shown that consumption of >5 servings/d of fruits and vegetables caused a stronger reduction in stroke (ischaemic and haemorrhagic stroke) compared to 3–5 servings, the latter consumption reducing stroke incidence significantly compared to <3 servings/d. A recommended portion is somewhat vaguely defined as 80–100 g (reviewed by Bouayed & Bohn, 2010).

In the world, certain dietary regimens are assumed to be healthier, providing high amounts of nutrients (vitamins, minerals) and non-nutrients (dietary fiber, carotenoids, polyphenols, monounsaturated fatty acids, etc), preventing the development (or the progression) of several human chronic diseases at epidemic level in several populations. For example, traditional Mediterranean regimens are based on diversity and high intake of plant-based foods such as olive oil, cereals, legumes, nuts and vegetables and also other food items such as honey, eggs and fish. Thus, moderate energy intake and limited animal fat are the landmarks of this regimen.

Several benefits have been attributed to Mediterranean diets, among them their ability to reduce several forms of cancer, cardiovascular diseases and related mortality (Psaltopoulou et al., 2004; Scarmeas et al., 2006; Lairon, 2007; Mekki et al., 2010; Fung et al., 2010). In this respect, a Greek prospective study has shown a negative relationship between Mediterranean regimen and hypertension (Psaltopoulou et al., 2004), which is an important precursor of other diseases such as renal insufficiency, stroke and especially several cardiovascular complications including atherosclerosis, myocardial infarction, congestive heart failure, peripheral vascular disease and sudden cardiac death. It has also been verified prospectively that a Mediterranean diet could reduce the risk of Alzheimer's disease (Scarmeas et al., 2006). In addition, dietary intervention studies have shown that the adoption of a Mediterranean diet reduced several cardiovascular risk factors in subjects at risk (primary prevention), and/or reduced cardiovascular events/mortality in patients following a first cardiac event (secondary prevention) (Reviewed by Lairon, 2007). It has also been recommended that the Mediterranean diet could improve dyslipidemia and prevent against lipid peroxidation and inflammation in chronic renal failure patients (Mekki et al., 2010).

Vegetarian diets could also be a good example to review the health promoting effects of plant food ingredients. Comparisons between vegetarians and non-vegetarians (omnivores) have shown that vegetarians have in general, lower risk of mortality from ischemic heart disease, hypertension, stroke, type 2 diabetes and certain cancers. Differences have also been noticed regarding body mass index, total serum LDL levels, and blood pressure. However, it seems that persons following a vegetarian regimen, especially vegans, possibly will need to include some fortified foods or supplements, providing e.g. vitamin D, vitamin B_{12}, ω-3 fatty acids, iron, calcium, zinc and iodine, to equilibrate their diet and to maintain optimal health (reviewed by Fang et al., 2009).

The use of antioxidants as antihypertensive agents in both preventive and therapeutic approaches is also emerging. The inhibition of vascular endothelial xanthine oxidase may result in antihypertensive effects favouring NO-induced vasorelaxation. It has been shown that i.v. injection of 4-amino-6-hydroxypyrazolo[3,4-*d*]pyrimidine, a synthetic xanthine oxidase inhibitor, reduced the blood pressure of hypertensive rats to 70% of the initial blood pressure (review: Fang et al., 2009). Several natural antioxidants, e.g. some polyphenols have the ability to inhibit xanthine oxidase and could be beneficial in the prevention or treatment of hypertension. Vascular oxidative stress plays a crucial role in vascular endothelial dysfunction by leading to disequilibrium between vasodilation and vasoconstriction. Besides impaired vasorelaxation, vascular endothelial dysfunction may also be related to platelet aggregation and monocyte adhesion (Guo & Bruno, 2011). Quercetin, a well-known antioxidant polyphenol, is regarded as a vasoprotective agent following its abilities to act as reducing, vasodilatory, anti-platelet, and anti-atherogenic compound. In addition, *in vitro* human cell studies have shown that quercetin, at physiological concentrations (0.001-1µM), is also able to increase the availability of •NO by increasing endothelial nitric oxide synthase (eNOS) mRNA expression. Human studies have shown that quercetin supplementation increased plasma •NO and decreased endothelin-1 within 2 h of ingestion in healthy participants. Although quercetin supplementation did not lower the rate of oxidized LDL (a risk factor of CVD) in healthy normal-weight persons, quercetin diminished oxidized LDL in obese patients. Quercetin supplementation also decreased blood pressures (systolic and diastolic) in hypertensive patients (reviewed by Guo & Bruno, 2011). As quercetin is a common polyphenol in plant foods such as apple, onion, plum, etc, it could be suggested that plant foods rich in quercetin may be beneficial for decreasing the risk of CVD.

3.4 Other chronic diseases and conditions

It has also been reported in several epidemiological studies, that antioxidant intake in elderly populations protected against several diseases and age related complications, which are more specific for the nervous system impairment, including dementia, cognitive deficiency and Alzheimer's disease (Grundman & Delaney, 2002). The brain is the most vulnerable organ to oxidative stress, especially during age, when the antioxidative system is prone to decline. The brain structure, which is rich in lipids, is very prone to lipid peroxidation resulting e.g. in decreased membrane fluidity and damage in membrane proteins, inactivating receptors, enzymes and ion channels, becoming a threat for neuronal function and even overall brain activity. Lipid peroxidation can disrupt membrane integrity, leading to neuronal cell death. The brain's high oxygen consumption and its modest antioxidant defenses constitute further reasons for the sensitivity of this vital organ to oxidative stress. Several food items have shown to protect aged animals against brain oxidation, improving cognitive function, and diminishing age-related anxiety. For example, aged rats fed for 10 weeks with a standard diet supplemented with fresh apple fruits have presented significantly lower oxidative stress and anxiety than aged rats fed with the standard diet. Interestingly, brain antioxidant status of aged rats fed with apple enriched diet, as assessed by SOD activity, was not different from young animals, fed with the standard diet with or without apples. Aged rats fed for one year with a diet containing olive oil naturally rich in antioxidants have displayed low anxiety and brain oxidative stress compared to aged rats fed either with a diet containing olive oil naturally low in antioxidants or with maize oil. Long-term intake of honey has also prevented rats from the side effects of the aging process, including high anxiety and spatial memory deterioration (Bouayed, 2011). From several studies, it appears that "anti-aging foods" including antioxidants will not allow increasing the life-span of species; however, they may permit to increase the quality of life by both retarding side effects related to the aging, and preventing (or retarding) diseases of the old age including Alzheimer's and Parkinson's diseases.

4. Conclusion

It has been estimated that healthy food choices, with regular physical activity and non smoking habits, can prevent over 80% of CHD, 70% of stroke, and 90% of type-2 diabetes (Willett, 2006). Epidemiological (retrospective and prospective) investigations, case-control studies and dietary intervention studies have strongly suggested or shown the importance of plant foods rich in antioxidants in the prevention against several chronic human diseases. It is well accepted that prevention is the most persistent, cost-effective strategy to deal with chronic diseases. Thus, natural foods rich in antioxidants could be employed as a strategy in the prevention of several chronic human diseases. Antioxidants in their natural matrices are generally assumed to be safe, and their concentration physiologic. Furthermore, it is thought that the advantageous effects of antioxidants in natural food sources are due to their additive and synergistic action. Among additional mechanisms independent from antioxidant properties, several antioxidative mechanisms are proposed as being responsible for maintaining optimal health and also preventing diseases. The protection against free radical-mediated lipid peroxidation, DNA and protein oxidation, and oxidative stress-related mitochondrial dysfunction constitutes the principal way of natural antioxidants for the prevention against several diseases including cancer, cardiovascular complications, neurodegenerative diseases and the side effects of aging. Other antioxidative protective mechanisms such as absorption of UV blue light by certain carotenoids could also play a role in prevention. Further prospective studies, both with whole food items and individual dietary constituents are warranted and needed.

5. Acknowledgements

The present project was supported by the National Research Fund, Luxembourg, and co-funded under the Marie Curie Actions of the European Commission (FP7-COFUND).

6. References

André, C.M. Larondelle, Y. Evers, D. (2010). Dietary antioxidants and oxidative stress from a human and plant perspective: A review. *Curr Nutr Food Sci*, 6, 2-12.

Atmani, D. Chaher, N. Atmani, D. Berboucha, M. Debbache, N. Boudaoud, H (2009). Flavonoids in human health: From structure to biological activity. *Curr Nutr Food Sci*, 5, 225-237.

Biehler, E. Bohn, T. (2010). Methods for assessing aspects of carotenoid bioavailability. *Curr Nutr Food Sci*, 6, 44-69.

Biehler, E. Mayer, F. Hoffmann, L. Krause, E. Bohn, T. (2010). Comparison of 3 spectrophotometric methods for carotenoid determination in frequently consumed fruits and vegetables. *J Food Sci*, 75, C55-C61.

Bieher, E. Alkerwi, A. Hoffmann, L. Krause, E. Guillaume, M. Lair, M.L. Bohn, T. (2011). Contribution of violaxanthin, neoxanthin, phytoene and phytofluene to total carotenoid intake: Assessment in Luxembourg. *J Fd Comp Anal*, doi:10.1016/j.jfca.2011.07.005

Bohn. T. (2008). Bioavailability of Non-Provitamin A Carotenoids. *Curr Nutr Food Sci*, 4, 240-258.

Bohn T. (2010). Isoflavone bioavailability from foods and supplements. Dietary factors impacting utilization. *Agro Food Industry Hi-Tech*; 21, 59-62.

Borel, P. Grolier, P, Armand, M. Partier, A. Lafont, H, Lairon, D. Azais-Braesco, V. (1996). Carotenoids in biological emulsions: solubility, surface-to-core distribution, and release from lipid droplets. *J Lipid Res*, 37, 250-261.

Bouayed, J. (2010). Polyphenols: a potential new strategy for the prevention and treatment of anxiety and depression. *Curr Nutr Food Sci*, 6, 13-18.

Bouayed, J. (2011). Relationship between oxidative stress and anxiety: emerging role of antioxidants within therapeutic or preventive approaches (2011), pp 27-38. Chapter invited for the book "Anxiety Disorder" (ISBN 978-953-307-592-1), V. Kalinin, Eds., Intech, Rijeka, Croatia.

Bouayed, J. Bohn, T. (2010). Exogenous antioxidants - Double-edged swords in cellular redox state: Health beneficial effects at physiologic doses versus deleterious effects at high doses. *Oxid Med Cell Longev*, 3(4), 228–237.

Bouayed, J. Hoffmann, L. Bohn, T. (2011a). Antioxidative mechanisms of whole-apple antioxidants employing different varieties from Luxembourg. *J Med Food*, doi:10.1089/jmf.2010.0260.

Bouayed, J. Hoffmann, L. Bohn, T. (2011b). Total phenolics, flavonoids, anthocyanins and antioxidant activity following simulated gastrointestinal digestion and dialysis of apple varieties: bioaccessibility and potential uptake. Food Chem, 128, 14–21.

Bouayed J, Deußer H, Hoffmann L, Bohn T. (2011c). Bioaccessible and dialyzable polyphenols in selected apple varieties following in vitro digestion *vs.* their native patterns. Food Chem, doi: 10.1016/j.foodchem.2011.10.030.

Bouayed, J. Rammal, H. Soulimani, R. (2009). Oxidative stress and anxiety: Relationship and cellular pathways. *Oxid Med Cell Longev*, 2(2), 63–67.

Bub, A. Watzl, B. Abrahamse, L. Delincée, H. Adam, S. Wever, J. Müller, H. Rechkemmer, G. (2000). Moderate intervention with carotenoid-rich vegetable products reduces lipid peroxidation in men. *J Nutr*, 130(9), 2200-2206.

Craig, W.J. (2010). Nutrition concerns and health effects of vegetarian diets. *Nutr Clin Pract*, 25, 613-620.

Delattre, J. Beaudeux, J.L. Bonnefont-Rousselot, D. (2005). Radicaux libres et stress oxydant. Aspects biologiques et pathologiques. Éditions Médicales internationals.

Dufresne, C.J. Farnworth, E.R. (2001). A review of latest research findings on the health promotion properties of tea. *J Nutr Biochem*, 12, 404-421.

Fang, J. Seki, T. Maeda, H. (2009). Therapeutic strategies by modulating oxygen stress in cancer and inflammation. *Adv Drug Deliver Rev*, 61 290–302.

Fiskum, G. (2000). Mitochondrial participation in ischemic and traumatic neural cell death. *J Neurotraum*, 17(10), 843-855.

Flora, S.J.S. (2009). Structural, chemical and biological aspects of antioxidants for strategies against metal and metalloid exposure. *Oxid Med Cell Longev*; 2, 63-67.

Franco, R. Sánchez-Oleab, R. Reyes-Reyesc, E.M. Panayiotidis, M.I. (2009). Environmental toxicity, oxidative stress and apoptosis: Ménage à Trois. *Mutat Res-Gen Tox En,* 674, 3–22.

Fung, T.T. Hu, B.H. Wu, K. Chiuve, S.E. Fuchus, C.S. Giovannucci, E. (2010). The Mediterranean and dietary approaches to stop hypertension (DASH) diets and colorectal cancer. *Am J Clin Nutr,* 92(6), 1429-1435.

Grundman, M. Delaney, P. (2002). Antioxidant strategies for Alzheimer's disease. *Proc Nutr Soc,* 61, 191-202.

Guo, Y. Bruno, R.S. (2011). Vasoprotective activities of quercetin. *Agro Food Industry Hi-tech,* 22, 16-19.

Gupta, A. Bhatt, M.L.B. Misra, M.K. (2009). Lipid peroxidation and antioxidant status in head and neck squamous cell carcinoma patients. *Oxid Med Cell Longev,* 2 (2) 68-72.

Halliwell, B. (1990). How to characterise a biological antioxidant. *Free Radic Res Commun,* 9, 1-32.

Halliwell, B. (2000). Lipid peroxidation, antioxidants and cardiovascular disease: how should we move forward? *Cardiovasc Res,* 47, 410–418.

Halliwell, B. (2002). Effect of diet on cancer development: is oxidative DNA damage a biomarker? *Free Radic Biol Med,* 32, 968-974.

Halliwell, B. (2006). Phagocyte-derived reactive species: salvation or suicide? *Trends Biochem Sci,* 31, 509-515.

Halliwell, B. Rafter, J. Jenner, A. (2005). Health promotion by flavonoids, tocopherols, tocotrienols, and other phenols: Direct or indirect effects? Antioxidant or not? *Am J Clin Nutr,* 81, 268S–276S.

Ho, Y.S. Xiong, Y. Ma, W. Spector, A. Ho, D.S. (2004). Mice lacking catalase develop normally but show differential sensitivity to oxidant tissue injury. *J Biol Chem,* 279, 32804–32812.

Harman D. (1956). Aging: a theory based on free radical and radiation chemistry. *J Gerontol* 11:298-300.

Jenab, M. Riboli, E. Ferrari, P. Sabate, J. Slimani, N. Norat, T. et al. (2006). Plasma and dietary vitamin C levels and risk of gastric cancer in the European prospective investigation into cancer and nutrition (EPIC-EURGAST). *Carcinogenesis,* 27, 2250-2257.

Johnson, E.J. (2002). The role of carotenoids in human health. *Nutr Clin Care,* 5(2), 56-65.

Key, T.J. (2011). Fruit and vegetables and cancer risk. *Brit J Cancer,* 104, 6 –11

Khachik, F. Carvalho, L. Bernstein, P.S. Muir, G.J. Zhao, D.Y. Katz, N.B. (2002a). Chemistry, distribution, and metabolism of tomato carotenoids and their impact on human health. *Exp Biol Med (Maywood),* 227(10), 845-851.

Khachik, F. de Moura, F.F. Chew, E.Y. Douglass, L.W. Ferris, F.L. 3rd, Kim, J. Thompson, D.J. (2006). The effect of lutein and zeaxanthin supplementation on metabolites of these carotenoids in the serum of persons aged 60 or older. *Invest Ophthalmol Vis Sci,* 47(12), 5234-5242.

Khachik, F. de Moura, F.F. Zhao, D.Y. Aebischer, C.P. Bernstein, P.S. (2002b). Transformations of selected carotenoids in plasma, liver, and ocular tissues of humans and in nonprimate animal models. *Invest Ophthalmol Vis Sci*, 43(11), 3383-3392.

Kondo, T. Sharp, F.R. Honkaniemi, J. Mikawa, S. Epstein, C.J. Chan, P.H. (1997). DNA fragmentation and prolonged expression of c-fos, c-jun and hsp70 in kainic acid induced neuronal cell death in transgenic mice overexpressing human CuZn-superoxide dismutase. *J Cereb Blood Flow Metab*, 17, 241-256.

Krinsky, N.I. Yeum, K-J. (2003). Carotenoid–radical interactions. *Biochem Bioph Res Co*, 305, 754–760.

Lairon, D. (2007). Intervention studies on Mediterranean diet and cardiovascular risk. *Mol Nutr Food Res*, 51, 1209-1214.

Lee, K.W. Lee, H.J. Lee, C.Y. (2004). Vitamins, phytochemicals, diets and their implementation in cancer chemoprevention. *Crit Rev Food Sci Nutr*, 44, 437-452.

Leopoldini, M. Russo, N. Toscano, M. (2011). The molecular basis of working mechanism of natural polyphenolic antioxidants. *Food Chem*, 125, 288–306.

Levine, M. Conry-Cantilena, C. Wang, Y. Welch, R.W. Washko, P.W. Dhariwal, K.R. Park, J. B. Lazarev, A. Graumlich, J.F. King, J. Cantilena, L. R. (1996). Vitamin C pharmacokinetics in healthy volunteers: evidence for a recommended dietary allowance. *Proc Natl Acad Sci USA*, 93, 3704–3709.

Maci, S. (2010). Lutein and zeaxanthin in the eye from protection to performance. *Agro Food Industry Hi-tech*, 21, 18-20.

Manach, C. Scalbert, A. Morand, C. Remesy, C. Jimenez, L. (2004). Polyphenols: food sources and bioavailability. *Am J Clin Nutr*, 79, 727-747.

Masaki, K.H. Losonczy, K.G. Izmirlian, G. Foley, D.J. Ross, G.W. Petrovitch, H. Havlik, R. White, R. White, L.R. (2000). Association of vitamin E and C supplement use with cognitive function and dementia in elderly men. *Neurology*, 54, 1265-1272.

Mekki, K. Bouzidi-bekada, N. Kaddous, A. Bouchenak, M. (2010). Mediterranean diet improves dyslipidemia and biomarkers in chronic renal failure patients. *Food Funct*, 1, 110-115.

Mirochnitchenko, O. Palnitkar, U. Philbert, M. Inouye, M. (1995). Thermosensitive phenotype of transgenic mice overproducing human glutathione peroxidases. *Proc Natl Acad Sci USA*, 92, 8120-8124.

Mortensen, A. Skibsted, L.H. Truscott, T.G. (2001). The interaction of dietary carotenoids with radical species. *Arch Biochem Biophys*, 385(1), 13–19.

National Academy of Sciences. (2000). Dietary Reference Intakes: Recommended Intakes for Individuals. Food and Nutrition Board, Institute of Medicine.

O'Neill, M.E. Carroll, Y. Corridan, B. Olmedilla, B. Granado, F., Blanco, I. Van den Berg, H. Hininger, I. Rousell, A.-M. Chopra, M. Southon, S. Thurnham, D.I.A. (2001). European carotenoid database to assess carotenoid intakes and its use in a five-country comparative study. *Brit J Nutr*, 85, 499-507

Palace, V.P. Khaper, N. Qin, Q. Singal, P.K. (1999). Antioxidant potentials of vitamin A and carotenoids and their relevance to heart disease. *Free Radic Biol Med*, 26(5/6), 746–761.

Pandey, K.B. Rizvi, S.I. (2009). Plant polyphenols as dietary antioxidants in human health and disease. *Oxid Med Cell Longev*, 2(5), 270-278.

Perrig, W.J. Perrig, P. Stahelin, H.B. (1997). The relation between antioxidants and memory performance in the old and very old. *J Am Geriatr Soc*, 45, 718-724.

Podmore, I.D. Griffiths, H.R. Herbert, K.E. Mistry, N. Mistry, P. Lunec, J. (1998). Vitamin C exhibits pro-oxidant properties. *Nature*, 392, 559.

Psaltopoulou, T. Naska, A. Orfanos, P. Trichopoulos, D. Mountokalakis, T. Trichopoulou, A. (2004). Olive oil, the Mediterranean diet, and arterial blood pressure: the Greek European Prospective Investigation into Cancer and Nutrition (EPIC) study. *Am J Clin Nutr*, 80, 1012–1018.

Rice-Evans, C. (2001). Flavonoid antioxidants. *Curr Med Chem*, 8, 797-807.

Rice-Evans, C.A. Miller, N.J. Paganga, G. (1996). Structure-antioxidant activity relationships of flavonoids and phenolic acids. *Free Radic Biol Med*, 20, 933-956.

Santos, I.M.S. Tomé, A. da R. Saldanha, G.B. Ferreira, P.M.P. Militão, G.C.G. de Freitas, R.V. (2009). Oxidative stress in the hippocampus during experimental seizures can be ameliorated with the antioxidant ascorbic acid. *Oxid Med Cell Longev*, 2(4), 214-221.

Sakakibara, H. Ogawa, T. Koyangi, A. Kobayashi, S. Goda, T. Kumazawa, S, Kobayashi, H, Shimoi, K. (2009). Distribution and excretion of bilberry anthocyanins in mice. *J Agric Food Chem*, 57, 7681–7686.

Scalbert, A. Morand, C. Manach, C. Rémésy, C. (2002). Absorption and metabolism of polyphenols in the gut and impact on health. *Biomed Pharmacother*, 56, 276–282.

Scarmeas, N. Stern, Y. Tang, M.X. Mayeux, R. Luchsinger, J.A. (2006). Mediterranean diet and risk for Alzheimer's disease. *Ann Neurol*, 59(6), 912-921.

Shils, M.E. Shike, M. Ross, A.C. Caballero, B. Cousins, R.J. (2006). Modern nutrition in health and disease. Lippincott Williams & Wilkins.

Silva, J.P. Gomes, A.C. Coutinho, O.P. (2008). Oxidative DNA damage protection and repair by polyphenolic compounds in PC12 cells. *Eur J Pharmacol*, 601, 50–60.

Splettstoesser, W.D. Schuff-Werner, P. (2002). Oxidative stress in phagocytes—"the enemy within". *Microsc Res Tech*, 57, 441-455.

Storey, K.B. (2004). Functional metabolism: regulation and adaptation. John Wiley & Sons Inc. New Jersey.

Souci, S.W. Fachmann, W. Kraut, H. (2000). Food Composition and Nutrition Tables. Stuttgart: CRC Press.

Sultana, R. Perluigi, M. Butterfield D.A. (2006). Protein oxidation and lipid peroxidation in brain of subjects with Alzheimer's disease: Insights into mechanism of neurodegeneration from redox proteomics. *Antioxid Redox Sign*, 8, 2021-2037.

Valko, M. Leibfritz, D. Moncol, J. Cronin, M.T.D. Mazur, M. Telser, J. (2007). Free radicals and antioxidants in normal physiological functions and human disease. *Int J Biochem Cell B*, 39, 44-84.

Wang, H. Cao, G. Prior, R.L. (1996). Total antioxidant capacity of fruits. J *Agric Food Chem*, 44, 701–705.

Wang, H. Gaok, J. Sun, X. Martinez-Wittinghan, F.J. Li, L. Varadaraj, K. Farrell, M. Reddy, V.N. White, T.W. Mathias, R.T. (2009). The effects of GPX-1 knockout on membrane transport and intracellular homeostasis in the lens. *J Membrane Biol*, 227, 25–37.

Willett, W.C. (2006). The Mediterranean diet: science and practice. *Public Health Nutr*, 9(1A), 105–110.

Yang, C.S. Sang, S. Lambert, J.D. Lee, M.-J. (2008). Bioavailability issues in studying the health effects of plant polyphenolic compounds. *Mol Nutr Food Res*, 52, S139 -S151.

2

Antioxidant and Pro-Oxidant Effects of Polyphenolic Compounds and Structure-Activity Relationship Evidence

Estela Guardado Yordi[1,*], Enrique Molina Pérez[1,*], Maria João Matos[2] and Eugenio Uriarte Villares[2]
[1]*University of Camagüey"Ignacio Agramonte y Loynaz"*
[2]*University of Santiago de Compostela*
[1]*Cuba*
[2]*Spain*

1. Introduction

Polyphenolic compounds are bioactive substances widely distributed in the vegetable kingdom. They act as natural antioxidants and their presence contributes to the color, flavor and aroma of food. Therefore, they are considered dietary antioxidants with interesting benefits to health.

The research and characterization of new bioactive substances, for the new design of functional foods, nutraceuticals, or drugs, has been intensified. Several studies showing controversial results of exogenous antioxidants, debating that the type, dosage and matrix of these antioxidants may be determining factors impacting the balance between beneficial and deleterious effects of these natural compounds (Bouayed & Bohn, 2010). There are also some proofs that they act as pro-oxidants, under certain conditions, such as high doses or the presence of metal ions (Decker, 1997; Raza & John, 2005; Watjen et al., 2005). The antioxidant or pro-oxidant activity intimately depends on their concentration (Bouayed & Bohn, 2010). The consequences of pro-oxidant activity could be the possible damage to the biomolecules such as DNA, proteins and lipids, and the consequent cellular death (Aruoma, 2003).

The relationship between the chemical structure and the anti-oxidant/pro-oxidant activities has been the main focus of important studies in the nineteenth century. Nevertheless, in general it was carried out without the use of bioinformatics. It was only observed in the last years that different research groups started to use quantitative structure-activity relationships (QSAR) methodologies, using different molecular descriptors.

In this chapter, the principal aim is to describe and analyze the potentialities of the QSAR methods, which has already been evaluated in the pharmaceutical sciences. These methodologies are going to be useful tools for the research and characterization of bioactive compounds, especially polyphenols, which can be used as functional foods in food science.

---

* Corresponding Authors

The possibilities of the *in silico* studies in a structural level are going to be discussed, since many of the antioxidant compounds have a dual behaviour by also displaying pro-oxidant activity.

2. Polyphenolic compounds and their presence in food items

2.1 Description of principal polyphenols, their presence in foods and the functions that they may play

Phenolic or polyphenolic compounds are plant metabolites widely spread throughout the plant kingdom. Phenolic compounds are essential for the growth and reproduction of plants, and are produced as a response for defending plants against pathogens and stress in general. Chemically, they are compounds with an aromatic ring linked to one or more hydroxyl groups. They are an important group with numerous substances, including structures from different moieties. They include simple compounds, such as phenolic acids and also more complex molecules such as hydrolysable and condensed tannins (Fig.1). The most prevalent group is the flavonoid group, including anthocyanins, flavonols, flavones, chalcones, dihydrochalcones, isoflavones and flavan-3-ols. They are composed of two aromatic rings (A and B) linked by an oxygenated heterocycle (C). Different subclasses depend on the degree of hydrogenation and substitution of the heterocycle (Fig.2). Another important subgroup is the one that include the phenylpropanoids, e.g. hydroxycinnamic acids (caffeic, ferulic and *p*-coumaric acids). Also important are the stilbenoids (resveratrol and piceatannol) and the benzoic acid derivatives (gallic and ellagic acids).

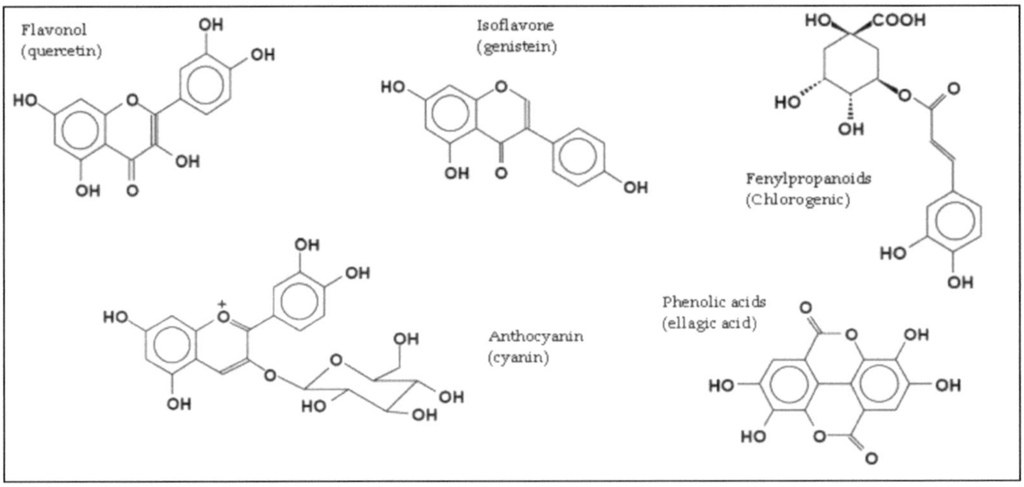

Fig. 1. Families and examples of polyphenols. Adapted from (Espín & Tomás-Barberán, 2005)

These compounds are present in important amounts in fruit and vegetables (Table 1) (Manach et al., 2004). They are considered as bioactive non-nutritional compounds, due to their properties, including antioxidant functions. The importance of antioxidant activities of phenolic compounds and their possible usage in processed foods as natural antioxidants have reached a new role in recent years.

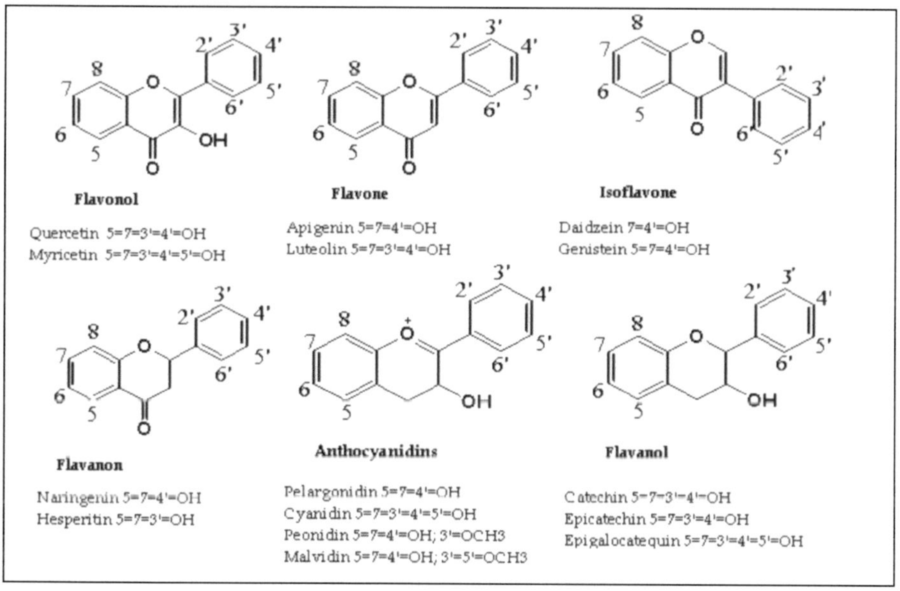

Fig. 2. Skeleton and some subclasses of flavonoids.

Antioxidant compounds are described as chemical structures that prevent the oxidation of a substrate at low concentrations (Halliwell, 1990). The type of substrates susceptible to this activity includes structures such as proteins, lipids, carbohydrates and DNA (Aruoma & Halliwell, 1995). They are also defined as compounds that protect the cellular system from potentially harmful processes that can cause excessive oxidation (Duthie et al., 2000).

Many phenolic compounds are responsible for some of the sensorial properties of food and therefore are important for their quality (Manach et al., 2004; Tomás-Barberán & Espín, 2001). There are bitter polyphenols, such as certain citrus flavanones (naringin from grapefruit and neohesperidine from the bitter oranges) or oleuropein present in olives. Among the polyphenols, there are pigments such as anthocyanins, responsible for the red, blue and violet characteristics of many fruits (strawberries, plums, grapes, etc.), vegetables (eggplant, red cabbage, radish, etc.) and red wine, or flavonols, with a cream-yellowish hue, which are found principally in the outer parts of fruits and vegetables (Tomás-Barberán et al., 2000). Proanthocyanidins (condensed tannins) and hydrolysable tannins confer astringency to fruits, and some simple phenols are important for the aroma of certain fruits, such as eugenol in bananas. Hydroxycinnamic acid derivatives such as caffeic, ferulic and sinapic acids are present in a number of fruits and vegetables and food products, cereals, and in some cases are the major polyphenols. Although they have not a direct impact on the organoleptic characteristics of foods containing them, they may indirectly negatively affect the quality if they are oxidized by oxidative enzymes found naturally in plant tissues. This leads to the formation of brown polymers (Tomás-Barberán & Espín, 2001).

The most relevant, in terms of abundance in the diet and due to their biological activity, are stilbenoids, hydroxytyrosol, ellagic acid and flavonoids (flavonols, flavan-3-ols and isoflavones). They are usually found in the form of glycosides. However, the human body cannot produce any of the polyphenols, therefore they must be obtained through the diet.

Chemical skeleton	Class	Example for each class	Sources
C6	Simple phenols	Hydroquinone	Gayuba
C6-C1	Benzoic acids	*p*-hydroxybenzoic acid	Coconut shell, white wine
C6-C3	Hydroxycinnamic acids	Caffeic acid	Apple, pear, red wine
C6-C3	Coumarins	Scopoletin	Cinnamon, mulberry from India, green tea
C6-C4	Naftoquinones	Vitamin K1	Spinach, cabbage, cereals
C6-C2-C6	Stilbenoids	*Trans*-resveratrol	Grapes, peanuts
C6-C3-C6	Flavonoids	Luteolin	Olive, grapes
(C6-C3-C6)	Condensed tannins	Procyanidins	Grapes, red wine, cherrys

Table 1. Classification of the phenolic compounds in relation to the number of carbon atoms and examples of their sources.

2.2 Food sources of flavonoids and phenolic acids

Flavonoids are widely distributed in plants, fruits and vegetables and represent substantial components, without significant contribution to the human diet (Aherne & O'Brien, 2002). Apart from being found in fruits and vegetables, they are present in seeds and flowers, as well as wine, green tea, black tea and soy, which are usually consumed in the human diet. Beer also contains significant amounts of flavonoids, mainly polyhydroxyflavanes (catechin and epicatechin), anthocyanidins (leucocyanidins or leucopelargonidine) and flavonols. Today more than 5.000 flavonoids have been identified (Ross & Kasum, 2002), among which are especially:

1. Citrus flavonoids: Quercetin, hesperidin, rutin, naranjin and limonene. Quercetin is a flavonoid present in green-yellow onions, apples, broccoli, cherries, grapes and red cabbage. Hesperidin is found in the peels of oranges and lemons. Narangin results in the bitter taste of many fruits such as orange, lemon and grapefruit. Limonene has been isolated from lemon and lime.
2. Flavonoids from soy or isoflavones (genistein and daidzein): they are present in soy foods such as beans, tofu, tempeh, soy milk, textured vegetable protein, flour and miso.
3. Anthocyanidins: they are plant pigments responsible for the colour of red and bluish-red cherries.
4. Proanthocyanidins are found in grape seed, red wine and pine bark sea extract.
5. Ellagic acid: it is a flavonoid found in fruits, such as grapes, and vegetables.
6. Catechins: green and black tea are good sources.
7. Kaempferol: is abundant in leek, broccoli, radish, endive and red beets.

Recently, the United States Department of Agriculture (USDA) published a "Database for the Flavonoid Content of Selected Foods". This database contains values for 385 food items with five subclasses of flavonoids: flavonols, flavones, flavanones, flavan-3-ols and anthocyanidins (United Stated Departament of Agriculture [USDA], 2007). Another important database is the "Phenol-Explorer" version 1.5.7 (database on polyphenol content in foods) (INRA & Wishart Research Group, 2009). It contains more than 35.000 content values for 500 different polyphenols in more than 400 foods. These data derive from the systematic collection of more than 60.000 original content values in more than 1.300 scientific publications. Data have been critically evaluated before inclusion in the database.

Phenolic acids, in general, describe phenols that possess a carboxylic acid function. However, when describing them as metabolites of the plant, they are referred to a different group of organic acids. These phenolic acids are distinguished according to two underlying structural compounds: the cinnamic and benzoic acids (Fig. 3). While for the rest of the phenolic acids the basic scheme is the same, the number and positions of hydroxyl and methoxy groups on the aromatic ring create variety. Hydroxybenzoic acids have a general structure C6-C1 (Macheix et al., 1990).

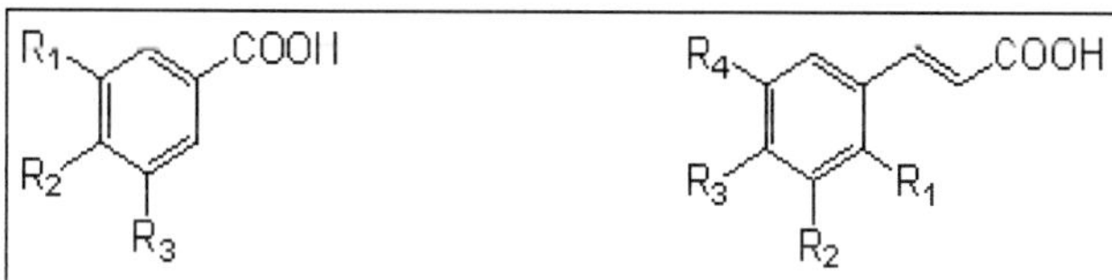

Benzoic acids

Acids	R1	R2	R3
Galic	OH	OH	OH
Vanillic	H	OH	OCH3
Syringic	OCH3	OH	OCH3

Cinnamic acids

Acid	R1	R2	R3	R4
Ferulic	H	H	OH	OCH3
p-cumaric	H	H	OH	H
Caffeic	H	H	OH	OH
Sinapic	H	OCH3	OH	OCH3

Fig. 3. Basic structure of benzoic and cinnamic acid derivatives.

Common benzoic acids are vanillic, *p*-hydroxybenzoic, syringic, protocatechuic, salicylic and gallic acids. They may be present in soluble form, conjugated with sugars or organic acids. Gallic acid is a trihydroxylic derivative, involved in the formation of gallotannins hydrolysates. The four cinnamic acid derivatives widely distributed in fruits are: *p*-coumaric, caffeic, ferulic and sinapic acids (Macheix et al., 1990). Cinnamic acid derivatives are mainly found in several conjugated forms, mainly esters of hydroxyl acids such as tartaric acid and sugar derivatives (Shahidi & Nacsk, 1995). Chlorogenic acid (5-caffeoylquinic acid) is one of the most important derivatives of cinnamic acid found in fruits, being sometimes the predominant phenolic compound. The wide distribution and high concentration of cinnamic acids in fruits may be due, in part, to the key role that they play in the biosynthesis of more complicated phenolic compounds such as shikimate and phenylpropanoid (Robards et al., 1999).

Plants contain a wide variety of phenolic compounds depending on the species (Bravo, 1998; Saraji & Mousavi, 2010). In general, plants of the *Solanaceae* family provide both chlorogenic acid and other hydroxycinnamic acids. In fruits, the content of hydroxybenzoic acid is generally low, with the exception of grapes, raspberries and strawberries. Usually, the content of hydroxycinnamic acid is higher. Caffeic acid is the predominant cinnamic acid in many fruits, constituting up to 75 % of phenolic acids. It is found in pears, apples, cherries and blueberries (Manach et al., 2004). However, *p*-coumaric acid is the major component within the cinnamic acid group of citrus and pineapple (Macheix et al., 1990). The presence of chlorogenic acid in fruits has also been widely studied (Martínez- Valverde et al., 2000).

Cinnamic acid derivatives are usually more abundant in fruits and vegetables than those of benzoic acid. Derivatives of cinnamic acid exist in all fruit parts, but concentrations are higher in the outer parts of ripe fruit. Concentrations generally decrease during the course of ripening, but total quantities increase as the size of the fruit increases (Hakkinen, 2000).

The content of polyphenols in fruit juices is generally between 2 and 500mg/L (Bravo, 1998). The highest concentrations were found in the pulp of oranges, with values around 31mg/100g of fresh weight (Justesen et al., 1998). Wine is a much studied source of phenolic compounds, especially of phenolic acids, anthocyanins, tannins and flavonoids. Within these food groups, the phenolic compounds of interest are the hydroxycinnamic (Buiarelli et al., 2010) and benzoic acids, along with flavan-3-ols and flavan-3-diols, anthocyanins, anthocyanidins, flavonols, flavones and condensed tannins. There are differences between the content of phenolic compounds in white wine and red wine. First, total content is lower for white wine. The average content of phenolic compounds in a typical white wine is about 250mg/L, although some types may contain up to 2000mg/L. By contrast, the content of total phenolic compounds in red wine typically ranges between 1000 and 4000mg/L. *p*-Coumaric, protocatechuic, syringic and caffeic acids are among the most abundant polyphenols in wines (Benassi & Cecchi, 1998).

Beer contains vanillic, *p*-coumaric and caffeic acids (Lunte et al., 1988). Tea is a beverage with a high content of phenolic compounds, noted for its high concentration of catechins, which constitute over 30 % of the dry weight of the leaf, and additional flavonols (quercetin, kaempferol and their glycosides), flavones and phenolic acids (gallic and chlorogenic acids).

In legumes and cereals, the main phenolic compounds are flavonoids, phenolic acids and tannins. Rice flour contains up to 86mg/100g of phenolic acids. This content is similar to those in wheat flour and oats. In both rice and wheat flour, the main phenolic component is ferulic acid, constituting at least 89 % of total phenolic acids (Sun et al., 2001). In addition, ferulic, sinapic, vanillic and syringic acids are among the most frequently found polyphenols in cereals (Lempereur et al., 1997).

3. Evidence for health benefits

The information linking diet to chronic diseases comes mainly from epidemiological studies and controlled assays in humans. These diseases are predominantly cardiac and vascular pathologies, cancer, atherosclerosis, diabetes type 2, Alzheimer's disease, hypertension, obesity, osteoporosis, chronic liver disease, nephritis or chronic renal and gastrointestinal diseases. These diseases are potentially preventable and are related to inappropriate lifestyles (Pasinetti, 2007).

According to WHO, in the early twentieth century around 20 % of the deaths were caused by cardiovascular or malignant tumors. The loss of quality of life becomes more critical in those over sixty years old, in whom these conditions predominate as causes of morbidity and mortality. There is an increasing epidemiological evidence associating diets rich in fruits and vegetables with lower rates of mortality from cardiovascular disease and some cancers. More limited evidences suggest that such diets may decrease the incidence of other chronic diseases, including diabetes, cataracts, macular degeneration, rheumatoid arthritis, neurodegenerative diseases and hypertension. Thus, the scientific community's attention has been focussing on the antioxidant components, which may provide protection against chronic diseases by decreasing oxidative damage in tissues. However, these antioxidant components are also involved in other cellular mechanisms such as apoptosis and have an important impact on intracellular signalling and gene regulation.

The so-called "antioxidant hypothesis" has been suggested based on the oxidative damage resulting from the action of reactive oxygen species (ROS) and nitrogen radicals, which are formed naturally in the body. ROS are known to be involved in pathogenic processes of numerous diseases described before (Diplock, 1994; Sies, 1997). When antioxidant defenses are insufficient, DNA, lipids, proteins and other molecules may be oxidatively damaged. Oxidation effects could be reduced by dietary antioxidants (e.g. polyphenolic compounds) (Ferrari & Torres, 2003). Fortunately, in the human body, there is a balance between oxidant and antioxidant species generation. When this equilibrium is broken, oxidative stress (OS) results. The balance between oxidation and antioxidation (redox balance) is critical in maintaining a healthy biological system (Bouayed, 2010; Bouayed & Bohn, 2010; Bouayed et al., 2009; Valko et al., 2007). However, despite that the role of the antioxidant defense mechanisms is to neutralize oxidant species, oxidative damages to proteins, lipids and DNA related to undetoxified ROS occur and accumulate during life, promoting aging process.

3.1 Cardiovascular and cerebrovascular disease prevention

In the last decades, several epidemiological studies have shown that dietary intake of foods rich in natural antioxidants was correlated with reduced risk of coronary heart disease. Particularly, a negative association between the consumption of polyphenol-rich foods and cardiovascular diseases has been demonstrated (Pokorny et al., 2001). This association has been partially explained based on the fact that polyphenols interrupt lipid peroxidation induced by ROS.

Epidemiological evidence for the importance of flavonoids in reducing mortality from coronary heart disease was provided by the Zutphen Elderly study (Anandh Babu & Liu, 2008; Hertog et al., 1993). These authors proved that the ingestion of flavonoids and other phenolic substances was inversely associated with mortality from myocardial infarction (Hertog et al., 1995).

Prospective studies have been conducted in order to analyze the effects of diet on heart disease on populations in the Netherlands (Hertog et al., 1993), USA (Rimm et al., 1996), UK (Hertog et al., 1997) and Finland (Knekt et al., 1996) and also on cerebrovascular disease in the Netherlands (Keli et al., 1996). In some of these studies there has been found a strong protective effect of flavonoids, from the group of polyphenols, against these diseases. However, in other studies the effect was less significant or even negative, as in the case of the British study (Hertog et al., 1997). These discrepancies may be due, among other factors, to defects in the evaluation methods of the ingestion of flavonoids and other polyphenols, the varieties of fruit and vegetable consumed, and differences postharvest such as cooking treatment effects, bioavailability and metabolism by colonic bacteria, and so on (Tomás-Barberán et al., 2000).

Knekt (2002) confirmed that people with higher intake of quercetin (a flavonol abundant in onions, apples, tea, wine and other fruits and vegetables) have lower myocardial infarction mortality rates. Also, cerebrovascular disease incidence was lower in those with higher intake of kaempferol, naringenin and hesperetin (very abundant in citrus) (Knekt et al., 2002). However, evidence on the role of flavonoids in the prevention of cardiovascular diseases is still a matter of discordance.

The oxidative modification of low density lipoprotein (LDL) is believed to have a crucial role in atherogenesis, and epidemiological studies have shown that consumption of fruits and vegetables and regular or moderate consumption of red wine is correlated with a reduced risk of cardiovascular disease (Renaud & Lorgeril, 1992). Other epidemiological studies have shown a direct relationship between tea consumption and cardiovascular disease. They hypothesized that the antioxidant effects of tea flavonoids may include prevention of oxidative damage to LDL (Kris-Etherton & Keen, 2002). Geleijnse et al. (2002) studied the association between intake of flavonoids in tea and the incidence of myocardial infarction in the Dutch population. The results indicated that the intake of flavonoids (quercetin + kaempferol + myricetin) was inversely associated with fatal myocardial infarction, when comparing individuals with higher compared to lower intake (Geleijnse et al., 2002).

3.2 Cancer prevention

Some of the most compelling evidence of a protective effect of diets against cancer, in recent years, is the evidence on the intake of fruits and vegetables (Block et al., 1992). A review of more than 160 epidemiological studies by Steinmetz & Potter (1996) showed that an increased consumption of fruits and vegetables was associated with a lower incidence of certain cancers, among which were the stomach, esophagus, lung, oral cavity and pharynx, endometrium, pancreas and colon (Steinmetz & Potter, 1996). This showed that the most important vegetables involved in this protective effect would be the ones consumed raw, followed by garlic and onions, cabbages, cauliflowers and broccoli, tomatoes and fruits in general. Among other protective components, the polyphenols have been generally cited. For example, it has been suggested that quercetin play an important role in the anticarcinogenity effects of plant foods such as apples and onions (Knekt et al., 1997) owing to its chemopreventive activity against carcinogens *in vitro* (Obermeier et al., 1995) and *in vivo* as suggested by animal studies (Deschner et al., 1991). EPIC is an important study that indicates that these retrospectively obtained results, at least respecting to cancer, might have been somewhat overestimated, however, still a significant reduction of consumption of fruits and vegetables on e.g. colorectal cancer was found (Bouayed & Bohn, 2010, van Duijnhoven et al., 2009).

Polyphenols can further act by inhibiting cell proliferation, which is deregulated in cancer. This inhibition has been demonstrated *in vitro* in many tumor cell lines. For example, Kuo (1996) published the antiproliferative effect of flavonoids on colon carcinoma cells through mechanisms of apoptosis induction (Kuo, 1996). Although the antiproliferative effects of polyphenols in general and in particular of flavonoids and isoflavonoids in cell cultures seems well established, there are relatively few data regarding the *in vivo* antiproliferative activity, and virtually nothing is known about the clinical relevance of this bioactivity (Birt et al., 2001). This antiproliferative effect suggests that polyphenols may have an effect via regulating the cell cycle or inducing apoptosis in tumor cells. In fact, many studies have shown the effect of polyphenols on the cell cycle of tumor cells in cultures in *in vitro* assays.

This has been demonstrated in cells of several types of leukemia, stomach cancer, lung, colon, bladder and prostate (Birt et al., 2001).

Polyphenols can also protect against cancer through their inhibition of oxidative DNA damage (Omenn, 1995). A mechanism of oxidation appears to be a major cause of mutations that could potentially be reduced by dietary antioxidants. Polyphenols, because of their amply demonstrated *in vitro* antioxidant activity, could prevent this oxidation if reaching those tissues where these oxidations occur. Again, relatively few data have been published about their role *in vivo*.

It has been suggested that flavonoids (flavones and isoflavones) can prevent cancer because they can be found in many plant foods that are associated with reduced cancer rates. In this sense, studies such as one following 10.054 people in Finland (Knekt et al., 2002) have shown that men with higher quercetin intake had a lower incidence of lung cancer and those who took more myricetin had a lower incidence of prostate cancer. Among the possible mechanisms of action were mentioned estrogenic/antiestrogenic activities, antiproliferative effects, induction of cell cycle block of cancer cells in a special phase and apoptosis, preventing oxidation, induction of detoxification enzymes (phase I enzymes and phase II), immune system regulation and changes in cell signalling (Birt et al., 2001). Recently, Banerjee et al. (2008) analyzed the use of genistein for cancer therapy (Banerjee et al., 2008), an isoflavonoid with potential estrogenic activity.

3.3 Neurodegenerative disease prevention

Neurodegenerative diseases are becoming increasingly prevalent with the aging of the general population. The twentieth century witnessed a significant demographic change in the human population of the industrialized world that is currently followed by a similar shift of life expectancy toward higher age ranges in Asia, Africa, and Middle and South America. Thus, neurodegenerative diseases are presently amongst the major contributors to disability and disease in human populations. Alzheimer's disease is the most prevalent of the neurodegenerative diseases followed by Parkinson's disease. Free radicals have been implicated in the development of neurodegenerative disorders such as Parkinson's and Alzheimer's diseases, among other multifactorial diseases such as diabetes, rheumatoid arthritis and chronic obstructive pulmonary diseases (Diplock, 1994; Diplock et al., 1998; Sies, 1997). These age-related diseases may, therefore, be beneficially influenced by antioxidant consumption. However, few epidemiological data exists on the association of antioxidants with the risk for these diseases (Diplock et al., 1998).

For instance studies suggest that OS may contribute to the pathogenesis of Alzheimer's disease. Although the etiology of this disease is not completely understood, deposits of aberrant proteins, namely β amyloid (Aβ) and τ-protein, OS dyshomeostasis of biometals and low levels of acethylcholine (ACh) seem to play significant roles. Monoamine oxidase B (MAO-B) activity is also increased in association with gliosis, which can result in higher levels of H_2O_2 and oxidative free radicals which are a possible source of OS for vulnerable neurons affected by Alzheimer's disease. The risk of this disease could be reduced by the consumption of antioxidants that counteract the negative effects of OS. Epidemiological studies have been carried out, such as the "Rotterdam Study", to evaluate the effect of diet on the prevention of neurodegenerative diseases and dementia. In this case, 5395 persons were followed during the nineties and it had been studied whether there was any relationship between diet and the development of Alzheimer's disease, especially following the consumption of antioxidants such as vitamins C and E, β-carotene and flavonoids. In this study, higher intakes of vitamins C and E were associated with a lower risk for developing Alzheimer's disease. This relationship was even more pronounced in the case of smokers, and was also observed in the case of β-carotene and flavonoids (Engelhart et al., 2002). However, epidemiological studies conducted in the USA found that only vitamin E, but not C or *β*-carotene may be associated with reduced risk of Alzheimer's disease (Morris et al., 2002). More research is needed regarding these data in order to demonstrate whether antioxidants in the diet can have an impact for the prevention of neurodegenerative diseases.

4. Polyphenolic compounds with antioxidant/pro-oxidant activities

In recent years, studies of natural antioxidants that are part of plant foods have become the focus of increased interest. Flavonoids and phenolic acids are receiving increased attention as potential antioxidants, primarily due to their wide presence in a large number of widely consumed foods. Different experimental methods developed for this purpose represent a major critical factor when it comes to strategies for developing fortified foods, or the characterization of a functional food and/or the formulation of an antioxidant supplement (Antolovich et al., 2002). There is a great diversity of variants and experimental assays designed for these studies (León et al., 2005).

Currently, the exogenous antioxidants, including polyphenols were considered "double-edged swords" in the cellular redox state. Several studies of exogenous antioxidants showed controversial results, especially when administered at high doses. The type, dosage and matrix of exogenous antioxidants may be determining factors impacting the balance between beneficial or deleterious effects of these natural compounds (Bouayed & Bohn, 2010). From epidemiological and dietary intervention studies, it appears, however, that exogenous antioxidants at physiologic (nutritional) doses play an important role in the maintenance or re-establishment of redox homeostasis, an essential state in maintaining healthy biological systems (Bouayed & Bohn, 2010; Valko et al., 2007).

4.1 Mechanisms involved in the antioxidant activity

The role of antioxidants in nutrition and health, as well as their mechanisms of action, have been extensively researched (Serafini, 2006). Although the biological functions of polyphenols and/or metabolism in the human body are not completely established, there is a consensus that the antioxidant activity of flavonoids could be a combination of metal chelating properties and free radical scavengers (Bohm et al., 1998; Bravo, 1998). Other authors also refer to the inhibition of oxidases, such as lipoxygenase (LO), cyclooxygenase (CO), myeloperoxidase (MPO), NADPH oxidase and xanthine oxidase (XO) (Groot & Rauen, 1998), as important mechanisms for avoiding the generation of higher amounts of ROS *in vivo*, as well as organic hydroperoxides. Moreover, they have been also known to inhibit enzymes indirectly involved in the oxidative processes, such as phospholipase A2 (FLA2) (Lindahl & Tagesson, 1997), while they stimulate others with recognized antioxidant activity, such as catalase (CAT) and superoxide dismutase (SOD) (Sudheesh et al., 1999). Therefore, the flavonoids interfere with the propagation reactions of free radicals and the radical formation itself (Van Acquire et al., 1996).

The chemical structure of polyphenols gives them the ability to act as free radical scavengers. The type of compound, the degree of methoxylation and the number of hydroxyl groups are some of the parameters that determine the antioxidant activity. As for phenolic acids, the oxidation inhibition is related to the chelation of metal ions via the *ortho*-dihydroxy phenolic structure, the scavenging of alkoxyl and peroxyl radicals, and the regeneration of α-tocopherol through reduction of the tocopheryl radical (Bors et al., 1990).

The structural features that have been associated with antioxidant activity are: a) a catechol group on the B-ring, which confers high stability to the radical formed after the capture reaction of the free radical, b) the 2,3-double bond in conjugation with a 4-oxofunction of a carbonyl group in the C-ring and c) the presence of hydroxyl groups at the 3 and 5 position (Fig. 4) (Bourne & Rice-Evans, 1998).

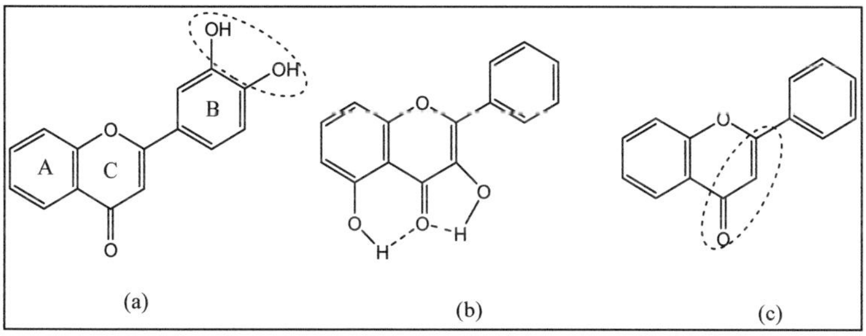

Fig. 4. Antioxidant activity-structure relationship of flavonoids. (a) a catechol moiety of the B-ring, (b) presence of hydroxyl groups at the 3 and 5 position, (c) the 2,3-double bond in conjugation with a 4-oxofunction of a carbonyl group in the C-ring.

Current studies on their metabolism established that flavonoids are located in the membrane at the interface lipid/water, being the first to react with the ROS formed in these areas. The type of conjugation during biotransformation and their location in the body, determines the ability of enzyme inhibition and antioxidant capacity. As part of this, they can act directly as "scavengers" of free radicals, by hydrogen or electron donation, leading to other more stable compounds, or compounds that can stabilize compounds obtained from free radicals or may have an additive effect on the endogenous antioxidant defense system by increasing or maintaining this antioxidant defence. These mechanisms are also the same for phenolic acids.

4.2 Evidence of pro-oxidant activity of polyphenols

4.2.1 Flavonoids and phenolic acids as pro-oxidant compounds

The need to clarify the safety aspects (McKevith et al., 2003), structure-activity (Yordi, 2010), bioavailability and metabolism of compounds with antioxidant activity when they become part of functional foods or nutraceuticals has been suggested (Fundación Española para la Ciencia y la Tecnología [FECYT], 2005). This is because many of the study results are inconclusive and sometimes contradictory. Examples of this are reported below, showing that some polyphenol antioxidants may have pro-oxidant activity under certain conditions such as at high doses or in the presence of metal ions (Azam et al., 2004; Bouayed & Bohn, 2010; Decker, 1997; Raza & John, 2005; Watjen et al., 2005) (Table 2).

Polyphenols and particularly flavonoids are examples of substances with such dual *in vitro* behaviour (Fukumoto & Mazza, 2000; Pérez-Trueba, 2003; Pérez-Trueba & Martínez, 2001; Sakihama et al., 2002). We recently published a list of such compounds and their dietary sources (Yordi, 2010). Phenolic acids have also been reported as pro-oxidants (Fukumoto & Mazza, 2000; Maurya & Devasagayam, 2010; Mozuraityte et al., 2009; Sakihama et al., 2002; Simiæ et al., 2007) (Table 2).

It is noteworthy that some of the most abundant flavonoids and phenolic acids present in foods were reported to act as pro-oxidants: quercetin, myricetin, kaempferol and caffeic, chlorogenic and ferulic acids (see Table 2). In the 2.2 section, some examples of the main food sources for these two groups of polyphenols are shown.

No.	Phenolic acids	CAS* Numbers	Pro-oxidant reference
1	*o*-coumaric acid	614-60-8	(Simiæ et al., 2007)
2	*p*-coumaric acid	501-98-4	(Fukumoto & Mazza, 2000; Mozuraityte et al., 2009; Simiæ et al., 2007)
3	*m*-coumaric acid	14755-02-03	(Simiæ et al., 2007)
4	ferulic acid	537-98-4	(Fukumoto & Mazza, 2000; Maurya & Devasagayam, 2010; Mozuraityte et al., 2009)
5	caffeic acid	331-39-5	(Fukumoto & Mazza, 2000; Maurya & Devasagayam, 2010; Mozuraityte et al., 2009)
6	salicylic acid	69-72-7	(Simiæ et al., 2007)
7	*p*-hydroxybenzoic acid	99-96-7	(Simiæ et al., 2007)
8	vanillic acid	306-08-1	(Fukumoto & Mazza, 2000) (Simiæ et al., 2007)
9	syringic acid	530-57-4	(Simiæ et al., 2007)
10	protocatechuic acid	99-50-3	(Fukumoto & Mazza, 2000)
11	gallic acid	149-91-7	(Fukumoto & Mazza, 2000)
12	chlorogenic acid	327-97-9	(Fukumoto & Mazza, 2000; Sakihama et al., 2002)
13	*m*-hydroxybenzoic acid	99-06-09	(Simiæ et al., 2007)
14	ellagic acid	476-66-4	(Fukumoto & Mazza, 2000)

* Chemical Abstracts Service

Table 2. Pro-oxidant activity of phenolic acids (benzoic and cinnamic acids derivatives).

4.2.2 New challenges in pro-oxidant activity

It is known that antioxidants should be present in the body in sufficient concentrations to prevent the accumulation of pro-oxidants (state of oxidative stress) (Sies & Jones, 2007). The human body has mechanisms that protect against the harmful effects of ROS that are continually being formed. These defense systems are antioxidant in nature and stabilize highly reactive compounds. The fact that some substances in foods have pro-oxidant activity is particularly interesting. Although there are few studies, this is due to the fact that OS can be induced from the pro-oxidant agents, either through the creation of ROS or inhibition of antioxidant systems (Puglia & Powell, 1984). This can generate oxidative damage to biomolecules such as proteins, DNA and lipids and eventually cells and tissues (Aruoma, 1999, 2003; James et al., 2003). It is recognized that the development of many chronic diseases may be due to OS (Espín & Tomás-Barberán, 2005; Halliwell & Whiteman, 2004), where an antioxidant/pro-oxidant balance is not achieved, resulting in a pathological process. The pro-oxidants catalyze then oxidative reactions of biomolecules, which may lead to cellular dysfunction, ending with cell death (Aruoma, 2003).

It is now recognized that the pro-oxidant action, of natural polyphenols, unlike their antioxidant properties, has a more specific preference against certain cellular targets,, since it appears to play an important role in the prevention of certain types of cancer (Lambert & Elias, 2010). In a recent review, the pro-oxidant activity of individual dietary polyphenols and their ability to induce mitochondrial dysfunction and consequently apoptosis has been suggested as a possible anticancer mechanism (Galati & O'Brien, 2004). Also, it was recently reported that dietary polyphenols could mobilize endogenous copper in humans, leading to oxidative DNA damage that could be responsible for inducing anti-cancer properties (Azmi et al., 2005).

We think it is interesting to predict, from the standpoint of polyphenolic structure-activity, the DNA damage leading to pro-oxidant substances (Yordi et al., 2011). Aruoma (2003) described that oxidative modifications to DNA are very important. These represent an early stage of carcinogenesis and may provide important OS biomarkers (Aruoma, 2003). It is known that one of the endpoints of the oxidation of this biomolecule are chromosomal aberrations (CA) which are changes in chromosome structure, visible by light microscopy (Astley & Lindsay, 2002). Physical or chemical agents capable of inducing these mechanisms are called clastogens (Bender et al., 1974; Galloway, 1994; Ishidate et al., 1988). The clastogen prediction, starting from a structural analysis, has been postulated by Estrada et al. (2006). These authors developed a quantitative structure-activity relationship (QSAR) model that allows the prediction of clastogens. They identified that one of the structural alerts which characterized clastogenic activity was the diphenylpyran group present in flavonoids (Estrada & Molina, 2006). These methodologies are going to be useful tools for the research and characterization of bioactive compounds, especially polyphenols, which can be used as functional foods, nutraceuticals or drugs. In the following section the possibilities of the *in silico* studies are going to be discussed.

5. Potentialities of bioinformatic methods approaches, such as QSAR studies, in structure-activity relationship studies

QSAR/QSPR (Quantitative structure-property relationship) methods are computational methods based on the physical, physicochemical, chemical and biological properties of organic compounds, depending ultimately on the molecular structure (Randic, 1998). This technique makes it possible to predict activities of new candidate compounds which are bioactive. These predictive methods require the existence of an experimental data set of chemicals, including activity/property relationship results. This allows for the development of mathematical models describing structure-activity/property relationships. This cheminformatics employs multivariate linear regression (MLR) and discriminant linear analysis (DLA) or nonlinear including neural networks. One advantage of these is to help identify the optimal structures for a given activity.

Chemical information of each molecule is encoded by molecular descriptors. These molecular descriptors such as CoMFA (Comparative Molecular Field Analysis) and graph-theoretic descriptors have become an interesting option for the generation of QSAR models (Devillers & Balaban, 1999). Nowadays, the number of papers in which graph-theoretic molecular descriptors are used in QSAR studies is increasing, especially the so-called topological indices. However, there are some misunderstandings on the role of such molecular descriptors in chemistry and drug investigation. These misunderstandings have produced severe criticisms to their use in QSAR, and sometimes excluded them from the pool of descriptors, which are traditionally used in structure-property-activity studies (Kubinyi, 1993).

The first serious attempt to establish an order in the definition of graph theoretical molecular descriptors was carried out by Randic in 1991 (Randic, 1991). This author gave a series of desired attributes that the topological indices need to have to be considered as useful molecular descriptors. In a more recent study, Milne in 1997 analyzed some aspects concerning the applications of graph theoretical molecular descriptors (Milne, 1997).

5.1 QSAR studies and antioxidant/pro-oxidant activities

Several studies of structure-activity relationship of polyphenolic compounds have been made on the basis of non-congeneric structures and empirically, which relate chemical structure with experimental results obtained, without using such graph-theoretic molecular descriptors or another topologic descriptor. In this sense, it is important to cite the paper by Cao et al. (1997) entitled "Antioxidant and pro-oxidant behavior of flavonoids: structure-activity relationships". An analysis by Fukumoto and Mazza (2000) related the antioxidant/pro-oxidant activities of phenolic compounds, comparing the results obtained by different established methods such as β-carotene bleaching and HPLC methods. For example this methods were used to assess activity of selected phenolic compounds, including several phenolic acids, flavonols, flavanols, flavanones, anthocyanidins/anthocyanins. Both pro-oxidant and antioxidant activities could be measured by the HPLC method. Comparisons in antioxidant activity of compounds with different structures could be made using the β-carotene bleaching and HPLC methods. Generally, antioxidant activity increased with an increase in hydroxyl groups and a decrease in glycosylation (Fukumoto & Mazza, 2000). Although these are no classical QSAR studies, they correlate the experimental results obtained with the chemical structure of compounds with antioxidant activity. Another more detailed analysis, but with the same characteristics, which established the relationship between the structure-antioxidant activity of polyphenolic compounds particularizing on possible mechanisms of action, was described by Perez-Trueba (2001) (Pérez-Trueba & Martínez, 2001). Regarding the analysis of the relationship between structure and pro-oxidant activity, there are few reports, among one of them the research by Gaspar et al. (1996) about flavonoid compounds with different hydroxyl substitutions.

Research teams such as Sergediene et al. (1999) have started to do QSAR studies, using different molecular descriptors, relating the influence of certain physical properties of polyphenols with pro-oxidant activity. The cytotoxicity of 13 polyhydroxybenzenes was described by the QSAR models crated by Nemeikaitė-Čėnienė et al. (2005). The correlations obtained quantitatively confirmed the parallelism between the polyphenol cytotoxicity and the rates of their single-electron oxidation, and point to the leading role of formation of the reactive oxygen species in their cytotoxicity. Depending on the examined system, this parallelism may be distorted due to the cytochrome P-450 and COMT-catalyzed transformation of polyphenols (Nemeikaitė-Čėnienėa et al., 2005). Selassie et al. (2005)

developed another QSAR study of phenolic compounds, based on physical properties such as hydrophobicity, partition coefficient o/w, etc., and the biological activities with respect to apoptosis and cytotoxicity in the cell line L1210, demonstrating the importance of hydrophobicity for these activities (Selassie et al., 2005). Later & Nandi (2007), developed another QSAR study using statistical techniques such as neural networks. More recently, Yuanqiang et al. (2010) related the measured caspase apoptotic activity of L1210 cell line analogues of phenolic compounds, obtaining good predictability from three dimension (3D), topographic descriptors used.

Studies have been focussing on the relationship between the chemical structure and the antioxidant activity of flavonoid compounds. These studies showed that the position of the hydroxyl groups, the dipole moment magnitude and the shape of the molecule play an important role in the inhibition of lipid peroxidation (Bakhtiyor et al., 2005). Amic, et al. (2007) created a more efficient model for the design of new flavonoids with antioxidant activity. More recently, Om & Kim (2008) focused their attention on the radical scavenging activity of flavonoids using 3D QSAR methods.

The understanding of the relationship between chemical structure and activity, and the ability to predict this, is a great advantage of such methods. We believe that these methods can be useful for predicting a specific activity of new polyphenolic compounds and/or synthezinging polyphenols with desired properties (Yordi et al., 2011).

5.2 QSAR studies under TOPS-MODE approach

In the last decade we have developed an approach to QSAR/QSPR and molecular design. It is known as the TOPS MODE approach, which is the acronym for topological substructural molecular descriptors/design (Estrada, 1996; Estrada, 1998; Estrada & Uriarte, 2001; Estrada et al., 2000). The TOPS-MODE approach is based on the calculation of spectral moments of molecular bond matrices appropriately weighted to account for hydrophobic, electronic and steric molecular features. Spectral moments are the trace of the *k*th power of a matrix, i.e. the sum of all entries in the main diagonal of such matrices. The reader is referred to (Estrada, 1996; Estrada, 1998) to obtain full details of this method.

A bond matrix is a square symmetric matrix in which non-diagonal entries are ones or zeroes if the corresponding bonds have a common atom or not, respectively (Estrada, 1995). These matrices represent the molecular skeleton without taking into account hydrogen atoms. For exemple, in fig. 5 it is shown the molecular graphe of the molecule 2,2-dimethylbutane and the adjacency matrix of corresponding link.

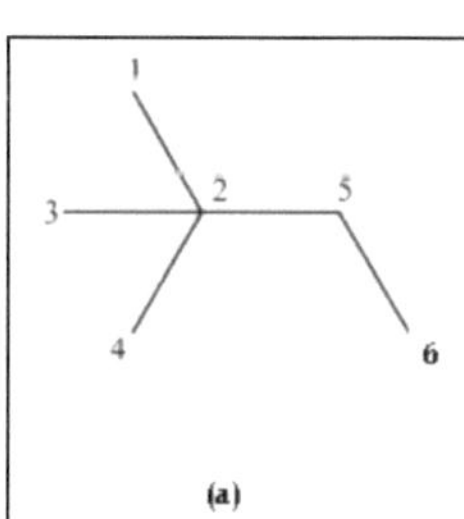

$$A = \begin{bmatrix} 0 & 1 & 0 & 0 & 0 & 0 \\ 1 & 0 & 1 & 1 & 1 & 0 \\ 0 & 1 & 0 & 0 & 0 & 0 \\ 0 & 1 & 0 & 0 & 0 & 0 \\ 0 & 1 & 0 & 0 & 0 & 1 \\ 0 & 0 & 0 & 0 & 1 & 0 \end{bmatrix}$$

(b)

Fig. 5. Molecular graphe (a) and its adjacency matrix link (b) corresponding to the 2,2-dimethylbutane molecule.

Bonds weights are placed as diagonal entries of such matrices and represent quantitative contributions to different physicochemical properties.

Among bond weights currently in use in our approach we have standard bond distance (SD), standard bond dipole moments (DM), hydrophobicity (H) (Wang et al., 2000), polar surface area (PS) (Ertl et al., 2000), polarizability (Pol) (Ghose & Crippen, 1987), molar refractivity (MR) (Ghose & Crippen, 1987), van der Waals radii (vdW) (Bondi, 1964), and Gasteiger-Marsilli charges (Ch) (Gasteiger & Marsilli, 1978).

The starting point for our approach is to calculate TOPS-MODE descriptors of the different types, e.g., H, PS, Pol, MR, vdW, and Ch, for the series of molecules under study. Then, we develop a quantitative model describing the property under study in terms of the spectral moments. In general this model can present the following form:

$$P = \sum_{j=1}^{n} b_j \mu_j + b_0 \tag{1}$$

where P is the property under study, b_j are the coefficients of the quantitative model (linear regression or discriminant analysis) and b_0 is the error.

The jth spectral moment of the bond matrix can be expressed as a sum of bond moments, which are simply the corresponding entries of the jth power of the bond matrix:

$$\mu_j = \sum_{i=1}^{m} \mu_j(i) \tag{2}$$

where $\mu_j(i)$ is the bond moment of the ith bond in a molecule with m bonds. Then, model (Trinajstic et al., 1986) can be written as:

$$P = \sum_{j=1}^{L} b_j \sum_{i=1}^{m} \mu_j(i) + b_0 = \sum_{i=1}^{m} \sum_{j=1}^{L} b_j \mu_j(i) + b_0 \tag{3}$$

where the right-hand side in (3) represents the contribution of bond i to the property P and is called the *bond contribution* and represented by $P(i)$:

$$P(i) = \sum_{j=1}^{L} b_j \mu_j(i) \tag{4}$$

and the property P can be expressed as an additive function of bond contributions:

$$P = \sum_{i=1}^{m} P(i) \tag{5}$$

5.2.1 Calculation of bond contributions

Bond contributions are numeric characterization of bonds which permit to identify some groups or regions of a molecular framework which can be responsible for a property/activity (Estrada & Molina, 2001). By carefully analyzing similar regions in different molecules we can obtain general rules about the contributions of molecular fragments to a particular property/activity. They are based on the substructural nature of TOPS-MODE. This procedure consists in transforming a QSPR or QSAR model into a bond additive scheme in which a property can be calculated as the sum of bond contributions for a molecule. Readers interested in learning more about the procedure for calculating work can be found in (Estrada & Molina, 2006). Yordi et al. (2011) calculated the bond contribution to the clastogenic activity of compounds with reported pro-oxidant activity. **Figure 6** shows one example of the calculated bond fragment. The selected compound (morin) was "active" in the QSAR model predictions. The hydroxyl group contribution is positive when C3 appears in ring C, corroborating the alert analysis. The 5,7-hydroxyl groups in ring A have 0.172 and 0.153 fragment contributions, respectively, and are more than the hydroxyl groups in ring C. It can be concluded that 5,7-hydroxyl substitutions are very important for the activity, so the methoxyl and hydroxyl groups have similar bond contributions. The 2:3 double bond at ring C is an example of a negative fragment contribution -0.103 of the "active" molecules. Only one fragment does not determine the biologic activity. Also, the additional collaborative effect at the molecule is needed

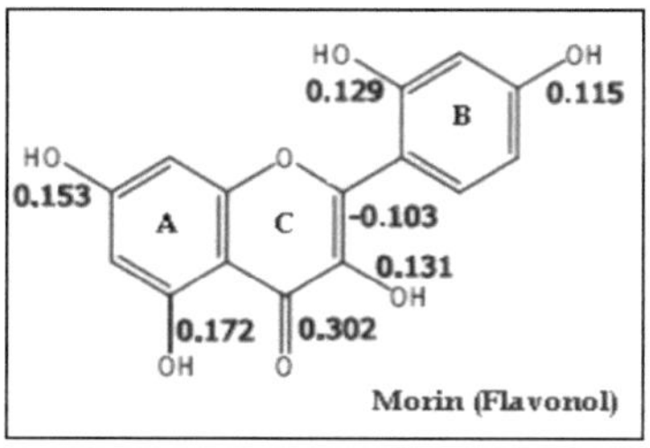

Fig. 6. Example of the Bond Contributions obtained from the TOPS-MODE classification model. Adapted from (Yordi et al., 2011).

5.2.2 Applications of the TOPS-MODE to the molecular design: MODESLAB

In recent years, the computer program TOPS-MODE (now called MODESLAB) was developed for Windows. This software facilitates the calculation of global spectral moments selecting any weight (weight) for different bonds in the molecule, while local spectral moments for single bonds or a defined fragment (contribution) in the molecule can be estimated in the same way after select the link/excerpt. The results are in file format, ready to be manipulated by statistical packages such as STATISTICA (StatSoft, 2001).

A novel application of TOPS-MODE was carried out for antibacterial drugs using computer-aided molecular design by Molina et al. in 2004. In this paper, the TOPS-MODE approach was largely probed to generate good predictive linear models in order to account for antimicrobial activity of a broader range of molecular structural patterns. Henceforth, we can assert that the TOPS-MODE approach may be used as an efficient alternative to massive screening of antimicrobial drugs (Molina et al., 2004).

This approach also allows formulating structural alert rules for chromosome aberration of organic compounds (Estrada & Molina, 2006). CA (clastogenic activity) are DNA changes generated by different repair mechanisms of DNA double strand breaks, which are microscopically visible. They are consequences of e.g. human exposure to ionising radiation or to mutagenic chemicals.

Speck et al. in 2009, used a mixed approach by employing atom-centered fragment, functional group counts, and TOPS-MODE descriptors in order to obtain fragment contributions that will provide a guide for the design of antituberculosis compounds (Speck et al., 2009).

Antytripanosomal agents (compounds agains a type of protozoa) were also studied. The model showed an accuracy of 100 % which means that the *in silico* methodology developed by our team is promising for the rational design of new antitrypanosomal drugs (Speck et al., 2009). In this sense, the molecules predicted can become or generate new leads for the development of more effective and less toxic compounds with antitrypanosomal activity. The methodology developed in this work provided in principle, the design, search and prediction of novel antitrypanosomal agents in a quantitative, rapid and easy way.

In this sense, the TOPS-MODE approach has been extended not only to the discovery of novel results but also toward the studying of the physicochemical and absorption properties of drugs (Cabrera et al., 2002; Estrada & González, 2003). On the other hand, most recently published papers in this area make use of reduced or homologous series of compounds to fit the QSAR for antimicrobial drug screening. This fact determines that these are not general models, which could be used to predict the biological activity of heterogeneous series of compounds. Approaches such as quantitative structure-activity relationships (QSAR) and molecular modeling are integrated with the study of complex networks to understand drug binding to human serum albumin (HSA)(Estrada & Molina, 2006). A topological substructural molecular design approach also has been used to formulate structural rules for binding of substrates of P-glycoprotein (P-gp). An important step in the development of any toxicological or metabolic activity of a chemical is the transport to the organs where the final effect takes place. One of the most important proteins in the transport of various molecules across extra- and intra-cellular membranes is P-glycoprotein (P comes from permeability) (Estrada et al., 2010).

The existence of a large number of polyphenols in different classes and the capacity to exercise different activities (estrogen, apoptotic, anti-depressant, anti-inflammatory, antioxidant, pro-oxidant etc) make this group of compounds very interesting to be studied using the TOPS-MODE approach, being able to predict not only the antioxidant activity, but any of the other properties mentioned. It is also possible to identify structural features and alerts that characterize a subclass of polyphenols that is likely to present a particular activity. It is also possible to do a substructural analysis, identifying how each fragment contributes to the activity. Allowing not only the selection of appropriate structures that are available but also the ability to synthesize, if they are not physically exist.

How does TOPS-MODE help in estimating antioxidant or another potential activity of polyphenols? For example, there are recent evidences about the potential use of dietary polyphenols as neuroprotective agents to reduce anxiety and to manage depression, analyzed by Bouayed (2010). In the paper "Polyphenols: A Potential New Strategy for the Prevention and Treatment of Anxiety and Depression" the pharmacological actions on the central nervous system of polyphenol compounds were analyzed (Bouayed, 2010). We can currently only imagine the possibilities offered by this approach in predicting antidepressant activity of polyphenol compounds. If we made a QSAR study using this approach, the first we need is to obtain a QSAR model. This model is building on a large number of molecules of different nature, which have proved to be antidepressant in different experimental models. Once obtained, the robust and validated model using multivariate statistical techniques, a virtual screening of it can be performed. For this purpose, data (Espín & Tomás-Barberán, 2005) from polyphenolic compounds of unknown activities are selected. The chemical information is encoded, molecular descriptors are calculated and the results interpreted. Then, the prediction is validated using the data of polyphenols analyzed by Bouayed (2010). Depending on the objectives of this *in silico* study, it is possible to identify/design polyphenolic structures suitable for the activity. It is also possible to identify the fragments that contribute positively to the activity.

6. Conclusion

In summary, the approach that encloses the calculation of the spectral moments of the bond adjacency matrix is known as TOPS-MODE approach. This has been applied for the description of some physicochemical properties of organic compounds, in quantitative structure toxicity relationship (QSTR), and has also been reported for the modelling of pharmacological activities (Estrada & Molina, 2001). The TOPS-MODE approach has been extended not only to the discovery of novel recoveries but also to the study of the absorption properties of drugs (Cabrera et al., 2002; Estrada & González, 2003).

The QSAR method is a proposal to be considered for the design of functional foods and nutraceuticals, particularly when analyzing the activities of polyphenolic compounds, taking into account the high structural diversity (Yordi, 2010). They can allow for predictions about the toxicity, absorption, metabolism, distribution of these compounds, and they can help to understand the dual nature (antioxidant/pro-oxidant) of these compounds from different matrices (food, medicine).

Studies on antioxidants should focus on their bioavailability (absorption, metabolism and cellular and tissue distribution), establishing whether the *in vitro* effects are applicable to the situation *in vivo*. Human trials should focus on specific populations with low intake of these substances (Cooper, 2004). It is also necessary to clarify certain aspects of security, for example, antioxidants may act as pro-oxidants under certain circumstances (McKevith et al., 2003). Assuming the possibility that polyphenols exert positive health effects, their bioavailability should be investigated. Also, considering epidemiological studies and trials on humans, it is evident that the health benefits of phytochemicals were observed predominantly when being consumed within their natural food matrices (fruits, vegetables, grain, etc.) (Bouayed & Bohn, 2010). Because the antioxidants are not occurring isolated in food items, other constituents may influence their activity. It should also be noted that their metabolites may have more relevance *in vivo* assays than the molecule itself, in terms of concentration or biological activity (Buttriss et al., 2002).

7. Acknowledgment

We are grateful to the Xunta de Galicia (09CSA030203PR, PGIDIT07PXIB, 10PXIB203303PR) and Ministerio de Sanidad y Consumo (FISPI09/00501) for financial support. M. João Matos thanks Fundação de Ciência e Tecnologia for the fellowship.

The authors thank the creator of the MODESLAB software PhD Ernesto Estrada from the University of Strathclyde, Glasgow, U.K.

8. References

Aherne, S. and O'Brien, N. (2002), "Dietary flavonols: chemistry, food content, and metabolism", *Nutrition Research Reviews,* Vol. 18, pp. 75-81.

Anandh Babu, P. and Liu, D. (2008), "Flavonoids and Cardiovascular Health", in Watson, R. (Ed.), *Complementary and Alternative Therapies and the Aging Population*. Academic Press., pp. 371-392.

Antolovich, M., et al. (2002), "Methods for testing antioxidant activity", *Analyst,* Vol. 127, pp. 183-198.

Aruoma, O. (1999), "Free radicals, antioxidant and international nutrition", *Asia Pacific J Clin Nutr,* Vol. 8 No. 1, pp. 53-63.

Aruoma, O. (2003), "Methodological considerations for characterizing potential antioxidant actions of bioactive components in plant foods", *Mutation Research,* Vol. 523-524, pp. 9-20.

Aruoma, O. and Halliwell, B. (1995), "DNA damage by free radicals: carcinogenic implications", in *Immunopharmacology Free Radical Species*. Academic Press, London, pp. 199-214.

Astley, S. and Lindsay, D. (2002), "European Research on the Functional Effects of Dietary Antioxidants", *Molecular Aspects of Medicine,* Vol. 23 pp. 1-3.

Azam, S., et al. (2004), "Prooxidant property of green tea polyphenols epicatechin and epigallocatechin-3-gallate: implications for anticancer properties.", *Toxicol In Vitro,* Vol. 18, pp. 555-561.

Azmi, A., et al. (2005), "Resveratrol-Cu(II) induced DNA breakage in human peripheral lymphocytes: implications for anticancer properties", *FEBS Lett.,* Vol. 579, pp. 3131-3135.

Bakhtiyor, R., et al. (2005), "A Quantitative Structure-Activity Relationship (QSAR) Study of the Antioxidant Activity of Flavonoids.", *QSAR & Combinatorial Science,* Vol. 24 No. 9, pp. 1056 - 1065.

Banerjee, S., et al. (2008), "Multi-targeted therapy of cancer by genistein", *Cancer Letters,* Vol. 269 No. 2, pp. 226-242

Benassi, M. and Cecchi, H. (1998), "Method development for the simultaneous determination of carboxylic acids, phenolic compounds and sorbic acid in white wines", *J. Liq. Chromatogr. Relat.Technol.* , Vol. 21, pp. 491-501.

Bender, M. A., et al. (1974), "Mechanisms of chromosomal aberration production III. Chemicals and ionising radiation", *Mutat. Res,* Vol. 23, pp. 197-212.

Birt, D. F., et al. (2001), "Dietary agents in cancer prevention: flavonoids and isoflavonoids", *Pharmacol.Ther,* Vol. 90, pp. 157-177.

Block, G., et al. (1992), "Fruit, vegetables, and cancer prevention: a review of the epidemiological evidence", *Nutr Cancer,* Vol. 18, pp. 1-29.

Bohm, H., et al. (1998), "Flavonols, Flavone and anthocyanins as natural antioxidants of foods and their possible role in the prevention of chronic disease", *Ernahrungswiss,* Vol. 2, pp. 147-163.

Bondi, A. (1964), "van der Waals volumes and radii", *J Phys Chem,* Vol. 68 No. 441-51

Bors, W., et al. (1990), "Flavonoids as antioxidants: determination of radical-scavenging efficiency", *Methods in Enzymology,* Vol. 186, pp. 343-355.

Bouayed, J. (2010), "Polyphenols: a potential new strategy for the prevention and treatment of anxiety and depression.", *Curr Nutr Food Sci,* Vol. 6, pp. 13-18.

Bouayed, J. and Bohn, T. (2010), "Exogenous antioxidants—Double-edged swords in cellular redox state.Health beneficial effects at physiologic doses versus deleterious effects at high doses", *Oxidative Medicine and Cellular Longevity* Vol. 3 No. 4, pp. 228-237.

Bouayed, J., et al. (2009), "Oxidative stress and anxiety: Relationship and cellular pathways. ", *Oxid Med Cell Longev* Vol. 2, pp. 63-67.

Bourne, L. and Rice-Evans, C. (1998), "Bioavailablity of ferulic acid ", *Biochem. Biophys. Res. Commun.*, Vol. 253, pp. 222-227.

Bravo, L. (1998), "Poliphenol: chemistry, dietary sources, metabolism and nutritional significance", *Nutr. Rev.*, Vol. 56 No. 11, pp. 317-333.

Buiarelli, F., et al. (2010), " Identification of hydroxycinnamic acid-tartaric acid esters in wine by HPLC-tandem mass spectrometry", *Food Chemistry*, Vol. 123 No. 3, pp. 827-833.

Buttriss, L., et al. (2002), "Antioxidants in food: a summary of the review conducted for the Food Standards Agency.", *Nutr. Bull.*, Vol. 27, pp. 227-236.

Cabrera, A., et al. (2002), "A ToSS MoDe Approach To Determine Physicochemical and Absorption Properties in 6-Fluoroquinolones Derivatives: Experimental Validation", *Eur. J. Pharm. Biopharm.*, Vol. 53, pp. 317-325.

Cooper, A. (2004), "Carotenoids in health and disease: recent scientific evaluations, research recommendations and the consumer.", *J. Nutr.*, Vol. 134, pp. 221S-224S.

Decker, E. (1997), "Phenolics: prooxidants or antioxidants?", *Nutr Rev*, Vol. 55, pp. 396-398.

Deschner, E., et al. (1991), "Quercetin and rutin as inhibitors of azoxymethanol-induced colonic neoplasia", *Carcinogenesis* Vol. 12, pp. 1193-1196.

Devillers, J. and Balaban, A. (1999), *Topological Indices and Related Descriptors QSAR and QSPR*, Gordon and Breach, Amsterdam.

Diplock, A. (1994), "Antioxidants and disease prevention", in Baum, H. (Ed.), *Molecular Aspects of Medicine*. Pergamon Press, Oxford and New York, pp. 295-376.

Diplock, A., et al. (1998), "Functional Food Science and defence against reactive oxidative species", *British Journal of Nutrition*, Vol. 80 No. 1, pp. S77-S112.

Duthie, G., et al. (2000), "Plant polyphenols in cancer and heart disease: implications as nutritional antioxidant", *Nutrition Research Reviews*, Vol. 13, pp. 79-106.

Engelhart, M., et al. (2002), "Dietary intake of antioxidants and risk of Alzheimer disease.", *J. Am. Med. Assoc.,*, Vol. 287, pp. 3223-3229.

Ertl, P., et al. (2000), "Fast calculation of molecular polar surface area as a sum of fragment-based contributions and its application to the prediction of drug transport properties", *J Med Chem*, Vol. 43, pp. 3714-3717.

Espín, J. and Tomás-Barberán, F. (2005), "Constituyentes bioactivos no-nutricionales de alimentos de origen vegetal y su aplicación en alimentos funcionales", in n.d. (Ed.), *Alimentos funcionales*. Fundación Española para la Ciencia y la Tecnologia [FECYT], Madrid, pp. 101-153.

Estrada, E. (1995), "Edge adjacency relationships and a novel topological index related to molecular volume", *J. Chem. Inf. Comput. Sci.*, Vol. 35, pp. 31-33.

Estrada, E. (1996), "Spectral moments of the edge adjacency matrix in molecular graphs. 1. Definition and applications to the prediction of physical properties of alkanes", *J. Chem. Inf. Comput. Sci.*, Vol. 36, pp. 844-849.

Estrada, E. (1998), "Spectral moments of the edge adjacency matrix in molecular graphs. 3. Molecules containing cycles. ", *J Chem Inf Comput Sci*, Vol. 38, pp. 23-27.

Estrada, E. and González, H. (2003), " What Are the Limits of Applicability for Graph Theoretic Descriptors in QSPR/QSAR? Modeling Dipole Moments of Aromatic Compounds with TOPS-MODE Descriptors", *J. Chem. Inf. Comput. Sci,* Vol. 43, pp. 75-84.

Estrada, E. and Molina, E. (2001), "Novel local (fragment-based) topological molecular descriptors for QSPR/QSAR and molecular design", *J Mol Graphics and Model,* Vol. 20 pp. 54-64.

Estrada, E. and Molina, E. (2006), "Automatic extraction of structural alerts for predicting chromosome aberrations of organic compounds", *J Mol Graphics and Model,* Vol. 25 pp. 275-288.

Estrada, E., et al. (2010), "Structural Contributions of Substrates to Their Binding to P-Glycoprotein. A TOPSMODE Approach", *Current Pharmaceutical Design,* Vol. 16, pp. 2676-2709.

Estrada, E. and Uriarte, E. (2001), "Recent advances on the role of topological indices in drug discovery research", *Curr. Med. Chem.,* Vol. 8, pp. 1573-1588.

Estrada, E., et al. (2000), "A novel approach to the rational selection and design of anticancer compounds. ", *J Med Chem* Vol. 43, pp. 1975-1985.

Ferrari, C. and Torres, E. (2003), "Biochemical pharmacology of functional foods and prevention of chronic diseases of aging", *Biomed. Pharmacother.,* Vol. 57, pp. 251-260.

Fukumoto, L. and Mazza, G. (2000), "Assessing Antioxidant and Prooxidant Activities of Phenolic Compounds", *J. Agric. Food Chem,* Vol. 48 No. 8

Fundación Española para la Ciencia y la Tecnología [FECYT]. (2005), "Alimentos funcionales.", in. Fundación Española para la Ciencia y la Tecnología, Madrid.

Galati, G. and O'Brien, P. (2004), "Potential toxicity of flavonoids and other dietary phenolics: significance for their chemopreventive and anticancer properties", *Free Radic Biol Med,* Vol. 37, pp. 287-303.

Galloway, S. (1994), "Chromosome aberrations induced in vitro: mechanisms, delayed expression, and intriguing questions", *Environ. Mol. Mutagen,* Vol. 23, pp. 44-53.

Gasteiger, J. and Marsilli, M. (1978), "A new model for calculating atomic charges in molecules. ", *Tetrahedron Lett,* Vol. 34, pp. 3181-3184.

Geleijnse, J., et al. (2002), "Inverse association of tea and flavonoid intakes with incident myocardial infarction: the Rotterdam study", *Am. J. Clin. Nutr,* Vol. 75, pp. 880-886.

Ghose, A. and Crippen, G. (1987), "Atomic physicochemical parameters for three-dimensional-structure-directed quantitative structure activity relationships. 2. Modeling dispersive and hydrophobic interactions.", *J Chem Inf Comput Sci,* Vol. 27, pp. 21-35.

Groot, H. and Rauen, U. (1998), "Tissue injury by reactive oxygen species and the protective effects of flavonoids", *Fundam Clin Pharmacol* Vol. 3, pp. 249-255.

Hakkinen, S. (2000), "Flavonols and Phenolic Acids in Berries and Berry Products", Kuopio University Publications Finland.

Halliwell, B. (1990), "How characterize a biological antioxidant", *Free Rad. Res.,* Vol. 9, pp. 1-32.

Halliwell, B. and Whiteman, M. (2004), "Measuring reactive species and oxidative damage in vivo and in cell culture: how should you do it and what do the results mean? ", *Br J Pharmacol.,* Vol. 142, pp. 231-255.

Hertog, M., et al. (1993), "Dietary antioxidant flavonoids and risk of coronary heart disease: the Zutphen elderly study", *The Lancet,* Vol. 342, pp. 1007-1011.

Hertog, M., et al. (1995), "Flavonoid intake and long-term risk of coronary heart disease and cancer in the Seven Countries Study", *Arch. Intern. Med.,* Vol. 155, pp. 381-386.

Hertog, M., et al. (1997), "Antioxidant flavonols and ischaemic heart disease in a Welsh population of men. The Caerphilly study", *Am. J. Clin. Nutr.,* Vol. 65, pp. 1489-1494.

INRA and Wishart Research Group. (2009), "Phenol-Explorer. Database on polyphenol content in foods ", available at: http://www.phenol-explorer.eu/ (accessed agust 2011).

Ishidate, M., et al. (1988), "A comparative analysis of data on the clastogenicity of 951 chemical substances tested in mammalian cell cultures", *Mutat. Res,* Vol. 195 pp. 151-213.

James, L., et al. (2003), "Acetaminophen-induced hepatotoxicity", *Drug Metab.Dispos.,* Vol. 31 No. 12, pp. 1499-1506.

Justesen, U., et al. (1998), "Quantilative analysis of flavonols, flavones and flavonones in fruits, vegetables and beverages by high-performance liquid chromatography with photo-diode array and mass spectrometric detection", *J Chromatogr A,* Vol. 799, pp. 101-110.

Keli, S., et al. (1996), "Dietary flavonoids, antioxidant vitamins, an incidence of stroke: the Zutphen study.", *Arch. Intern. Med,* Vol. 156, pp. 637-642.

Knekt, P., et al. (1996), "Flavonoid intake and coronary mortality in Finland: a cohort study", *Br Med J.,* Vol. 321, pp. 478-481.

Knekt, P., et al. (1997), "Dietary flavonoids and the risk of lung cancer and other malignant neoplasms", *Am J Epidemiol,* Vol. 146, pp. 223-230.

Knekt, P., et al. (2002), "Flavonoid intake and risk of chronic diseases", *Am. J. Clin. Nutr.,* Vol. 76, pp. 560-568

Kris-Etherton, P. and Keen, C. (2002), "Evidence that the antioxidant flavonoids in tea and cocoa are beneficial for cardiovascular health", *Curr. Opin. Lipidol,,* Vol. 13, pp. 41-49.

Kubinyi, H. (1993), "QSAR Hansch Analysis and related Approaches", in *Parameters in Methods and Principles in Medicinal Chemistry* Mannhold, pp. 21.

Kuo, S. (1996), "Antiproliferative potency of structurally different dietary flavonoids on human colon cancer cells", *Cancer Lett.,* Vol. 110, pp. 41-48.

Lambert, J. and Elias, R (2010), "The antioxidant and pro-oxidant activities of green tea polyphenols: A role in cancer prevention", *Archives of Biochemistry and Biophysics,* Vol. 501 No. 1, pp. 65-72.

Lempereur, I., et al. (1997), "Genetic and agronomic variation in arabinoxylan and ferulic acid contents of durum wheat (Triticum durum L.) grain and its milling fractions", *J Cereal Sci.* , Vol. 25, pp. 103-110.

León, O., et al. (2005), *Balance antioxidante-prooxidante: salud y enfermedad.*

Lindahl, M. and Tagesson, C. (1997), "Flavonoids as phospholipase A2 inhibitors: importance of their structure for selective inhibition of group II phospholipase A2", *Inflammation,* Vol. 21 No. n.d., pp. 347-356.

Lunte, C., et al. (1988), "Determination of selected phenolic acids in beer extract by liquid chromatography with voltammetric -amperometric detection", *Analyst* Vol. 113, pp. 95-98.

Macheix, J., et al. (1990), *Fruit Phenolics*, Boca Raton, Florida.
Manach, C., et al. (2004), "Polyphenols: food sources and bioavailability", *American Journal of Clinical Nutrition*, Vol. 79, pp. 727-747.
Martínez- Valverde, I., et al. (2000), "Significado nutricional de los compuestos fenólicos de la dieta", in. Campus Universitario de Espinardo, Universidad de Murcia.
Maurya, D. and Devasagayam, T. (2010), "Antioxidant and prooxidant nature of hydroxycinnamic acid derivatives ferulic and caffeic acids", *Food Chem. Toxicol.*, Vol. 48, pp. 3369-3373.
McKevith, B., et al. (2003), "The Food Standards Agency's antioxidants in food programme - a summary.", *J. Hum. Nutr. Dietet.* , Vol. 16, pp. 257-263.
Milne, G. (1997), "Mathematics as a basis for chemistry", *Chem. Inf. Comput. Sci*, Vol. 37, pp. 639-644.
Molina, E., et al. (2004), "Designing Antibacterial Compounds through a Topological Substructural Approach", *J. Chem. Inf. Comput. Sci*, Vol. 44, pp. 515-521.
Morris, M., et al. (2002), "Dietary intake of antioxidant nutrients and the risk of incident Alzheimer disease in a biracial community study.", *J. Am. Med. Assoc.,*, Vol. 287, pp. 3230-3237.
Mozuraityte, R., et al. (2009), "Antioxidant Activity of Phenolic Acids in Lipid Oxidation Catalyzed by Different Prooxidants", *J. Agric. Food Chem.*, Vol. 57 No. 21, pp. 10377-10385.
Nemeikaitė-Čėnienėa, A., et al. (2005), "Quantitative structure-activity relationships in prooxidant cytotoxicity of polyphenols: Role of potential of phenoxyl radical/phenol redox couple", *Archives of Biochemistry and Biophysics*, Vol. 441 No. 2, pp. 182-190
Obermeier, M., et al. (1995), "Effects of bioflavonoids on hepatic P450 activities", *Xenobiotica*, Vol. 25, pp. 575-584.
Omenn, G. (1995), "GAT accounts for the association of vegetables and fruits with lower incidence of cancers and coronary heart diseases", *Ann. Epidemiol.*, Vol. 5, pp. 333-335.
Pasinetti, G. (2007), "Lifestyle factors and Alzheimer's disease: Experimental approaches and clinical implications", *Alzheimer's and Dementia*, Vol. 3 No. 3, pp. S198.
Pérez-Trueba, G. (2003), "Los flavonoides antioxidantes o prooxidantes", *Rev Cubana Invest Biomédicas*, Vol. 22 No. 1
Pérez-Trueba, G. and Martínez, G. (2001), "Los flavonoides como antioxidantes naturales", *Acta Farmacéutica Bonaerense* Vol. 20 No. 4
Pokorny, J., et al. (2001), *Antioxidants in food Practical applications*, Woodhead Publishing Ltd, Cambridge.
Puglia, C. and Powell, S. (1984), "Inhibition of cellular antioxidants: a possible mechanism of toxic cell injury. ", *Environ. Health Perspect.* , Vol. 57, pp. 307-311.
Randic, M. (1991), *J. Math. Chem.*, Vol. 7, pp. 171.
Randic, M. (1998), *Encyclopedia of Computational Chemistry*, John Wiley & Sons., New York:.
Raza, H. and John, A. (2005), "Green tea polyphenol epigallocatechin-3-gallate differentially modulates oxidative stress in PC12 cell compartments", *Toxicol Appl Pharmacol*, Vol. 207, pp. 212-220.
Renaud, S. and Lorgeril, M. (1992), "Wine, alcohol, platelets, and the french paradox for coronary heart disease", *Lancet*, Vol. 339, pp. 1523-1527.

Rimm, E., et al. (1996), "Relation between intake of flavonoids and risk for coronary heart disease in male health professionals", *Ann Intern Med,* Vol. 125, pp. 384-389.

Robards, K., et al. (1999), "Phenolic compounds and their role in oxidative processes in fruits", *Food Chemistry,* Vol. 66, pp. 401-436.

Ross, J. and Kasum, C. (2002), "Dietary flavonoids: Bioavailability, metabolic effects, and safety", *Annu Rev Nutr,* Vol. 22, pp. 19-34.

Sakihama, Y., et al. (2002), "Plant phenolic antioxidant and prooxidant activities: phenolics-induced oxidative damage mediated by metals in plants", *Toxicology,* Vol. 177, pp. 67-80.

Saraji, M. and Mousavi, F. (2010), "Use of hollow fibre-based liquid-liquid-liquid microextraction and high-performance liquid chromatography-diode array detection for the determination of phenolic acids in fruit juices", *Food Chemistry,* Vol. 123 No. 4, pp. 1310-1317

Selassie, C., et al. (2005), "Cellular apoptosis and cytotoxicity of phenolic compounds: a quantitative structure-activity relationship study", *J Med Chem,* Vol. 48 No. 23, pp. 7234-7242.

Serafini, M. (2006), "The role of antioxidants in disease prevention", *Medicine* Vol. 34 No. 12, pp. 533-535.

Shahidi, F. and Nacsk, M. (1995), *Food Phenolics: Sources, Chemistry, Effects, and Application,* Technomic Publishing Company, Lancaster.

Sies, H. (1997), *Antioxidants in Disease Mechanisms and Therapy,* Academic Press, London.

Sies, H. and Jones, D. (2007), "Oxidative Stress", in Fink, G. (Ed.), *Encyclopedia of Stress.* Academic Press.

Simiæ, A., et al. (2007), "Electrochemical Behavior and Antioxidant and Prooxidant Activity of Natural Phenolics", *Molecules,* Vol. 12, pp. 2327-2340.

Speck, A., et al. (2009), "Design of novel antituberculosis compounds using graph-theoretical and substructural approaches", *Mol Divers,* Vol. 13, pp. 445-458.

StatSoft, I. (2001), "STATISTICA (data analysis software system)", in.

Steinmetz, K. and Potter, J. (1996), "Vegetables, fruit, and cancer prevention: a review", *J Am Diet Assoc,* Vol. 96 No. n.d., pp. 1027-1039.

Sudheesh, S., et al. (1999), "Antioxidant activity of flavonoids from Solanum melongena", *Phytother Res,* Vol. 13, pp. 393-396.

Sun, R., et al. (2001), "Quantitative determination of hydroxycinnamic acids in wheat, rice, rye, and barley straws, maize stems, oil palm frond fiber, and fast-growing poplar wood", *J. Agric. Food Chem.,* Vol. 49, pp. 5122-5129.

Tomás-Barberán, F. and Espín, J. (2001), "Phenolic compounds and related enzymes as determinants of quality in fruits and vegetables", *J. Sci. Food Agric.,* Vol. 81, pp. 853-876.

Tomás-Barberán, F., et al. (2000), "Antioxidant phenolic metabolites from fruit and vegetables and changes during postharvest storage and processing", *Studies in Natural Products Chemistry,* Vol. 23, pp. 739-795.

Trinajstic, N., et al. (1986), "On the quantitative structureactivity relationships in drug research", *Acta Pharm. Jugosl.,* Vol. 36, pp. 267-279.

United Stated Departament of Agriculture [USDA]. (2007), "USDA Database for the Flavonoid Content of Selected Foods, Release 2.1 ", available at: http://www.ars.usda.gov/SP2UserFiles/Place/12354500/Data/Flav/flav02-1.mdb (accessed April 2011).

Valko, M., et al. (2007), "Free radicals and antioxidants in normal physiological functions and human disease", *Int J Biochem Cell Biol,* Vol. 39, pp. 44-84.

Van Acquire, S., et al. (1996), "Structural aspects of antioxidant activity of flavonoids", *Free Radic Biol Med,* Vol. 20, pp. 331-342.

Wang, R., et al. (2000), "Calculating partition coefficient by atomadditive method", *Perspect Drug Discovery Des,* Vol. 19, pp. 47-66.

Watjen, W., et al. (2005), "Low concentrations of flavonoids are protective in rat H4IIE cells whereas high concentrations cause DNA damage and apoptosis. ", *J Nutr* Vol. 135, pp. 525-531.

Yordi, E. G. (2010), "Sistema de cribado computacional para la identificación de compuestos prooxidantes, presentes en alimentos, que dañan al ADN: flavonoides", Universidad de La Habana, Cuba.

Yordi, E. G., et al. (2011), "Structural Alerts for Predicting Clastogenic Activity of Pro-oxidant Flavonoid Compounds: Quantitative Structure—Activity Relationship Study ", *J Biomol Screen* DOI: 10.1177/1087057111421623

3

Whole Grain Consumption and Health of the Lower Gastrointestinal Tract: A Focus on Insoluble-Bound Phenolic Compounds

Neal Okarter
Cornell University
USA

1. Introduction

A whole grain is the intact, ground, cracked or flaked caryopsis, whose principal anatomical components - the endosperm, germ and bran - are present in the same relative proportions as they exist in the intact caryopsis (AACC, 1999). The endosperm, the largest component of the caryposis contains starchy carbohydrates, proteins, vitamins and minerals and provides energy for the rest of the plant. The bran, the multi-layered outer skin of the grain, protects the germ and the endosperm from damage from sunlight, pests, water, and diseases. The germ or embryo is the part of the grain that becomes a new plant when fertilized by pollen.

Whole grain food products can be defined by one of two definitions. An intact whole grain food product is a product that has the original composition of bran, germ, and endosperm throughout the entire lifetime of the product, from field to consumption. A reconstituted whole grain food product is a product that has the original components of a whole grain recombined to the relative proportion naturally occurring in the grain kernel. Due to advances in food processing and the commonplace nature in which these processes take place, the bulk of the whole grain food products would be considered reconstituted whole grain products.

Many grains are consumed on a daily basis in a number of products from around the world. Wheat, *Triticum aesitivum*, has become the prominent grain based on total consumption. Corn, *Zea mays L.*, is another commonly eaten grain and is consumed as tortilla, popcorn, or corn cakes. Rice, *Oryza sativa L.*, is the major staple for a majority of the world's population, especially in Asian countries. Rice is rarely eaten as a whole grain. Generally, the endosperm fraction, i.e. the polished rice, without the bran and germ fractions, is eaten. Rice can also be parboiled, incorporating B vitamins into the endosperm of the grain following heating of the whole grain. White rice is not considered a whole grain. Oats, *Avena sativa L.*, are almost always eaten whole since their bran and germ fractions are rarely removed. They also tend to have a sweet taste, making good breakfast cereals and beers. Millet, *Panicum miliaceum L.*, is rarely consumed by humans in North America, but is very common in Asia and also in Africa. Sorghum, *Sorghum bicolor*, is also rarely consumed in North America. Barley, *Hordom vulgare L.*, has a tough hull that is difficult to remove and therefore requires long cooking times. Rye, *Secale cereale L.*, has high fiber content in its endosperm and is consumed with highest frequency in parts of Scandinavia and Russia. Pseudocereals are plants with seeds that can be milled and used much the same way as cereal flours (Brady *et al.*, 2007). These pseudocereals, in addition to rice, millet and maize, do not contain the protein gluten, intolerance to which is known as Celiac disease. Pseudocereals include buckwheat, quinoa, and amaranth. A list of common grains and grain-based food products is provided below (**Table 1**).

Species	**Common Name**	**Common Food Products**
Zea mays	Corn, Maize	Corn cakes, tortilla, popcorn, hominy
Orize sativa	Rice	White rice, brown rice, parboiled rice
Triticum aestivum	Wheat	Breads, flours, pasta, baked goods
Secale cereal	Rye	Breads
Avena sativa	Oats	Oatmeal, flour
Triticum aestivum spelta	Spelt	Breads, baked goods
Hordom vulgare	Barley	Hulled barley
Sorghum bicolor	Sorghum	Couscous, porridge, molasses
Panicum miliaceum	Millet	Porridge, millet
Chenopodium quinoa	Quinoa	Cooked quinoa, pasta,
Amaranthus caudatus	Amaranth	Breads, pasta
Fagopyrum esculentum	Buckwheat	Porridge, pasta, pancakes, breads

Table 1. Common grains and grain-based food products

2. The health benefits of whole grain consumption

Whole grain consumption has been associated with reduced risk of chronic diseases including cardiovascular disease, type 2 diabetes, obesity and some cancers (Okarter & Liu, 2010). The results from select epidemiological studies that investigated the association between increased whole grain consumption and reduced risk of chronic diseases are reported below.

2.1 Cardiovascular disease

Jacobs *et al* reported results from the Iowa Women's Health Study. There was an inverse association between whole grain intake and risk of death from ischemic heart disease (IHD), after adjustment for potentially confounding factors and dietary fiber intake (RR = 0.70) (Jacobs *et al.*, 1998). After adjustment for total dietary fiber intake, Jacobs *et al* found there was still an inverse association between whole grain intake and IHD across all intakes of whole grain, RR = 0.77; 95% CI 0.54 - 1.10 for highest quintile of whole grain intake, highlighting that other factors in addition to dietary fiber may have contributed to the beneficial health effects.

Liu *et al* reported results from the Nurse's Health Study (NHS). There was an inverse association between whole grain intake and risk of coronary heart disease (CHD) (RR = 0.51; 95% CI 0.41 - 0.64), for women in the highest quintile for whole grain consumption when compared to the lowest (Liu *et al.*, 1999).

Lockheart *et al* investigated the association between dietary patterns and risk of first myocardial infarction using the data from a case-control study performed in Norway. After adjusting for family history of heart disease, smoking, energy intake, and other possible confounding factors, consumption of whole grain breakfast cereals was inversely associated with risk of first myocardial infarction (RR = 0.64; 95% CI = 0.45 - 0.90) (Lockheart *et al.*, 2007) when comparing the group with the highest level of whole grain breakfast cereal intake to the group with the lowest.

Flint *et al* investigated the association between whole grain consumption and hypertension in men using data from the Health Professionals Follow-Up Study. After adjusting for fruit and vegetable consumption, smoking, family history of hypertension, physical activity and other possible confounding factors, whole grain consumption was associated with reduced incident of hypertension, when comparing the highest quintile of whole grain consumption to the lowest (RR = 0.81, 95% CI = 0.75-0.87) (Flint *et al.*, 2009).

Wang *et al* investigated the association between whole grain consumption and hypertension using data from the US Health Professional's Follow-Up Study. Whole grain consumption was inversely associated with hypertension when comparing the highest quintile of whole grain intake to the lowest, after adjusting for possible confounding lifestyle, clinical, and dietary factors (RR = 0.89; 95% CI = 0.82 - 0.97) (Wang *et al.*, 2007). Contrarily, no significant association was seen between refined grain consumption and hypertension.

The positive effects of whole grain consumption on cardiovascular parameters are thought to be mediated by improvements in body weight, dyslipidemia, and insulin resistance (Harris & Kris-Etherton, 2010). However, the Dietary Guidelines Advisory Committee 2010 rated the evidence for the protective relationship between whole grains and cardiovascular disease as "moderate" (Harris & Kris-Etherton, 2010). It is important to understand that causality cannot be assigned in observational studies such as the ones mentioned above. It therefore remains unknown whether or not whole grains are protective against cardiovascular disease. Whole grain consumption may be a marker of a healthy lifestyle (Harris & Kris-Etherton, 2010). It is also becoming more difficult to determine the amount of whole grains consumed due to the increasing number of new food products containing varying amounts of whole grain.

2.2 Type 2 diabetes and obesity

Obesity has been linked to the development of type 2 diabetes and cardiovascular diseases. Increased whole grain consumption has been associated with reduced risk of type 2 diabetes and obesity.

Meyer *et al* investigated the association between whole grain consumption and the relative risk of type 2 diabetes. Whole grain consumption was inversely associated with risk of type 2 diabetes when comparing the highest quintile of whole grain intake to the lowest (RR = 0.79; 95% CI = 0.65 - 0.96) (Meyer *et al.*, 2000).

Fung *et al* investigated the association between whole grain consumption and risk of type 2 diabetes, using the data from the US Health Professionals Follow-Up Study. After adjusting for confounding factors including fruit and vegetable consumption, whole grain consumption was inversely associated with risk of type 2 diabetes when the highest quintile of whole grain intake was compared to the lowest (RR = 0.70; 95% CI = 0.57 - 0.85) (Fung *et al.*, 2002).

Montonen *et al* investigated the association between whole grain intake and risk of type II diabetes using data from the Finnish Mobile Clinic Health Examination Survey (Montonen *et al.*, 2003). Whole grain consumption was inversely associated with type 2 diabetes when comparing the highest quartile of whole grain consumption to the lowest, after adjusting for fruit, berry, and vegetable consumption and other confounding factors, (RR = 0.65; 95% CI = 0.36 - 1.18).

Bazzano *et al* investigated the association between consumption of whole breakfast cereals and weight gain in men using data from the Physician's Health Study. After adjusting for baseline BMI, physical activity, age, and other possible confounding factors, whole grain breakfast cereal consumption was inversely associated with risk of having a BMI greater than 25 (RR = 0.83; 95% CI = 0.71 - 0.98) and body weight gain of more than 10 kg (RR = 0.78; 95% CI = 0.64 - 0.96), 8 years after initial subject evaluation (Bazzano *et al.*, 2005).

Newby *et al* investigated the association between whole grain consumption and BMI, weight, and waist circumference using data from the Baltimore Longitudinal Study on Aging. Consumption of whole grain was inversely associated with BMI, weight, and waist circumference when the highest quintile of whole grain consumption was compared to the lowest after adjusting for refined grain intake, total energy intake, and other possible confounding factors (Newby *et al.*, 2007).

Munter *et al* investigated the association between whole grain, bran, and germ intake using data from the first and second trials of the Nurse's Health Study. Whole grain intake was inversely associated with risk of type 2 diabetes in the first trial of the Nurse's Health Study (RR = 0.63; 95% CI = 0.57 - 0.69) and in the second trial of the Nurse's Health Study (RR = 0.68; 95% CI = 0.57 - 0.86) after adjusting for physical activity, total energy intake, and other possible confounding factors (Munter *et al.*, 2007).

The positive association between increased whole grain consumption and reduced risk of type 2 diabetes, obesity, and major weight gain is most likely due to the presence of complex carbohydrate and the slower release of sugars into the blood. For this reason, the inclusion of whole grain has been recommended as a preventative measure from type 2 diabetes.

2.3 Health of the lower gastrointestinal tract

Colorectal cancer is the third most commonly diagnosed cancer in males and the second in females. In 2008, there were an estimated 608,700 deaths due to the disease (Jemal *et al.*, 2011). Incidence rates of colorectal cancer have been increasing in countries in East Asia, Eastern Europe, and Western Europe such as Spain (Center *et al.*, 2009a, Center *et al.*, 2009b). However, incidence rates of colorectal cancer have been declining in the United States, Canada, and Australia (Center *et al.*, 2009a, Center *et al.*, 2009b).

The declining trend in incidence rates of colorectal cancer seen in the United States, Canada, and Australia may be due to a variety of modifiable factors including dietary patterns, smoking, and physical inactivity. Population-based screening in economically developed countries has played a role in reducing incidence rates. The United States is the only country with significantly decreasing incidence rates of colorectal cancer in both males and females in the last few years, possibly as a result of early detection and removal of precancerous lesions through colorectal cancer screening (Center *et al.*, 2009b). Rates of colorectal cancer continue to increase in many countries with limited resources and health infrastructure (Center *et al.*, 2009a).

Data regarding the potential health benefits of whole grain consumption in the lower gastrointestinal tract, specifically with regards to colorectal cancer, vary between studies. This chapter will focus on the potential health benefit of whole grain consumption on the risk of colorectal cancer. Based on frequency of consumption, this chapter will also focus on the potential health benefit of wheat and rice consumption.

3. Proposed mechanisms for whole grain consumption and protection from colorectal cancer

Whole grains contain many phytochemicals; naturally occurring, non-nutrient compounds found in plants. These phytochemicals are generally thought to be responsible for the proposed association between increased whole grain consumption and reduced risk of colorectal cancer. One of the most studied class of whole grain phytochemicals is phenolic compounds. Phenolic compounds contain one or more aromatic rings and one or more hydroxyl groups (**Figure 1**). The predominant phenolic compounds found in whole grains and whole wheat are phenolic acids. Phenolic acids are hydroxybenzoic-acid and hydroxycinnamic-acid derivatives. Phenolic acids are generally found esterified or bound to cell wall polymers and are therefore insoluble when extraction solvents are used to extract phenolic compounds from whole or refined grains (Sosulski *et al.*, 1982). These compounds can be released from the cell wall by alkaline or acidic hydrolysis, or enzymatic activity. Andreasen *et al.* proposed a mechanism by which insoluble-bound phenolic compounds may be released in the lower gastrointestinal tract by gut microflora where they may exhibit potential health benefits (Andreasen *et al.*, 2001). For these reasons, the content of phenolic compounds in the insoluble-bound fraction of whole grains deserves the greatest attention when elucidating a mechanism for the association between whole grain consumption and reduced risk of colon cancer.

a. Flavonoids b. Hydroxycinnamic acids c. Hydroxybenzoic acids

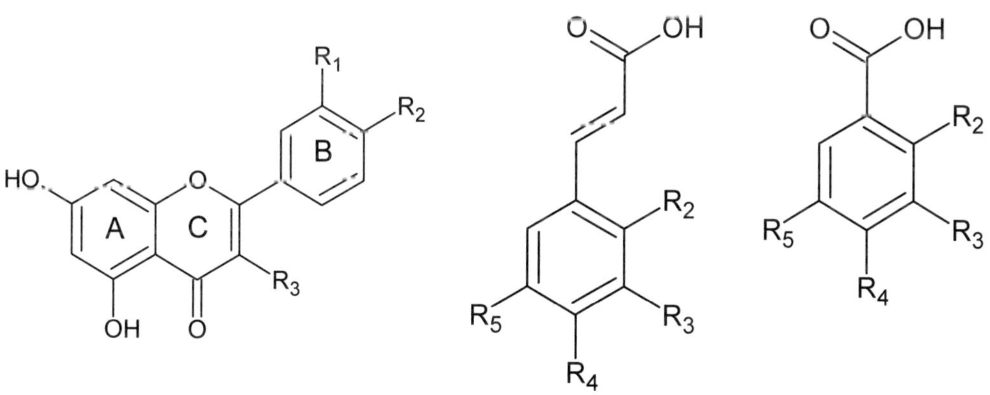

Fig. 1. Structure of phenolic acids and flavonoids

The multistage model for carcinogenesis involves the initiation of a normal cell, the promotion of an initiated cell to a preneoplasia, and finally the progression of the preneoplasia to an invasive tumor. The initiation of a normal cell can be caused by reactive oxygen species. Reactive oxygen species can oxidize biologically important molecules such as DNA.

3.1 Phenolic composition of the insoluble-bound fraction of whole grains

Many studies have determined the total phenolic content of the insoluble-bound fraction of whole grains. This is done using the total phenolic assay, which uses Folin-Ciocalteu Reagent (FCR). Initially, this assay was intended for the analysis of proteins taking advantage of the FCR's activity toward the amino acid tyrosine which has a phenol group. Singleton then used the assay to determine the content of total phenols in wine. The assay has since been used to determine the total phenolic content of fruits, vegetables, and grains. Phenolic compounds react with FCR under basic conditions only. Disassociation of the phenolic proton results in a phenolate anion, which is capable of reducing FCR (Huang *et al.*, 2005). The absorbance of the resulting solution is then measured at 750 nm and compared to the absorbance of various concentrations of a standard such as gallic acid to express the total phenolic content in terms of equivalents. Despite the undefined chemical nature of FCR, the total phenolic assay is convenient and reproducible.

The total phenolic content and phenolic composition of the insoluble-bound fraction of whole grains were determined in several studies. In these studies, the insoluble-bound phenolic compounds were extracted using solvents following alkaline hydrolysis. In one study investigating the total phenolic content and phenolic composition of grains, the total phenolic content of the insoluble-bound fraction of whole grains ranged from 24 (amaranth and buckwheat) to 255 (corn) mg gallic acid equivalents (GAE)/100 g. Ferulic acid and *p*-coumaric acid were found in the insoluble-bound fraction of all grain samples. Caffeic acid was only detected in the insoluble-bound fraction of barley and corn (4.2 and 1.8 μmol/100 g, respectively). Vanillic acid was only detected in the insoluble-bound fraction of quinoa. The insoluble-bound *p*-hydroxybenzoic acid content of amaranth and quinoa was 11.2 and 15.2 μmol/100 g, respectively. No flavonoids (quercetin, kaempferol, catechin, or rutin) or syringic acid were detected in the insoluble-bound fraction of whole grains.

The phenolic content and composition of the insoluble-bound fraction of grains is summarized below (**Table 2**).

The phenolic composition of a wide variety of grains and grain-based products was determined using reversed phase (RP)-high performance liquid chromatography (HPLC) with a diode array detector (DAD), (Mattila *et al.*, 2005). High levels of total phenolic acids were found in wheat bran (4527 mg/kg) and whole wheat flour (1342 mg/kg). The total phenolic acid content of refined wheat flour (167 mg/kg) was much lower than that of whole wheat flour (1342 mg/kg). White and brown rice samples had small amounts of free phenolic acids *p*-coumaric acid and ferulic acid. Parboiled rice contained mainly ferulic acid and *p*-coumaric acid (120 and 38 mg/kg, respectively). Roughly twice the amounts of ferulic acid and *p*-coumaric acid were found in brown rice (240 and 76 mg/kg, respectively).

The total phenolic content and phenolic acid composition of eight Maryland-grown varieties of soft wheat was determined using an RP-HPLC-DAD method (Moore *et al.*, 2005). Total

	Insoluble-Bound Phenolic Compounds					
	TPC†	**FA***	***p*-CA***	***p*-HBA***	**VA***	**CA***
Corn	255 ± 5.6	558 ± 8.8	70.2 ± 2.4	nd	nd	1.8 ± 0.03
Wheat	122 ± 0.9	192 ± 15.7	12.1 ± 0.3	nd	nd	nd
Barley	94.1 ± 4.7	133 ± 12.2	19.0 ± 1.3	nd	nd	4.2 ± 1.6
Oats	79.5 ± 3.9	70.2 ± 7.4	26.4 ± 3.0	nd	nd	nd
Brown Rice	65.8 ± 0.7	88.6 ± 6.5	32.7 ± 0.8	nd	nd	nd
Quinoa	39.4 ± 2.9	35.5 ± 5.1	13.1 ± 1.0	15.2 ± 1.2	19.2 ± 11.2	nd
Amaranth	23.8 ± 0.3	6.5 ± 0.6	6.8 ± 0.2	11.2 ± 0.9	nd	nd
Buckwheat	23.5 ± 0.8	5.3 ± 0.2	6.3 ± 0.4	nd	nd	nd

†Total phenolic content values expressed as mg GAE/100 g.
*Phenolic compound content reported as μmol/100 g.
nd: not detected (detection limit of 160 μg/g of grain).
TPC, total phenolic content; FA, ferulic acid; *p*-CA, *p*-coumaric acid; *p*-HBA, *p*-hydroxybenzoic acid; VA, vanillic acid; CA, caffeic acid.

Table 2. The total phenolic content and phenolic acid content of the insoluble-bound fraction of whole grains.

phenolic content ranged from 40-80 mg GAE/100 g wheat. Vanillic, syringic, *p*-coumaric, and ferulic acids were found in all soft wheat samples. However, no *p*-hydroxybenzoic acid was detected in any of the soft wheat samples. Ferulic acid was the predominant phenolic acid found in all soft wheat samples. Similar to findings from other studies, most of the ferulic acid was found in the insoluble-bound fraction, ranging from 407-588 μg/g of grain, contributing between 89-95% of total ferulic acid content (84-90% of the total identified phenolic acids). The insoluble-bound fraction contributed the greatest proportion of *p*-coumaric acid for all soft wheat samples. Total phenolic content and contents of phenolic acid were not significantly correlated with antioxidant activity, suggesting that phenolic compounds were not predominantly responsible for the observed antioxidant activity. From this study, it was unclear which compounds were responsible for the observed antioxidant activity.

The total phenolic content and phenolic acid composition of the insoluble-bound fraction of six diverse varieties of whole wheat was determined using an RP HPLC-DAD method (Okarter *et al.*, 2010). Ferulic acid was the predominant phenolic acid found in the insoluble bound fractions. The insoluble-bound ferulic acid content ranged from 272 (Caledonia) to 482 (KanQueen) μmol ferulic acid/100 g dry weight (DW). The bound ferulic acid content of KanQueen was significantly different ($p < 0.05$) from all other bound ferulic acid contents. The percentage of ferulic acid found in the insoluble-bound fraction ranged from 87.4 (Caledonia) to 97.2 % (KanQueen). *p*-Coumaric acid was also found in the insoluble-bound fraction of whole wheat. Insoluble-bound *p*-coumaric acid content ranged from 15.9 (Cham1) to 29.0 (KanQueen) μmol /100 g DW. The percentage of *p*-coumaric acid found in the insoluble bound fraction ranged 32.3 (Caledonia) to 63.4% (KanQueen). Syringic acid was found in the insoluble-bound fraction. Insoluble-bound syringic acid content ranged from 3.1 (Caledonia) to 9.8 (KanQueen) μmol /100 g DW. The insoluble-bound syringic acid content of KanQueen was significantly different ($p < 0.05$) from all other bound syringic acid contents. No insoluble-bound syringic acid was detected in the Foster variety. Caffeic acid was only found in the insoluble bound fraction. Caffeic acid contents ranged from 3.2 (Caledonia) to 7.4 (KanQueen) µmol /100 g DW.

The total phenolic content and phenolic composition of the insoluble-bound fraction of two commercial blends of whole wheat and their refined flours were determined using an RP-HPLC-DAD method (Okarter, 2010). The total phenolic content of the insoluble-bound fraction of Magnolia and Barretta whole wheat was 95.8 and 97.5 mg GAE/100 g, respectively. The total phenolic content of the insoluble-bound phenolic fraction of Magnolia and Barretta refined wheat was 12.8 and 13.8 mg GAE/100 g, respectively. The insoluble-bound ferulic acid content of whole wheat ranged from 297-320 µmol/100 g. The insoluble-bound ferulic acid content of refined wheat ranged from 27-37 µmol/100 g. The insoluble-bound *p*-coumaric acid content of whole wheat ranged from 17-18 µmol/100 g. The insoluble-bound *p*-coumaric acid content of refined wheat ranged from 7-8 µmol/100 g. Caffeic acid was detected in the insoluble-bound fraction of whole wheat, but not refined wheat. The caffeic acid content of whole wheat ranged from 6-8 µmol/100 g.

3.2 Antioxidant activity of whole grains

Phenolic compounds are generally regarded as antioxidants. An antioxidant is a compound that can prevent or greatly retard the oxidation of easily oxidizable materials (Chipault, 1962). Antioxidants can transfer a hydrogen atom from a phenolic compound to the reactive oxygen species, preventing the oxidation of biologically important molecules. After transfer of the hydrogen atom to the reactive oxygen species, the antioxidant remains a stable compound by delocalizing the unpaired electron amongst the alternating single and double bonds. This process is known as hydrogen atom transfer. Antioxidants can also transfer an unpaired electron to an oxidant, resulting in the reduction of the oxidant. This process is known as single electron transfer. However, for phenolic compounds, free radical reduction by hydrogen atom transfer is preferred, as this requires lower energy (Leopoldini *et al.*, 2011). Antioxidant activity assays have been developed to determine the antioxidant activity of phenolic compounds using both processes.

3.2.1 Chemical antioxidant activity

Generally, antioxidant activity is assessed using a number of antioxidant activity assays. These assays involve the use of a free radical generator, a probe, and a standard. Though there is a large number of chemical antioxidant activity assays, the most common antioxidant activity assays are described below.

The Oxygen Radical Absorbance Capacity (ORAC) assay assesses the antioxidant activity of hydrophilic antioxidants (Cao *et al.*, 1993). Samples, controls, and a standard (e.g. Trolox, a hydrophilic analogue of vitamin E) are mixed with fluorescein solution and incubated at 37 °C before addition of 2,2′-azobis (2-amidinopropane) dihydrochloride (AAPH) solution to initiate the reaction. The fluorescence intensity (excitation, 485 nm; emission, 525 nm) is measured every minute at ambient conditions (pH 7.4, 37 °C). As the reaction progresses, fluorescent intensity decreases. In the presence of phenolic compounds or phenolic extracts, the decrease in fluorescent intensity is prolonged. The decrease in fluorescent intensity is compared to that of Trolox and the antioxidant activity of the sample is then expressed as Trolox equivalents.

The Total Peroxyl Radical-Trapping Antioxidant Parameter (TRAP) assay uses a fluorescent probe (R-phycoerythrin) and free radical initiator (AAPH) to assess the antioxidant activity of phenolic compounds and phenolics extracts (Wayner *et al.*, 1985). As with the ORAC assay, the decrease in fluorescent intensity is prolonged in the presence of phenolic compounds or phenolic extracts. The decrease in fluorescence is monitored (excitation, 495 nm; emission, 575 nm) over time and compared to that of Trolox. The decrease in fluorescent intensity is compared to that of Trolox and the antioxidant activity of the sample is then expressed as Trolox equivalents.

The 2,2′-Azinobis(3-ehtyl benzothiazoline-6-sulfonic acid) (ABTS) assay uses the ABTS•- radical, which is dissolved in an aqueous potassium persulfate solution, resulting in a dark blue solution (Re *et al.*, 1999). This solution is then diluted with ethanol or buffer (pH 7.4) until a specific absorbance at 734 nm is reached. After addition of a standard, phenolic compounds or phenolic extracts, ABTS•- is converted to $ABTS^{2-}$ by single electron transfer resulting in a colorless solution. The decrease in absorbance is compared to that of Trolox, and the antioxidant activity of the sample is then expressed as Trolox equivalents. In this assay, Trolox or vitamin C can be used as a standard. When Trolox is used, the assay is referred to as the Trolox Equivalent Antioxidant Capacity (TEAC) assay. When vitamin C is used as the standard, the assay is referred to as the Vitamin E C Equivalent Antioxidant Capacity (VCEAC) assay. The concentration of phenolic compound or phenolic extracts giving the same percentage change of absorbance of the ABTS•- as that of 1 mM Trolox is regarded as TEAC.

The Ferric Ion Reducing Antioxidant Power (FRAP) assay assesses the ability of Trolox, phenolic compounds, or phenolic extracts to reduce an Fe (III) salt to an Fe (II) salt (Benzie & Strain, 1996). After addition of Trolox, phenolic compounds or phenolic extracts, the Fe (III) is reduced to an Fe (II) salt. The absorbance of the resulting solution is measured at 593 nm. The resulting absorbance is compared to that after addition of Trolox and the antioxidant activity of the sample is then expressed as Trolox equivalents.

The antioxidant activity of grains was assessed using the TEAC, FRAP, and TRAP assays (Pellegrini *et al.*, 2006). Whole barley had the highest total antioxidant activity when using the TEAC and FRAP assays (obtained by the sum of soluble and bound compounds). Brown rice had the second highest total antioxidant activity when using the FRAP assay and spelt had the second highest total antioxidant activity when using the TEAC assay. However, spelt and brown rice had the highest total antioxidant activity when using the TRAP assay. White rice exhibited the lowest total antioxidant activity when using any of the three antioxidant activity assays. For all the grains analyzed the insoluble-bound fraction contributed to the majority of the total antioxidant activity. In the case of barley, brown rice, and spelt, the insoluble bound fraction contributed 50% of the total antioxidant activity when using the TEAC assay. When using the TRAP assay, the insoluble-bound fraction contributed to all of the antioxidant activity of white rice.

The antioxidant activity of the insoluble-bound fraction of eight whole grains was determined using the ORAC assay (Okarter, 2010). The antioxidant activity of the insoluble-bound fraction ranged from 748 (amaranth) to 10089 (corn) µmol Trolox equivalents/100 g grain. The three pseudocereals had similar antioxidant activities in the insoluble-bound fraction. In this study, the total phenolic content of grains was correlated with antioxidant activity ($R^2 = 0.880$. $p < 0.001$), suggesting that phenolic compounds were responsible for the observed antioxidant activity.

The total antioxidant activity of rice (white, red, and black) extracts was determined using the TEAC assay (Shen *et al.*, 2009). In this study, antioxidant activity was expressed as mM TEAC. Antioxidant activity ranged from 0.01 to 5.53 mM TEAC among the total rice accessions. Among the white rice, antioxidant activity ranged from 0.01 to 0.41 mM TEAC. Among the red rice, antioxidant activity ranged from 0.29 to 2.96 mM TEAC. The antioxidant activity of the black rice samples from this study was approximately three times that of the red rice. These data suggest that the phenolic compounds, which were responsible for the color of the rice, were also responsible for the observed antioxidant activity.

The antioxidant activity of eight Maryland-grown varieties of soft wheat was assessed using the ORAC and TEAC assay (Moore *et al.*, 2005). When using the TEAC assay, antioxidant activities ranged from 14.3 to 17.6 μmol of Trolox equivalents/g of soft wheat grains. The highest antioxidant activity was observed with the SS560 soft wheat line, and the least effective variety was Vigoro Tribute. There was significant variation between the wheat varieties when using the TEAC assay. Extracts from all soft wheat varieties or experimental lines exhibited significant ORAC values. When using the ORAC assay, antioxidant activity ranged from 32.9 μmol Trolox equivalents/g (Vigoro Tribute) to 47.7 μmol Trolox equivalents/g (Choptank).

The antioxidant activity of the insoluble-bound fraction of six diverse varieties of whole wheat was assessed using the ORAC assay (Okarter *et al.*, 2010). The ORAC ranged from 3190 (KanQueen) to 5945 (Roane) μmol Trolox equivalents/100 g DW. Total phenolic content was correlated with ORAC ($R^2 = 0.810$; $p < 0.001$).

The data regarding the antioxidant activity of the insoluble-bound fraction of whole grains varies between studies and are difficult to interpret due to the use of different antioxidant activity assays between studies and the lack of standardization between these antioxidant activity assays, the use of different grain varieties, and the different origin of the grain varieties between studies. This was clearly illustrated in the study that investigated the antioxidant activity of grains using three different antioxidant activity assays (Pellegrini *et al.*, 2006).

Antioxidant activity is correlated with total phenolic content in the vast number of studies investigating the antioxidant activity of grains. However, the correlation of total phenolic content with antioxidant activity may not be a meaningful correlation. Various phenolic compounds have varying antioxidant activities when assessed using the different antioxidant activity assays. For example, ferulic acid and *p*-coumaric acid have similar TEAC values (1.90 and 2.00, respectively). However, caffeic acid has a TEAC value of 1.00 even though its structure is similar to that of ferulic and *p*-coumaric acids. Further, the antioxidant activity of derivatives of hydroxycinnamates can vary significantly (Shahidi & Chandrasekara, 2010). Therefore, determining the phenolic composition of the phenolic extracts is also important.

3.2.2 Cellular antioxidant activity

The relevance of chemical antioxidant activity assays has been questioned due to the use of non-physiological temperature and/or pH, not accounting for bioavailability, uptake, or metabolism (Fardet *et al.*, 2008, Wolfe & Liu, 2007). Wang *et al* (1999) developed a cell-based antioxidant activity assay, in which 2′,7′-dichlorofluorosceine diacetate (DCFH-DA) is used to measure the loss of fluorescence upon quenching of the AAPH-induced reactive oxygen species by the pure compound or sample extract in HepG2 cell cultures (Wang & Joseph, 1999).

In this assay, HepG2 cells are on a 96-well microplate in growth medium/well. After seeding, the growth medium is removed and triplicate wells are treated for 1 h with 100 µL treatment medium containing various concentrations of pure phenolic compound plus DCFH-DA. Following treatment, AAPH is applied to cells in 100 µL Hank's Balanced Salt Solution (HBSS). Emission at 538 nm is measured with excitation at 485 nm every 5 min for 1 hour at 37°C. The concentrations used to determine the cellular antioxidant activity did not impact cellular viability, i.e. did not reduce the number of HepG2 cells by more than 10% compared to the control after 24 hours.

The cellular antioxidant activity of the phenolic compounds found in the insoluble-bound fraction of whole grains employing the DCFH-DA is reported below (**Figure 2**). Ferulic acid, the predominant phenolic acid found in whole grains, does not show any cellular antioxidant activity (Wolfe & Liu, 2007). Of the phenolic acids found in the insoluble-bound fraction of whole grains, only caffeic acid had any cellular antioxidant activity. All other phenolic acids did not show cellular antioxidant activity.

Another study aimed to complete the data regarding the cellular antioxidant activity of phenolic compounds found in the insoluble-bound fraction of whole grains (Okarter, 2010). None of the phenolic acids tested resulted in any cellular antioxidant activity. The cellular antioxidant activity of phenolic compounds is summarized below (**Figure 2**). Caffeic acid

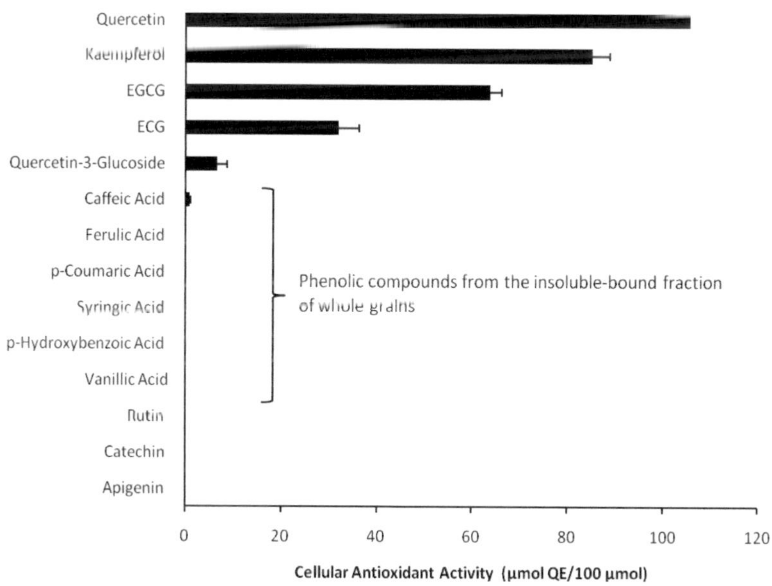

Fig. 2. Cellular antioxidant activity of phenolic compounds found in the insoluble-bound fraction (mean ± standard deviation, n = 3) using the DCFH-DA assay. These data were pooled from several sources (Okarter, 2010, Wolfe & Liu, 2007, Wolfe & Liu, 2008). QE, quercetin equivalents; EGCG, epigallocatechin gallate; ECG, epicatechin gallate

was the only phenolic acid that had any cellular antioxidant activity. This may be due to the structure of caffeic acid. Caffeic acid is the only phenolic acid that has two hydroxyl groups located next to each other on the aromatic ring. Other phenolic acids may not have shown any cellular antioxidant activity due to the fact that these compounds were test at concentrations that did not reduce the number of HepG2 cells by more than 10% compared to the control after 24 hours. At higher concentrations, these compounds may have measureable cellular antioxidant activity. However, the number of HepG2 cells remaining may not be enough to obtain a fluorescence signal. It may also be possible that phenolic compounds are absorbed to varying degrees by the HepG2 cell, further affecting their cellular antioxidant activity. The absorption of phenolic compounds by HepG2 cells was not assessed in any of these studies.

3.3 Antiproliferative activity of phenolic compounds and phenolic extracts from whole grains

Cell counting and cell proliferation assays are used to assess the ability of phenolic compounds or phenolic extracts to prevent or alter the proliferation of cell cultures *in vitro*.

The Trypan Blue Stain Assay is the most common cell counting assay. This assay involves direct cell counting after staining cells with trypan blue staining followed by microscopic quantification using a hemacytometer. This process involves trypsinizing adherent cells, removing cells from culture, centrifugation and re-suspension, staining with trypan blue, and cell counting. Loss of cells due to trypsinization and the resulting degradation of chromatin are sources of underestimated cell numbers.

The Methylene Blue Stain Assay is another common cell counting assay (Oliver *et al.*, 1989). Methylene Blue is a basic dye that is positively charged at pH 5-8. It binds electrostatically to negatively charged groups within cells, predominately phosphate moieties of nucleic acids and some charged groups in proteins. This explains the variation in dye bound by cells from different species, and also by different cell types within the same species. Lowering the pH below 2 with HCl causes acidic groups to be protonated, liberating the Methylene Blue into the elution solvent. The absorbance of the resulting solution is assessed at 570 nm and fit to a standard curve to determine cell number.

The Microculture Tetrazolium Salt (MTS) assay is based on the conversion of a tetrazolium salt into a colored, aqueous soluble formazan product by mitochondrial activity of viable cells at 37°C. The amount of formazan produced by dehydrogenase enzymes is directly proportional to the number of living cells in culture and can be measured at 492 nm.

The antiproliferative activity of phenolic acids and phenolic extracts from brown rice was determined in three different colon cell lines using the MTS assay (Hudson *et al.*, 2000). Caffeic acid (50 μM) inhibited the proliferation of all three colon cell lines (HT29, SW480, and HCEC) used in the study. Methoxycinnamic acid inhibited the proliferation of two of the colon cell lines used in the study. Ferulic acid and sinapic acid inhibited the proliferation of only one cell line (HCEC and SW480, respectively). *p*-Coumaric acid did not have any effect on cell proliferation in any of the three colon cell lines. Phenolic extracts from brown rice inhibited the proliferation of two out of the three colon cell lines used in the study (SW480 and HCEC).

The antiproliferative activity of phenolic acids and phenolic extracts from the insoluble-bound fraction of two commercial blends of whole wheat and their refined flours was determined using the Methylene blue stain assay (Okarter, 2011). *p*-Coumaric acid (50 μM) significantly ($p < 0.05$) reduced the number of Caco-2 cells at 96 hours post-treatment compared to the number of cells when cells were grown under control growth conditions (**Figure 3**). Ferulic acid and caffeic acid significantly ($p < 0.05$) reduced the number of Caco-2 cells at 96 hours post-treatment at a concentration of 500 μM compared to the number of cells when cells were grown under control growth conditions (**Figure 3**). The concentrations of phenolic acids used in this study are achievable through diet (Janicke *et al.*, 2005).

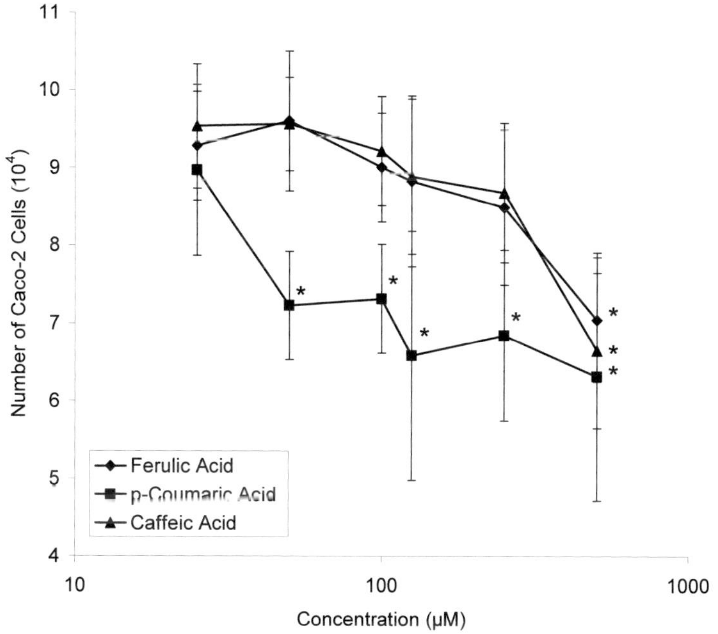

* indicates significant difference from the number of Caco-2 cells when cultured under control growth conditions (no phenolic acids added; $p < 0.05$).

Fig. 3. Number of Caco-2 cells after 96 hours of treatment with phenolic acids (mean ± standard deviation, n = 3)

Phenolic extracts from the insoluble-bound fraction of whole wheat but not refined wheat significantly inhibited the proliferation of Caco-2 cells at 24 and 96 hours post-treatment, compared to cells cultured under control growth conditions (**Table 3**).

In another study, the effect of ferulic and *p*-coumaric acids on Caco-2 cell proliferation was assessed using the Trypan Blue stain assay (Janicke *et al.*, 2005). Ferulic acid and *p*-coumaric acid treatment significantly inhibited the proliferation of Caco-2 cells at 72 hours post treatment at a concentration of 1500 μM.

Results from these studies suggest that the phenolic compounds found in the insoluble-bound fraction of whole grains protect against colon cancer by inhibiting the proliferation of cells.

	24 hours	96 hours
	Cell number (10^4)	Cell number (10^4)
Medium Control	7.3 ± 0.7	9.0 ± 0.8
Solvent Control ‡	7.6 ± 0.6 (104.2%)	9.5 ± 0.9 (104.2%)
Refined Wheat †		
Barretta	7.1 ± 1.3 (96.6%)	8.8 ± 1.1 (97.8%)
Magnolia	7.0 ± 1.6 (96.3%)	8.9 ± 0.9 (99.6%)
Whole Wheat †		
Barretta	4.6 ± 0.2 (59.9%) *	7.0 ± 2.1 (76.6%) *
Magnolia	4.5 ± 1.2 (59.0%) *	7.1 ± 1.4 (78.2%) *

‡ 10% v/v sterile water in growth medium; † Phenolic extracts were delivered at a grain sample concentration of 100 mg/mL

*Within each column indicates a significant difference from the medium control at $p < 0.05$

Table 3. Effects of phenolic extracts from the insoluble-bound fraction of whole and refined wheat on Caco-2 cell number 24 and 96 hours post-treatment. Values are reported as number of Caco-2 cells (mean ± standard deviation, n = 3). Percent proliferation of the medium alone (control) is in parentheses.

4. *In vivo* data regarding whole grain consumption and health of the lower gastrointestinal tract

Given the amount of *in vitro* data, supported by proposed mechanisms for the association between whole grain consumption and cancer development, we do not see these clear associations with colon cancer development *in vivo*. Gene activation or changes in intracellular signaling cascades may partially explain why we do not see these associations *in vivo*.

4.1 Animal studies

One study investigated the bioavailability of ferulic acid in rats fed a ferulic acid-supplemented diet (Adam *et al.*, 2002). The amount of ferulic acid in the feces was negligible regardless of the amount of ferulic acid that was added to the diet suggesting that ferulic acid is absorbed by the gastrointestinal tract or metabolized by gut microflora in the cecum. In plasma, ferulic acid was detected only after enzymatic treatment (5 x 10^6 units/L β-glucoronidase and 2.5 x 10^5 units/L sulfatase), suggesting that ferulic acid is circulated only as conjugates. Ferulic acid was not detected in the plasma 18 h after consuming a meal supplemented with ferulic acid suggesting it is either poorly absorbed or quickly eliminated following absorption via urinary excretion. Fecal excretion of ferulic acid was enhanced following its addition to a high wheat semi-purified diet, suggesting that the absorption of this phenolic acid is decreased when present in a complex matrix such as cereals. These data suggest that the cereal matrix severely limits the bioavailability of ferulic acid in rats, and imply a significant role for gut microflora in ferulic acid bioavailability.

Another study investigated the effects of brown rice, rice bran, and polished rice on preneoplastic lesions of the colon in rats (Li *et al.*, 2011). Consumption of brown rice, rice bran, or polished rice had no effect on plasma or hepatic thiobarbituric acid reactive substances (TBARS), a measure of antioxidant activity. Further, consumption of brown rice,

rice bran, or polished rice did not affect the number of mucin-depleted foci (MDF), preneoplastic lesions of colon cancer. Consumption of rice bran significantly reduced the number of 2-crypt aberrant-crypt foci (ACF), another preneoplastic lesion of colon cancer, compared to rats consuming the control diet ($P<0.05$). Most ACF were observed in the middle colon. Consumption of rice bran (AIN-93G containing 2.2% rice bran) significantly reduced the number of ACF in the middle colon ($P<0.05$). Interestingly, the number of ACF in the distal colon was significantly increased in rats fed a diet containing 17.8% polished rice compared to rats fed a control diet ($P<0.05$). COX-2 protein expression in the middle colon was significantly lower in all rats fed experimental diets compared to rats fed a control diet ($P<0.05$). There were no significant differences in COX-2 protein expression among all treatment groups in the proximal and distal colon.

The data from these animal trials suggest that ferulic acid and perhaps other hydroxycinnamic acids obtained from food have no effect on antioxidant measures and the formation of other markers of colon cancer. This may be because the food matrix affects the bioavailability of these hydroxycinnamic acids, and phenolic acids are metabolized by gut microflora.

4.2 Human studies

Many clinical studies have investigated the effects of whole grain consumption on various health-related outcomes or incidence of colorectal cancer. These studies are summarized below.

One study concluded that absorption of hydroxycinnamic acids present in cereals is limited (Kern *et al.*, 2003a). Six healthy volunteers underwent a 2-day low phenolic compound diet prior to the study day that avoided bran cereals, whole grain products, seeds and nuts, fruits and vegetables, herbs and spices, and other products. The volunteers were allowed to eat foods low in phenolic compounds (white bread, white pasta, white rice, etc). On the study day and immediately before eating the test meal a blood sample (30 mL) was taken as a baseline control. Following blood sampling, volunteers consumed 100 g of a commercial breakfast cereal (85% wheat bran) with skimmed milk. Subsequent blood samples (30 mL each) were taken at various times after the test meal. Low-phenolic compound meals and water were served during the study day. Volunteers collected urine for 24 h on the day prior to the study day and throughout the study day. The researchers found that ferulic acid and sinapic acid were the major hydroxycinnamic acids taken up in humans after the consumption of a high-bran cereal. Further, maximum levels of these compounds reached in plasma were in the nanomolar range, though the content in food was in the micromolar range. The researchers concluded that absorption of the compounds occurred mostly from the small intestine and that the bulk of ester-linked dimeric compounds were excreted in feces or further metabolized by colonic microflora.

Another study reported that there were no effects on antioxidant measures (oxygen radical absorbance capacity in blood, and isoprostane and thiobarbituric acid reactive substances in urine) after 14 days of consumption of whole grain or refined grain food products in healthy subjects (mean age, 27.1 years; mean BMI, 23.9 kg/m^2) (Enright & Slavin, 2010). The study was a randomized, crossover design with two 14-day intervention periods (whole grain or refined grain) with no washout period in between. Subjects were assigned to either a diet containing eight servings for men and six servings for women of whole grain foods in addition to their regular diet or a diet containing eight servings for men and six servings for women of refined grain foods in addition to their regular diet. After 14 days, they were switched to the other group. The results from this study suggest that higher total phenolic content of test meals, higher *in vitro* antioxidant activity, and the presence of more phenolic compounds, as found in whole grains compared to refined grains, do not correlate with increased antioxidant activity *in vivo.*

Epidemiological studies have also investigated the association between whole grain consumption and reduced risk of colorectal cancer. Results from these studies are described below (**Figure 4**), suggesting a slight but often not significant positive health effect on rectal and colorectal cancer upon consumption of whole grains.

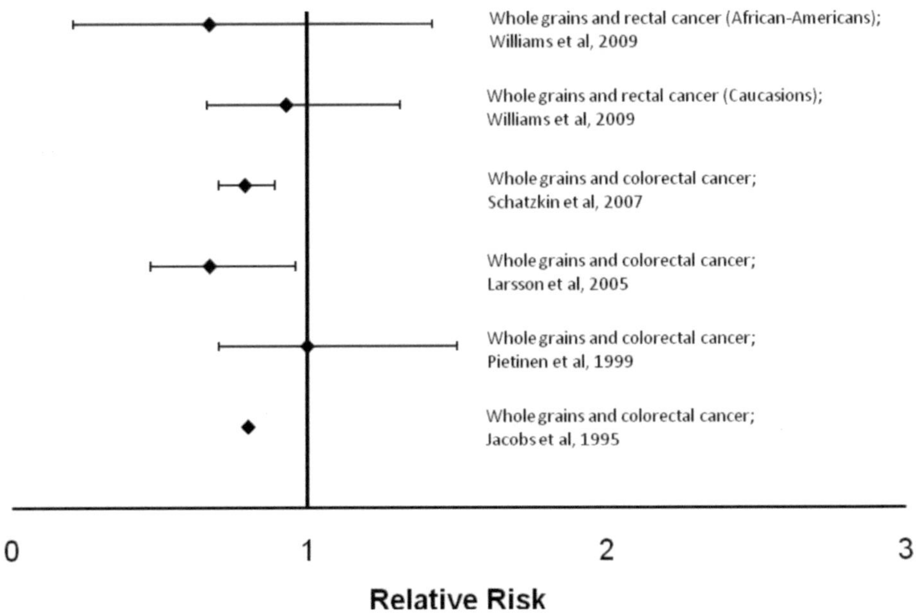

Fig. 4. The association between whole grain consumption and reduced risk of colorectal cancer according to several human studies (relative risk ± 95% confidence interval)

Researchers from Scandinavia investigated the association between cereal consumption and risk of colorectal cancer using data from the Alpha-Tocopherol, Beta-Carotene Lung Cancer Prevention Study (Pietinen *et al.*, 1999). Subjects were male smokers aged 50-69 years who were recruited from the total male population of the same age range in southwestern Finland. Diet was assessed at baseline using a self-administered modified dietary questionnaire. The questionnaire included 276 food items and a portion size picture booklet with photographs of foods, each with 3-5 different portion sizes. Food consumption data were processed using the software and food composition database provided by the National Public Health Institute. After adjusting for confounding factors including years of smoking, BMI, physical activity level and education, there was no significant association between whole grain cereal consumption and risk of colorectal cancer, when comparing the highest quartile of whole grain cereal consumption to the lowest.

One study investigated the association between whole grain consumption and colon cancer using data from the Swedish Mammography Cohort. Dietary information was obtained from a 67-item food-frequency questionnaire administered at baseline. Incidence of colorectal cancer was obtained after a mean of 14.8 years. After adjusting for red meat, fruit, and vegetable consumption and other possible confounding factors, whole grain consumption was inversely associated with risk of colon cancer (RR = 0.67; 95% CI = 0.47 - 0.96) when comparing the highest quintile of whole grain intake (≥4.5 servings/day) to the lowest (<1 serving/day) (Larsson *et al.*, 2005).

Another study investigated the association between whole grain consumption and colorectal cancer using data from the NIH-AARP Diet and Health Study. The diet of nearly 490,000 subjects aged 50-71 years was assessed with a self-administered food-frequency questionnaire at baseline in 1995–1996. During the 5 year follow up, 2974 incident colorectal cancer cases were identified. The researchers found that there was an inverse association between whole grain consumption and risk of colorectal cancer (RR = 0.79; 95% CI = 0.70 - 0.89) when comparing the highest quintile of whole grain intake (1.3 servings/1000 kcal/day) to the lowest (0.2 servings/1000 kcal/day) after multivariate analysis (Schatzkin *et al.*, 2007). This association was stronger for men (RR = 0.79; 95% CI = 0.68 - 0.91) than for women (RR = 0.87; 95% CI = 0.70 - 1.07). The association between whole grain consumption and reduced risk of site specific tumors was strongest for the rectum (RR = 0.64; 95% CI = 0.51 - 0.81).

Researchers from the United States investigated the association between whole grain consumption and risk of rectal cancer in both Whites and African-Americans using data from the North Carolina Cancer Study - Phase II. The dietary intake of 1520 Whites and 384 African-Americans was assessed using the Diet History Questionnaire, which consisted of 124 separate food items and assessed the frequency of consumption and portion size consumed for each food item. Participants estimated their food and beverage intake from the 12 months prior to the study. After adjusting for age, non-steroidal anti-inflammatory drug use, total energy, and other possible confounding factors, whole grain consumption was not associated with reduced risk of rectal cancer in White-Americans (RR = 0.93; 95% CI = 0.66-1.31) or African-Americans (RR = 0.67; 95% CI – 0.71-1.42) when comparing the highest quartile of whole grain intake to the lowest (Williams *et al.*, 2009).

Thus, the findings regarding the potential health benefits of whole grain consumption in the lower gastrointestinal tract are mixed. Some studies show an association between whole grain consumption and reduced risk of colorectal cancer while some studies show no association. The difference in findings may be due to the study design, sample population, type of whole grains consumed from baseline to follow-up, and/or other possible factors.

5. Barriers to potential health benefits in the lower gastrointestinal tract

There are many possible factors that may affect the ability of phenolic compounds from the insoluble-bound fraction of whole grains to provide any health benefits in the lower gastrointestinal tract. These factors are discussed in further detail below.

5.1 Effect of processing on phenolic composition

It is important to recognize that human beings rarely eat whole grains unprocessed. Human beings eat food products made from whole and/or refined grain flours. The various food processing methods that occur prior to consumption of grain-based food products (kneading, baking, puffing, fermenting, extruding, boiling, nixtalmizing, etc.) may alter the total phenolic content and/or phenolic composition of the insoluble-bound fraction of the grain.

One study investigated the effect of processing on phenolic acid content of grain-based food products using an RP-HPLC-DAD method (Mattila *et al.*, 2005). Baking, cooking, or other processing did not destroy phenolic acids. The total phenolic acid content of traditional and precooked oat flakes was identical. The contents of phenolic acids in rye and wheat flours and corresponding products made from these flours (bread and pasta) were found to be similar as well. There were only minor differences when comparing the phenolic acid contents of organic and conventional rye and wheat flours.

Another study investigated the effect of cooking and processing on the phenolic acid profile of bran enriched pastas (Fares *et al.*, 2010). Processing decreased the content of free phenolic acids by nearly 50%. However, the phenolic acid content of the insoluble-bound fraction did not change after processing. The total phenolic acid content decreased after processing, possibly as a result of the oxidative degeneration that occurs due to the addition of water, heat treatment, and kneading during processing. Cooking generally increased the content of phenolic acids found in the insoluble-bound fraction of all samples to varying degrees (36-87%). Pasta had significantly enhanced *in vitro* antioxidant properties after cooking, most likely because cooking releases esterified phenolic compounds from the plant cell wall and leads to the formation of Maillard reaction products, which have measurable antioxidant activity.

The effect of fermentation on the phenolic content and antioxidant activity of cereals (buckwheat, barley, wheat, and rye) was assessed using the FRAP assay (Dordevic *et al.*, 2010). Fermentation led to increase in the total phenol content of all four cereals by up to 39%. Antioxidant activity was increased when samples were fermented with *L. rhamnosus* compared to samples fermented with *S. cerevisiae*. Fermentation did not have any effect on antioxidant activity assessed using the FRAP assay.

Another study investigated the difference in antioxidant activity between pasta made from whole wheat compared to pasta made from refined wheat (Hirawan *et al.*, 2010). The total phenolic content of refined wheat and whole wheat spaghetti were significantly different before and after cooking. Cooking significantly decreased the total phenolic content of refined wheat and whole wheat spaghetti by 22-53%. The average antioxidant activity, assessed using the ORAC assay, was not significantly different between refined wheat spaghetti and whole wheat spaghetti.

5.2 Metabolism of phenolic acids

One study investigated the formation of conjugates and metabolites from hydroxycinnamics (Kern *et al.*, 2003b). The human small intestine epithelium may contribute to the metabolism and bioavailability of hydroxycinnamates. Methyl-cinnamate-sulfate and methyl-cinnamate glucuronide conjugates were the main metabolites formed after metabolism of hydroxycinnamaic acids in Caco-2 cells. These results suggest that sulfation may be the preferred metabolic pathway for hydroxycinnamic acids in the small intestinal epithelium. The glucuronide derivatives of hydroxycinnamic acids may be products of liver metabolism.

The products of microfloral metabolism of dietary phenolic compounds were previously summarized (Aura, 2008). Ferulic acid can be transformed to 3-(3-hydroxyphenyl) proprionic acid (Aura, 2008). Caffeic acid can be transformed either to 4-ethylcatechol or to 3-(3-hydroxyphenyl) propionic acid by human fecal microflora (Peppercorn & Goldman, 1971). Quercetin can be transformed to 2-(3-hydroxyphenyl) acetic acid by a process called ring fission (Aura *et al.*, 2002). Other hydroxycinnamates and flavonoids may be metabolized by the intestinal microflora to similar structures.

5.3 Uptake of phenolic acids from the intestinal lumen

The monocarboxylic acid transporter (MCAT) is an active transporter located on the apical and basolateral sides of the Caco-2 cell. Transport of hydroxycinnamates from the intestinal lumen to the plasma involves the simultaneous transport of hydrogen ions. Studies have shown that the MCAT is involved in the transport of hydroxycinnamates and their metabolites into epithelial cells (Konishi *et al.*, 2003, Konishi & Kobayashi, 2004, Konishi, 2005, Konishi & Kobayashi, 2005).

Other research in this field suggests that the MCAT may not be involved in the uptake of phenolic acids from the lumen of the gastrointestinal tract (Watanabe *et al.*, 2006). The uptake of phenolic acids from the intestinal lumen may occur via the nateglinide/H+ active transport system (Itagaki *et al.*, 2005). One study showed that phenolic acids (caffeic acid, *p*-coumaric acid, and chlorogenic acid) have different affinities for this transporter as demonstrated by the ability to inhibit nateglinide uptake (Saito *et al.*, 2005).

Research data regarding the uptake of phenolic acids from the intestinal lumen are mixed. However, the existing data do show that phenolic acids have varying affinities for the transporters used to transport phenolic acids into the intestinal epithelium. It is important to know and understand the varying affinities of hydroxycinnamate metabolites and conjugates in order to understand the potential of phenolic acids from the insoluble-bound fraction to impart potential health benefits in the lower gastrointestinal tract.

6. Conclusion and future research

Whole grain consumption has been associated with reduced risk of cardiovascular disease, type 2 diabetes, and obesity. However, data regarding the association between increased whole grain consumption and reduced risk of colorectal cancer are mixed. The potential health benefits of whole grain consumption in the lower gastrointestinal tract may be due to the content of phenolic compounds; compounds with one or more aromatic ring and one or more hydroxyl group. The phenolic content and composition of whole grains have been reported on several occasions. Generally, phenolic compounds are found esterified to cell wall polymers, which enables them to survive digestion in the upper gastrointestinal tract, and allows them to impart their potential health benefit in the lower gastrointestinal tract.

Phenolic compounds are believed to impart health benefit, at least in part, due to their antioxidant activity. Generally, antioxidant activity is assessed using a number of *in vitro* chemical antioxidant activity assays. However, these assays do not take into account factors such as bioavailability and metabolism. When *in vitro* cellular antioxidant activity assays are used to assess the antioxidant activity of phenolic compounds found in grains, data show that these compounds do not have any antioxidant activity, except for caffeic acid. These data suggest that any potential health benefit of whole grain phenolic compounds is independent of antioxidant activity.

Phenolic compounds found in whole grains have the ability to inhibit the proliferation of colon cells *in vitro*. Further, phenolic extracts from whole wheat and brown rice inhibit the proliferation of colon cells *in vitro*. Scientific evidence suggests that other pathways such as intracellular signaling, cell cycle arrest, and apoptosis are the major reasons for the observed effect of treatment with phenolic compounds on cell proliferation (Hou *et al.*, 2004, Janicke *et al.*, 2005, Romier *et al.*, 2008).

There are several factors to take into account when considering the potential health benefits of whole grain phenolic compounds in the lower gastrointestinal tract. The processing of whole grains affects the content of phenolic compounds found in the insoluble-bound fraction of whole grains. Further, intestinal microflora metabolize these phenolic compounds, further reducing their contents. Remaining phenolic compounds and their metabolites must then compete for transport into the cell via the MCAT, the nateglinide/H+ transport system, or other similar transport systems.

Further research in the field of whole grain consumption and health of the lower gastrointestinal tract must consider all the factors that affect the content of phenolic compounds including food processing, microfloral metabolism, uptake into the intestinal epithelium, and the effect of these compounds on the proliferation of colon cells. Other phytochemicals and nutrients found in grains (carotenoids, vitamin E, alkylresorcinols, and γ-oryzanols) and dietary fiber and soluble phenolic compounds that survive digestion in the upper gastrointestinal tract also contribute to the health benefits of whole grain consumption. The use of animal models provides valuable insights but must also consider the composition of the diet given to test animals as well as the microfloral content, composition, and activity of the animals' gastrointestinal tract. Epidemiological studies should focus more on and report more detailed descriptions of the grains consumed by subjects. Prospective or intervention studies could answer some aspects on the benefits of grain on the colon.

Whole grains play an important role in reducing the risk of chronic disease. More work is needed to understand the potential health benefits of whole grains in the lower gastrointestinal tract.

7. References

AACC. (1999). AACC Members Agree on Definition of Whole Grain, St. Paul, MN

Adam, A, Crespy, V, Levrat-Verny, MA, Leenhardt, F, Leuillet, M, Demigne, C, & Remesy, C. (2002). The bioavailability of ferulic acid is governed primarily by the food matrix rather than its metabolism in intestine and liver in rats. *Journal of Nutrition*, 132 (7), 1962-1968

Andreasen, MF, Kroon, PA, Williamson, G, & Garcia-Conesa, MT. (2001). Intestinal release and uptake of phenolic antioxidant diferulic acids. *Free Radical Biology & Medicine*, 31 (3), 304-314

Aura, A. (2008). Microbial metabolism of dietary phenolic compounds in the colon. *Phytochemical Reviews*, 7 407-429

Aura, AM, O'Leary, KA, Williamson, G, Ojala, M, Bailey, M, Puupponen-Pimia, R, Nuutila, AM, Oksman-Caldentey, KM, & Poutanen, K. (2002). Quercetin derivatives are deconjugated and converted to hydroxyphenylacetic acids but not methylated by human fecal flora in vitro. *Journal of Agricultural and Food Chemistry*, 50 (6), 1725-1730

Bazzano, LA, Song, YQ, Bubes, V, Good, CK, Manson, JE, & Liu, SM. (2005). Dietary intake of whole and refined grain breakfast cereals and weight gain in men. *Obesity Research*, 13 (11), 1952-1960

Benzie, IF, & Strain, JJ. (1996). The ferric reducing ability of plasma (FRAP) as a measure of "antioxidant power": the FRAP assay. *Anal Biochem*, 239 (1), 70-76

Brady, K, Ho, CT, Rosen, RT, Sang, SM, & Karwe, MV. (2007). Effects of processing on the nutraceutical profile of quinoa. *Food Chemistry*, 100 (3), 1209-1216

Cao, GH, Alessio, HM, & Cutler, RG. (1993). Oxygen-Radical Absorbency Capacity Assay for Antioxidants *Free Radical Biology and Medicine*, 14 (3), 303-311

Center, MM, Jemal, A, Smith, RA, & Ward, E. (2009a). Worldwide variations in colorectal cancer. *CA: A Cancer Journal for Clinicians*, 59 (6), 366-378

Center, MM, Jemal, A, & Ward, E. (2009b). International trends in colorectal cancer incidence rates. *Cancer Epidemiology Biomarkers & Prevention*, 18 (6), 1688-1694

Dordevic, TM, Siler-Marinkovic, SS, & Dimitrijevic-Brankovic, SI. (2010). Effect of fermentation on antioxidant properties of some cereals and pseudo cereals *Food Chemistry*, 119 957-963

Enright, L, & Slavin, J. (2010). No effect of 14 day consumption of whole grain diet compared to refined grain diet on antioxidant measures in healthy, young subjects: a pilot study. *Nutrition Journal*, 9 12-19

Fardet, A, Rock, E, & Remesy, C. (2008). Is the in vitro antioxidant potential of whole-grain cereals and cereal products well reflected in vivo? *Journal of Cereal Science: 48 (2) 258-276*, 48 (2), 258-276

Fares, C, Platini, C, Baiano, A, & Menga, V. (2010). Effect of processing and cooking on phenolic acid profile and antioxidant activity of durum wheat pasta enriched with debranning fractions of wheat *Food Chemistry*, 119 1023-1029

Flint, AJ, Hu, FB, Glynn, RJ, Jensen, MK, Franz, M, Sampson, L, & Rimm, EB. (2009). Whole grains and incident hypertension in men. *American Journal of Clinical Nutrition*, 90 (3), 493-498

Fung, TT, Hu, FB, Pereira, MA, Liu, S, Stampfer, MJ, Colditz, GA, & Willett, WC. (2002). Whole-grain intake and the risk of type 2 diabetes: a prospective study in men. *American Journal of Clinical Nutrition*, 76 (3), 535-540

Harris, KA, & Kris-Etherton, PM. (2010). Effects of whole grains on coronary heart disease risk. *Current Atherosclerosis Reports*, 12 (6), 368-376

Hirawan, R, Ser, WY, Arntfield, SD, & Beta, T. (2010). Antioxidant properties of commercial, regular- and whole-wheat spaghetti *Food Chemistry*, 119 258-264

Hou, YZ, Yang, J, Zhao, GR, & Yuan, YJ. (2004). Ferulic acid inhibits endothelial cell proliferation through NO down-regulating ERK1/2 pathway. *Journal of Cellular Biochemistry*, 93 (6), 1203-1209

Huang, D, Ou, B, & Prior, RL. (2005). The chemistry behind antioxidant capacity assays. *Journal of Agricultural and Food Chemistry*, 53 (6), 1841-1856

Hudson, EA, Dinh, PA, Kokubun, T, Simmonds, MSJ, & Gescher, A. (2000). Characterization of potentially chemopreventive phenols in extracts of brown rice that inhibit the growth of human breast and colon cancer cells. *Cancer Epidemiology Biomarkers & Prevention*, 9 (11), 1163-1170

Itagaki, S, Otsuka, Y, Kubo, S, Okumura, H, Saito, Y, Kobayashi, M, Hirano, T, & Iseki, K. (2005). Intestinal uptake of nateglinide by an intestinal fluorescein transporter. *Biochimica et Biophyssica Acta*, 1668 (2), 190-194

Jacobs, DR, Jr., Meyer, KA, Kushi, LH, & Folsom, AR. (1998). Whole-grain intake may reduce the risk of ischemic heart disease death in postmenopausal women: the Iowa Women's Health Study. *Am J Clin Nutr*, 68 (2), 248-257

Janicke, B, Onning, G, & Oredsson, SM. (2005). Differential effects of ferulic acid and p-coumaric acid on S phase distribution and length of S phase in the human colonic cell line Caco-2. *Journal of Agricultural and Food Chemistry*, 53 (17), 6658-6665

Jemal, A, Bray, F, Center, MM, Ferlay, J, Ward, E, & Forman, D. (2011). Global Cancer Statistics. *CA: A Cancer Journal for Clinicians*, 61 (2), 69-90

Kern, SM, Bennett, RN, Mellon, FA, Kroon, PA, & Garcia-Conesa, MT. (2003a). Absorption of hydroxycinnamates in humans after high-bran cereal consumption. *Journal of Agricultural and Food Chemistry*, 51 (20), 6050-6055

Kern, SM, Bennett, RN, Needs, PW, Mellon, FA, Kroon, PA, & Garcia-Conesa, MT. (2003b). Characterization of metabolites of hydroxycinnamates in the in vitro model of human small intestinal epithelium caco-2 cells. *Journal of Agricultural and Food Chemistry*, 51 (27), 7884-7891

Konishi, Y, Kobayashi, S, & Shimizu, M. (2003). Transepithelial transport of p-coumaric acid and gallic acid in Caco-2 cell monolayers. *Bioscience, Biotechnology, and Biochemistry* 67 (11), 2317-2324

Konishi, Y, & Kobayashi, S. (2004). Microbial metabolites of ingested caffeic acid are absorbed by the monocarboxylic acid transporter (MCT) in intestinal Caco-2 cell monolayers. *Journal of Agricultural and Food Chemistry*, 52 (21), 6418-6424

Konishi, Y. (2005). Transepithelial transport of microbial metabolites of quercetin in intestinal Caco-2 cell monolayers. *Journal of Agricultural and Food Chemistry*, 53 (3), 601-607

Konishi, Y, & Kobayashi, S. (2005). Transepithelial transport of rosmarinic acid in intestinal Caco-2 cell monolayers. *Bioscience, Biotechnology, and Biochemistry*, 69 (3), 583-591

Larsson, SC, Giovannucci, E, Bergkvist, L, & Wolk, A. (2005). Whole grain consumption and risk of colorectal cancer: a population-based cohort of 60 000 women. *British Journal of Cancer*, 92 (9), 1803-1807

Leopoldini, M, Russo, N, & Toscano, M. (2011). The molecular basis of working mechanism of natural polyphenolic antioxidants. *Food Chemistry*, 125 (2), 288-306

Li, SC, Chou, TC, & Shih, CK. (2011). Effects of brown rice, rice bran, and polished rice on colon carcinogenesis in rats *Food Research International* 44 209-216

Liu, S, Stampfer, MJ, Hu, FB, Giovannucci, E, Rimm, E, Manson, JE, Hennekens, CH, & Willett, WC. (1999). Whole-grain consumption and risk of coronary heart disease: results from the Nurses' Health Study. *American Journal of Clinical Nutrition*, 70 (3), 412-419

Lockheart, MS, Steffen, LM, Rebnord, HM, Fimreite, RL, Ringstad, J, Thelle, DS, Pedersen, JI, & Jacobs, DR, Jr. (2007). Dietary patterns, food groups and myocardial infarction: a case-control study. *British Journal of Nutrition*, 98 (2), 380-387

Mattila, P, Pihlava, J-m, & Hellstrom, J. (2005). Contents of phenolic acids, alkyl- and alkenylresorcinols, and avenanthramides in commercial grain products. *Journal of Agricultural and Food Chemistry*, 53 (21), 8290-8295

Meyer, KA, Kushi, LH, Jacobs, DR, Jr., Slavin, J, Sellers, TA, & Folsom, AR. (2000). Carbohydrates, dietary fiber, and incident type 2 diabetes in older women. *American Journal of Clinical Nutrition*, 71 (4), 921-930

Montonen, J, Knekt, P, Jarvinen, R, Aromaa, A, & Reunanen, A. (2003). Whole-grain and fiber intake and the incidence of type 2 diabetes. *American Journal of Clinical Nutrition*, 77 (3), 622-629

Moore, J, Hao, Z, Zhou, K, Luther, M, Costa, J, & Yu, LL. (2005). Carotenoid, tocopherol, phenolic acid, and antioxidant properties of Maryland-grown soft wheat. *Journal of Agricultural and Food Chemistry*, 53 (17), 6649-6657

Munter, JSLd, Hu, FB, Spiegelman, D, Franz, M, & Dam, RMv. (2007). Whole grain, bran, and germ intake and risk of type 2 diabetes: a prospective cohort study and systematic review. *PLoS Medicine*, 4 (8), e261

Newby, PK, Maras, J, Bakun, P, Muller, D, Ferrucci, L, & Tucker, KL. (2007). Intake of whole grains, refined grains, and cereal fiber measured with 7-d diet records and associations with risk factors for chronic disease. *American Journal of Clinical Nutrition*, 86 (6), 1745-1753

Okarter, N. (2010). Whole wheat phytochemicals and potential health benefit. Cornell University Department of Food Science

Okarter, N, Liu, CS, Sorrells, ME, & Liu, RH. (2010). Phytochemical content and antioxidant activity of six diverse varieties of whole wheat. *Journal of Food Chemistry* 119 249-257

Okarter, N, & Liu, RH. (2010). Health benefits of whole grain phytochemicals. *Critical Reviews in Food Science and Nutrition* 50 (3), 193-208

Oliver, MH, Harrison, NK, Bishop, JE, Cole, PJ, & Laurent, GJ. (1989). A rapid and convenient assay for counting cells cultured in microwell plates: application for the assessment of growth factors. *Journal of Cell Science* 92 (Pt 3), 513-518

Okarter, N. (2011) Phenolic extracts from the insoluble-bound fraction of whole wheat inhibit the proliferation of colon cancer cells. Life Sciences and Medicine Research Vol. 2011 LSMR-38

Pellegrini, N, Serafini, M, Salvatore, S, Del Rio, D, Bianchi, M, & Brighenti, F. (2006). Total antioxidant capacity of spices, dried fruits, nuts, pulses, cereals and sweets consumed in Italy assessed by three different in vitro assays. *Molecular Nutrition and Food Research* 50 (11), 1030-1038

Peppercorn, MA, & Goldman, P. (1971). Caffeic acid metabolism by bacteria of the human gastrointestinal tract. *Journal of Bacteriology*, 108 (3), 996-1000

Pietinen, P, Malila, N, Virtanen, M, Hartman, TJ, Tangrea, JA, Albanes, D, & Virtamo, J. (1999). Diet and risk of colorectal cancer in a cohort of Finnish men. *Cancer Causes Control*, 10 (5), 387-396

Re, R, Pellegrini, N, Proteggente, A, Pannala, A, Yang, M, & Rice-Evans, C. (1999). Antioxidant activity applying an improved ABTS radical cation decolorization assay. *Free Radical Biology & Medicine*, 26 (9-10), 1231-1237

Romier, B, Van De Walle, J, During, A, Larondelle, Y, & Schneider, YJ. (2008). Modulation of signalling nuclear factor-kappaB activation pathway by polyphenols in human intestinal Caco-2 cells. *British Journal of Nutrition*, 100 (3), 542-551

Saito, Y, Itagaki, S, Otsuka, Y, Kobayashi, Y, Okumura, H, Kobayashi, M, Hirano, T, & Iseki, K. (2005). Substrate specificity of the nateglinide/H(+) cotransport system for phenolic acids. *Journal of Agricultural and Food Chemistry*, 53 (15), 6100-6104

Schatzkin, A, Mouw, T, Park, Y, Subar, AF, Kipnis, V, Hollenbeck, A, Leitzmann, MF, & Thompson, FE. (2007). Dietary fiber and whole-grain consumption in relation to colorectal cancer in the NIH-AARP Diet and Health Study. *American Journal of Clinical Nutrition*, 85 (5), 1353-1360

Shahidi, F, & Chandrasekara, A. (2010). Hydroxycinnamates and their in vitro and in vivo antioxidant activities *Phytochemical Reviews* 9147-170

Shen, Y, Jin, L, Xiao, P, Lu, Y, & Bao, J. (2009). Total phenolics, flavonoids, antioxidant capacity in rice grain and their relations to grain color, size and weight. *Journal of Cereal Science*, 49 (1), 106-111

Sosulski, F, Krygier, K, & Hogge, L. (1982). Free, Esterified, and Insoluble-Bound Phenolic-Acids .3. Composition of Phenolic-Acids in Cereal and Potato Flours. *Journal of Agricultural and Food Chemistry*, 30 (2), 337-340

Wang, H, & Joseph, JA. (1999). Quantifying cellular oxidative stress by dichlorofluorescein assay using microplate reader. *Free Radical Biology & Medicine*, 27 (5-6), 612-616

Wang, L, Gaziano, JM, Liu, SM, Manson, JE, Buring, JE, & Sesso, HD. (2007). Whole- and refined-grain intakes and the risk of hypertension in women. *American Journal of Clinical Nutrition*, 86 (2), 472-479

Watanabe, H, Yashiro, T, Tohjo, Y, & Konishi, Y. (2006). Non-involvement of the human monocarboxylic acid transporter 1 (MCT1) in the transport of phenolic acid. *Bioscience, Biotechnology, and Biochemistry* 70 (8), 1928-1933

Wayner, DD, Burton, GW, Ingold, KU, & Locke, S. (1985). Quantitative measurement of the total, peroxyl radical-trapping antioxidant capability of human blood plasma by controlled peroxidation. The important contribution made by plasma proteins. *FEBS Letters*, 187 (1), 33-37

Williams, CD, Satia, JA, Adair, LS, Stevens, J, Galanko, J, Keku, TO, & Sandler, RS. (2009). Dietary patterns, food groups, and rectal cancer risk in Whites and African-Americans. *Cancer Epidemiol Biomarkers Prev*, 18 (5), 1552-1561

Wolfe, KL, & Liu, RH. (2007). Cellular Antioxidant Activity (CAA) Assay for assessing antioxidants, foods, and dietary supplements. *Journal of Agricultural and Food Chemistry*, 22 8896-8907

Wolfe, KL, & Liu, RH. (2008). Structure/activity relationships of flavonoids in the Cellular Antioxidant Activity Assay. *Journal of Agricultural and Food Chemistry*, 56 (18), 8404-8411

4

Nutrition and Bone Health in Old Age

Manuel Díaz-Curiel[1], María J. Moro-Álvarez[2] and Susana Sanz-Baena[2]
[1]*Department of Internal Medicine, Metabolic Bone Disease Unit, Fundación Jiménez Díaz, Madrid*
[2]*Department of Internal Medicine, Metabolic Bone Disease Unit, Hospital Central Cruz Roja, Madrid, Spain*

1. Introduction

Osteoporosis is a progressive skeletal disease characterized by loss of bone mass and density. It is the second leading healthcare problem in the world after cardiovascular diseases, according to the World Health Organisation. The prevalence of osteoporosis is increasing due to progressive aging of the population.

Bone mass loss from both the spine and hip, starting at the post-menopause in women and around 60 years of age in men, continues during aging. It is also at old age when the vast majority of fractures due to bone fragility occur and when measures intended to minimize such losses, including the diet and other aspects related to nutritional status, are therefore particularly important.

The role of nutritional deficiency in the production of bone growth changes and promotion of osteoporosis in the elderly is well known today. Multiple factors are involved in bone mass loss in these stages of life, including decreased bone formation, decreased calcium absorption by the intestine, changes in regulation by calciotropic hormones, physical activity, nutritional status, and so on.

While calcium has been shown to have a beneficial effect on bone health at all ages, individual nutritional intake of calcium is below the recommended levels in all countries and at all ages. The best way to achieve an adequate calcium intake is through a balanced diet that should include dairy products.

Vitamin D deficiency is common in many elderly populations and increases fracture risk. Regular exposure to sunlight, taking vitamin D-rich food and vitamin D supplements when required, helps maintain adequate levels of the vitamin.

Bone is a live and active tissue that is continuously being remodeled by the formation of new tissue and elimination of old tissue. This remodeling may be affected by nutritional deficiencies and dietary excesses. This article analyzes how main food items in our diet could influence bone formation and health.

2. Aging and bone mass. The role of nutrition

2.1 Importance of bone mass

Bone mass is clearly the most significant factor related to fracture pathogenesis. Experimental studies have found a high *in vitro* correlation between bone resistance to fracture and bone mineral density (BMD) in both the lumbar spine and femoral neck. Bone mass has been reported to have a different relative importance in fracture incidence depending on the type of bone and bone structure. The contribution of bone mass to femoral neck fracture represented 85% of all factors, while in other bones the relative importance ranged from 67% and 82% (1).

2.2 Changes in bone mass over time

Bone mass at any age is the result of two variables: the amount of bone accumulated during growth, known as peak bone mass, and the subsequent proportion of bone lost.

In humans, bone density increases during the growth period and continues to increase even after growth, when height stays constant, reaching a peak at 25-30 years of age for bone with a mainly trabecular composition (vertebra) and at 35-40 years of age for bones with a mainly cortical composition (femur and radius). It is estimated that 90% of bone mass accumulates up to 20 years of age, and an additional 10% between age 20 and 35 (2). It is obvious that an inadequate accumulation of skeletal mass during youth increases the risk for a fracture incidence at subsequent stages of life (3).

Age is the most important determinant in BMD, and it is known that an overall decrease in the amount of cortical bone, and more specifically of trabecular bone, occurs from a certain age, leading to the so-called age-related physiological osteopenia (4,5). This decrease, measured using densitometric techniques, has been estimated to start from approximately 40 years of age, and cumulative losses at 80-90 years of age are 30% in men and 45%-50% in women. This loss is not linear and is more significant in the early post-menopausal years in women, after which an overall annual decrease by 0.5%-1% occurs. The loss affects trabecular bone and, less markedly, cortical bone (6,7).

In women, the significance of the relationship of bone mass loss to estrogen deficiency is well known (8). It is also known that while in elderly men bone mass has no relationship to androgen levels, a close relationship exists in women between BMD and estrogen levels. It is therefore concluded that estrogens may play a significant role in the maintenance of skeletal mass, at least in the elderly (9,10). Moreover, BMD improvements have been seen in patients, both men and women over 70 years of age, treated with low androgen doses (11). Table 1 summarizes all other factors.

2.3 Pathophysiology of bone mass loss in the elderly

Multiple factors are involved in bone mass loss, including decreased bone formation, decreased calcium absorption by the small intestine, changes in the regulation of calciotropic hormones, physical activity, overall nutritional status, and so on. Each of these factors are discussed in more detail below.

1. Decreased osteoblast function
2. Decreased intestinal calcium absorption
3. Changes in calciotropic hormones
4. Other factors

Table 1. Aging and bone mass loss. Most common causes for bone mass loss.

2.3.1 Decreased osteoblast formation

A decreased bone formation occurs with increasing age, partly related to overall osteoblast aging. However, the fact that these cells are able to respond to adequate stimuli, as occurring in fractures in the elderly, where callus develops, suggests that the mechanism actually subsiding is the lack of action of local factors which are mediators in this action, such as insulin-like growth factor-1 (IGF-1), also called somatomedin C (12).

It is known that with age a decrease occurs in the endogenous function of the growth hormone-releasing hormone (GHRH) that contributes to reduced growth hormone (GH) secretion (13). The formation stage of the remodeling cycle would also be decreased. This stage, which normally lasts 80 days at 40 years of age, is reduced to 60 days in subjects older than 60 years. (14)

2.3.2 Decreased intestinal calcium absorption

Decreased calcium absorption with age may contribute to the development of osteopenia. This has been shown with techniques that measure intestinal calcium absorption using isotopically labeled calcium, and could be related to either a vitamin D deficiency that may lead to osteomalacia when this deficiency is very pronounced or to a decreased conversion of vitamin D into 1,25$(OH)_2$ vitamin- D in the kidney. It is known that overall kidney function is physiologically impaired with age, and renal synthesis of this hormone is therefore decreased. Finally, some authors suggest that intestinal vitamin D receptors decline with age, which causes a relative resistance to calcitriol and a resulting impairment in intestinal calcium absorption (15).

The demonstrated therapeutic effects of vitamin D administration for fracture prevention (16) are in line with this hypothesis. Several studies conducted in both men and women reported that low vitamin D levels in the elderly were related to both estrogen deficiency in women and low androgen levels in men (17). Thus, at advanced ages, calcium and vitamin D supplements given to women with a low calcium intake and vitamin D deficiency had a significant effect on bone mineral density, with increases by 0.5±4.8% ($p=0.02$) in the hip and 2.12±4.06% ($p=0.04$) in the lumbar spine (18,19,20, 21).

2.3.3 Changes in calciotropic hormones

There may also be hormonal factors related to age-related osteopenia. Vitamin D is a steroid prohormone that is obtained either by photosynthesis in the plant, through its conversion in the human skin, or through nutritional intake. As already discussed, a great number of changes in vitamin D metabolism have been reported in the elderly. Exposure to sunlight typically decreases with age because of limitations of elderly people to walk, due to institutionalization, etc. However, a decrease also occurs in the capacity of the skin conversion of the vitamin (22). In addition, a decrease occurs in renal production of calcitriol that is attributed to aging itself. While parathormone (PTH) levels increase with age, PTH is not able to increase the synthesis of 1,25$(OH)_2$vitD (23). Higher PTH levels would cause an increased bone resorption and, thus, osteopenia.

Calcitonin (CT) levels, both basal and in response to calcium infusion, have also been shown to gradually decrease in elderly subjects of both sexes. This is due to either a decrease in the functional capacity of thyroid C cells or a decreased capacity of such cells to respond to natural secretagogues, such as calcium or pentagastrin.(24)

2.3.4 Decreased physical activity

Aging leads to a decrease in physical activity. As this is a significant factor in bone formation, a decreased physical activity would involve a reduced bone formation in elderly subjects (25).

2.3.5 Nutritional changes

There are various nutritional changes in the elderly that may contribute to bone mass reduction. A deficient calcium intake at this age has been shown, due to either a decreased dietary supply or gastrointestinal problems related to aging (26). Wasting diseases, eating alone, taste loss, swallowing problems, unmet special dietary requirements, and even economical factors are factors contributing to the weight loss and malnutrition occurring in elderly people, leading to a decreased calcium intake. Elderly people with chronic diseases with or without disability and those with acute conditions have considerable changes in markers of nutritional status.

The reduced calcium intake, is a risk for increased negative calcium balance which is attempted to be balanced by an increased calcium passage from bone stores to plasma, causing bone demineralization, albeit to some extent, calcium homoeostasis can be achieved by increased tubular re-absorption.

Structured questionnaires to assess nutritional risk are now available, including Mininutritional assessment (MNA) and its short form, MNA-SF (27,28).

3. Special nutritional needs in the elderly: Energy, protein, minerals, vitamins

3.1 Energy

At this stage of life, an adequate, balanced diet that allows elderly subjects for duly carrying out their daily activities and for maintaining a satisfactory state of health is very important. However, studies suggest that nutritional and energy deficits frequently exist in the elderly (29,30).

The limitations of basal energy expenditure and expenditure caused by thermogenesis provide a first guideline for assessing energy intake in elderly people. (31) A decrease in intake by up to 600 calories between 30 and 80 years of age has been reported. Loss of physical activity alone would account for a reduction of virtually 400 calories. In the Baltimore longitudinal study in elderly subjects, requirements decreased from 2,700 cal/day at 30 years to 2,100 cal/day at 80 years.

In the NHANES study, men in the 24-34 year age group had an intake of 2,700 cal/day. From 65 years of age, mean intake was 1,829 cal/day. In women, the same age groups showed a reduction from 1,800 cal/day to 1,259 cal/day (32).

The RDAs are 2,300 cal/day for men weighing 65 kg and 1,900 cal/day for women of the same weight, both over 51 years of age. In both cases, intake is approximately 30 cal/kg body weight/day (33). Energy reduction may involve the loss of essential nutrients. Iron, thiamine, riboflavin, niacin, vit. A, vit. C, and calcium deficiencies are most common. The decreased efficiency by which the elderly absorb essential nutrient some factors such as iron and vitamin C should also be considered.(34)

3.2 Protein

Protein requirements in healthy elderly subjects are highly controversial, not only with respect to the amount of protein, but also the amount of required individual amino acids, especially for the essential amino acids.

The nutritional recommendation is to provide the same amount of proteins as for young adults, approximately 20% of energy intake, but since muscle mass is decreased, with this protein recommendation the amount of protein per kg of lean fraction is higher. This recommended intake is not intended to increase muscle mass, but rather to provide an amount of protein sufficient to avoid a deficient intake.

Protein supply is related to energy supply. The recommended dietary allowance (RDA) of protein is 0.8 g/kg/day. This may be adequate when the diet provides adequate energy. However, when the limits are at 30 cal/kg body weight/day, this RDA may be low for achieving positive nitrogen balances, as in the case of low energy consumption, amino acids could be converted into glucose and keto-bodies for energy delivery, and are therefore lost for more specific functions. It has therefore been suggested that such an amount may be inadequate for maintaining nitrogen balance in healthy, ambulatory elderly subjects because the efficacy with which elderly people use these proteins may be decreased during the aging process. An intake of 1 to 1.25 g/kg/day, adequately meeting the requirements in healthy elderly people, is thus recommended (35). This amount should be even further increased under conditions of disease, stress, or lesion to prevent malnutrition. The heterogeneity of the elderly population should however be considered, and these recommendations should be adapted to each individual and situation. Protein requirements should be increased (up to 1.5 g/kg/day) in people under stress conditions such as infection, fracture, surgery and specially healing of bedsores, and are reduced in renal and/or liver insufficiency.

Protein should represent at least 12% of the diet's calories, and two thirds of the amount provided should be in the form of proteins of animal origin with a high biological value. Higher intakes are not recommended because they do not slow muscle catabolism and may impose an excess overload on the liver and kidney (36). Protein deficiencies in elderly people are not rare. Signs of protein deficiency include fatigue, delayed healing, and decreased physical resistance (30,37).

Proteins of a high biological value, i.e. rich in essential amino acids, should be taken. Milk, cheese, and eggs may therefore be preferred to meat, meat products, fish, etc. because the former are easier to obtain, store, and prepare.

3.3 Lipids

Approximately 30% of total calories should be provided by fat. Absorption of lipid-soluble vitamins is very much reduced without fat. It has not been clearly elucidated to what extent fat intake should be restricted in the elderly, as severe restrictions could affect several of their functions and may also cause nutritional deficiencies. Low fat diets are unpalatable and usually poorly accepted by the population. These diets may enhance development of osteoporosis because of a decreased intestinal absorption and an increased renal loss of calcium and vitamin D.

Recent studies conducted in subjects over 65 years have noted that a high intake of polyunsaturated fat is associated with an increased risk of osteoporotic fracture in the elderly, while a diet rich in monounsaturated fat decreases such risk. An inverse relationship was also seen between HDL cholesterol levels and risk of fracture (38).

3.4 Calcium

Most calcium salts or compounds need hydrochloric acid for the conversion into soluble Ca^{2+} and fractional absorption of calcium from the diet in general diminishes in patients with reduced gastric acid secretion. The importance of gastric secretion for calcium absorption takes on clinical relevance with aging. Hypochlorhydria, achlorhydria and atrophic gastritis all occur as a result of aging(39).

It is admitted that intestinal calcium absorption decreases by 30% to 50% in adult age (40). Approximately 99% of body calcium is in the skeleton; the remaining 1% is distributed among extracellular fluids and cell membranes.

On the other hand, age-related bone mass loss involves high risk of deformity, fracture, and disability. Variations in serum calcium levels (4.5-5.5 mmol/L) are maintained by the interaction of PTH, estrogens, CT, and vitamin D [1,25$(OH)_2$D]. If calcium intake or absorption is inadequate, serum levels may be maintained at the expense of bone mineral. The recommended calcium (RDA ??) intake for men and women is 800 mg/day, and 1,500 mg/day for people over 65 years of age (41). The actual intake tends to be below those recommendations (42)

Milk and its products are the richest sources of calcium. Some people however avoid dairy products, do not take animal products, or do not tolerate lactose. Yogurt may be a good source of calcium for people who do not tolerate lactose because the bacterial cultures used for yogurt production contain some lactase and this fact promotes its digestion. It is known that with lactose, efficacy in calcium absorption increases to 60% of intake; without lactose, absorption capacity decreases to 30%.(43). A way to receive adequate amounts of calcium while maintaining a low fat diet is to take partially or totally skimmed products, such as skimmed yogurt or milk or partially skimmed milk. Partially or completely skimmed milk contains almost the same amount of calcium as whole milk but much less fat.(usually, ca. 1.5%). Milk and dairy products however are not the only sources of calcium. There are foods

VEGETABLES	
• Cabbage	400
• Dry soya	226
• Swiss chard	110
• Pinto beans	106
• Lentils	79
CEREALS	
• Enriched white bread	84
NUTS	
• Hazelnuts	250
• Dry almonds	234
• Walnuts	99
• Peanuts	74
DAIRY PRODUCTS	
• Whole milk powder	909
• Pasteurized whole milk	123
• Skimmed milk	133
• Yogurt	150
Cheese	
• Emmentaler	1.180
• Edamer	900
• Cabrales	700
• Gruyère	700
• Roquefort	700
• Sheep cheese	400
• Cream cheese	300
• Burgos	210
• Camembert	162
• Cottage cheese	100
FISH	
• Sardines in oil (with spines)	354
• Sole	70
MEAT	8 to 12
CONFECTIONERY	
• Molasses	273
• Milk chocolate	228
OTHER: Eggs (without shell)	40

Table 2. Calcium content of some foods (mg/100 g)

with little fat but which are rich in calcium, such as oranges and cabbage. Calcium-enriched food may represent an alternative. When supplemental calcium is given, vitamin D should be administered concurrently to improve absorption. Supplements should preferably be given with meals and separate from treatments indicated for osteoporosis, such as bisphosphonates, because they may cause interference and the desired result is not achieved.

Certain characteristics of the diet may influence effective intestinal absorption of calcium. Protein-rich food contributes to create a slightly acidic urinary pH, which promotes calcium excretion in urine and contributes to bone demineralization. Fat, particularly saturated, decreases calcium absorption and enhances calcium excretion in urine (44). Fat intake should therefore be monitored. Phytic acid in whole meal flours and oxalic acid and uronic acid in vegetables may combine with calcium ions, forming salts and preventing their absorption. However, 50 g/day of fiber should be exceeded for these mechanisms to seriously compromise calcium supply (45).

Due to the above reasons, the National Osteoporosis Foundation (NOF) recommends a calcium intake of approximately 1,200 to 1,500 mg/day in elderly people. Higher intakes do not provide additional benefits and may increase the risk of renal stones or cardiovascular disease (46,47).

3.5 Phosphorus

Phosphorus is essential for the structural integrity of the cell and for metabolic and catabolic reactions; it regulates a great number of enzymes and controls energy storage in the body, as well as energy transformations. Phosphorus plays a significant role in oxygen supply to tissues through 2,3- diphosphoglycerate and ATP levels in red blood cells. It is also part of the buffer systems in urine and plasma, and its presence may possibly be critical in the defense against infection (48). The body of a 70 kg adult contains ca.670 g of phosphorus, and the main sources of the mineral are milk and dairy products. Meat, fish, eggs, and cereals also provide phosphorus. The daily RDA for adults is 800 mg/day.

Excess dietary phosphate, associated with low calcium supply, causes hypocalcaemia, stimulation of PTH secretion, and bone mass loss. A dietary calcium:phosphorus ratio of 1 or higher and a maximum ?? total amount of phosphorus of 800 mg/day are therefore recommended. When this ratio is not met, even if the recommended calcium intake is met, demineralization may occur (49). Carbonated beverages (with gas) may contain high amounts of phosphorus, which alter bone remodeling. (50)

By contrast, except for diseases leading to hypophosphatemia, particularly renal and gastrointestinal and exceptionally congenital diseases, defect in phosphorus supply are rarely found in humans.

3.6 Magnesium

This is an indispensable element required for biochemical processes affecting energy metabolism and neuromuscular transmission. It is also an essential cation in the control of calcium/phosphorus metabolism through the hormonal action of vitamin D, parathormone, and calcitonin (51). Magnesium absorption disorders and deficiencies related to advanced age do not apparently occur. Significant dietary sources of magnesium include nuts, cereals, legumes, bananas, vegetables, and dairy products, however, its intake is well distributed between the various sources.

Over fifty percent of the human body magnesium is found in bone. When magnesium deficiency occurs, the mineral is mobilized from bone, which may have an impact on bone health. As occurring with other minerals, the recommended daily requirements vary with age and physiological needs. The recommended magnesium supply is 350 mg/day in adults and 300 mg/day in women (52).

Plasma magnesium homeostasis is achieved through changes in both intestinal absorption and renal excretion, more concrete renal re-absorption of the mineral. Only in cases of severe deficiency (lack of supply, malabsorption states, renal losses, etc.), magnesium is mobilized at the expense of its bone deposits together with increased magnesium absorption. This occurs through PTH and, to a lesser extent, vitamin D.

3.7 Vitamins

The risk of restrictive diets is an inadequate micronutrient supply. The deficiency takes some time to manifest, but may have serious consequences. Table 3 provides some international recommendations for vitamin dietary intake.

Supply/day	Males	Females
Thiamine	1.2 mg	1.0 mg (0.5 mg/1,000 kcal)
Riboflavin	1.4 mg	1.2 mg
Ascorbic acid	60 mg	60 mg
Vitamin A *	1,000 mg	800 mg
Vitamin D	700-800 IU	700-800 IU
Vitamin K	80 ng	60 ng
Niacin	15 mg	13 mg
Folate	200 mg	180 mg
Vitamin B_6	2.0 mg	2.0 mg
Vitamin B_{12}	2.0 ng	2.0 ng
Vitamina E	10 mg	8 mg
(Tocopherol equivalent:	0.4 mg of Vit. E/1 g	polyunsaturated fatty acids)
Biotin	30-100 ng	30-100 ng

* 1 ng of retinol = 0.3 IU = equivalent to 6 ng of β-carotene in terms of vitamin A activity.

Table 3. RDAs (recommended dietary allowances) of micronutrients potentially related to reduced bone mass density.

It should be considered that some factors, such as smoking, emotional stress, high alcohol intake, drugs, etc. may strongly modify bioavailability and vitamin requirements in the elderly. Some gastrointestinal changes, such as atrophic gastritis, may lead to folate malabsorption due to pH changes in the proximal bowel. Intrinsic factors and vitamin B_{12} absorption are also affected by these changes (53). Homocysteine levels increase with age and are inversely related to levels of folates and vitamin B_{12}, two essential cofactors for remethylation to methionine. There are studies reporting a change in BMD related to vitamin B_{12} levels (54), and it has been suggested that folate supplementation may prevent fractures in elderly women, although it has not been shown to decrease bone metabolism. Lack of vitamin B_{12} has been proposed as an independent risk factor for osteoporosis and fractures (55,56).

Normal conversion of β-carotenes into vitamin A is usually decreased in the elderly, and it is therefore advisable to increase dietary intake of vegetables and fruit. Nevertheless, both low and excess intake of vitamin A increases fracture risk. Independently, beta-carotene is by itself has been hypothesized to be associated with decreased risk of developing osteoporosis. Vitamin C helps in the synthesis and repair of bone collagen, and vitamin K is required for adequate formation of bone structure.

Among all vitamins, a special notion should be given to vitamin D, one of the most important nutrients related to osteoporosis. Its better known functions are to promote calcium and phosphorus absorption, and ensuring adequate bone mineralization. Vitamin D occurs in common foods such as dairy products (many milks are enriched in vitamin D), some fortified cereals, blue fish (including salmon, tuna, sardine, or anchovy), and eggs. Vitamin D may also be synthesized in our skin due to the action of sunlight upon it, although the amount synthesized depends on many other factors, such as the time of day or year or the latitude.

Studies show that vitamin D production is decreased in elderly or institutionalized persons (57-60). It has also been shown that, assuming a similar exposure time, vitamin D synthesis is lower in old age as compared to young adults or children. Skin conversion appears to be less effective in the elderly. Renal synthesis of 1,25-dihydroxyvitamin D is also impaired in this population group due to a decreased response to PTH. Finally, some resistance to the action of 1,25-dihydroxyvitamin D appears to exist at target organs (61). All of this results in a lower calcium absorption and a PTH elevation, which leads to an increased remodelling and mineralization loss.

For all of these reasons, men and women over 70 years of age are less likely to cover their vitamin D requirements with sun exposure. Elderly people need a greater intake of food containing vitamin D or vitamin D supplementation.

Vitamin D supplementation may prevent bone loss in the elderly (62). Moreover, studies show that the use of calcium and vitamin D decreases fracture risk. In a placebo-controlled study on 1634 women receiving 1200 mg/day of calcium and 800 U/day of vitamin D3 for 18 months, Chapuy et al. noted a 25% reduction in hip fractures ($p=0.043$) and a 15% reduction in non-vertebral fractures ($p=0.015$) (63). In another study by Dawson-Hugues et al, supplementation for 3 years with 500 mg of Ca and 700 IU of vitamin D decreased incidence of non-vertebral fractures by 50% ($p=0.02$) (64).

In addition, most studies showing antifracture efficacy with the different antiresorptive and bone-forming agents have included calcium and vitamin D supplementation, and the different antiosteoporotic drugs have been shown to be less effective when administered to patients with inadequate calcium and vitamin D levels. Thus, calcium and vitamin D supplementation should be given as an adjuvant therapy, added to any other anti-osteoporotic treatment (65).

Vitamin D doses currently recommended by the guidelines for elderly subjects are 700-800 UI/day (46,47). The available evidence also alerts about the importance of detecting vitamin D deficiency to be treated as part of a fall prevention program in the elderly (66).

3.8 Flavonoids and others

Bone morphogenetic proteins (BMPs) stimulate bone formation, and the BMP2 gene has been found to be related to osteoporosis. The possibility of positive effects through dietary sources, such as those rich in polyphenols that stimulate the BMP2 promoter and its effects on bone formation is currently being investigated. Flavonoids contained in certain food products may therefore promote bone formation It could thus be possible to improve bone mass by such dietary means and to decrease the risk of osteoporosis at these stages of life (67).

4. Deficiencies and conditions affecting bone in the elderly

4.1 Energy balance and body weight in the elderly

Intake and energy expenditure gradually decrease with age. Changes also occur in body composition, including loss of lean mass and increased fat. These modifications may be considered adjustments with the intention of prolonging life, including changes occurring in middle and old age.

From 30 years of age, the basal metabolic rate (BMR) has been reported to decrease by approximately 5 kcal/day/year, and energy intake by approximately 12 kcal/day/year. Lean body mass decreases by 2%-3% each year, and this decrease is mainly related to muscle mass loss (sarcopenia) (68). Loss of strength, and the resulting increase in propensity to fall, is much faster than the concomitant muscle mass loss. This suggests a difficult to reverse impairment in muscle quality. A mean 15% decrease occurs in total body water between the middle age and old age. Fat is redistributed around the trunk, being mainly deposited in the abdominal region, while subcutaneous and limb fat is decreased (69). Weight usually changes little, but may decrease by 10% between 70 and 80 years of age. It remains to be established whether lean body mass reduction is an adaptive mechanism to decrease activity level or is the consequence of the age-related decrease in physical activity. From the functional viewpoint, lean mass loss is associated to changes in many physiological functions (respiratory, gastrointestinal, immune, etc.). The etiology and purpose of changes in body composition in the elderly remain obscure and highly controversial.

15% to 20% of elderly people experience a weight loss defined as a 5% loss of their usual weight. This proportion reaches 27% in selected high-risk populations (old people with Psychosocial determinants and the higher prevalence of acute and chronic illnesses). In a study conducted in Spain on 450 subjects older than 65 years, 20% showed a weight loss greater than 4% at one year of follow-up (70).

The aging process involves changes in organs and systems that may influence intake decrease and weight loss in the elderly to a greater or lesser extent, thereby contributing to an increased bone demineralization and development of osteoporosis. Some of these changes and their relationship to nutritional status are discussed in detail below.

4.2 Changes in special senses

Taste and smell start to decline at about 60 years of age, with losses being more severe at 70 years (71). Not only aging itself may affect these senses, but also some drugs commonly used in the elderly (digoxin and theophylline), nutritional deficits (zinc, niacin, etc.), associated diseases such as Bell paralysis, multiple sclerosis, Sjögren syndrome, glossitis, etc. Loss or decrease in these senses is important, because when stimulated they induce increases in salivation and gastric and pancreatic secretion.

Decreased hearing, vision, and coordination may result in a decreased intake by causing loss of appetite and food recognition and impair the intake process. Senile cataract is also a significant problem as a cause of vision loss in elderly people. Antioxidant nutrients including vitamin C, carotenoids, and vitamin E and their potential relationship to senile cataract have been studied (72). High carotenoid levels in plasma have been reported to be related to a delayed cataract formation. It is thought that vitamin E could play a significant role in the maintenance of the integrity of cell membranes of the lens of the eye. Vitamin C levels are 30 times higher in the lens as compared to plasma; however, subjects with cataract or elderly people have low vitamin C levels in the lens, particularly in the nucleus, which is where senile cataract starts (73). Some studies found a lower incidence of cataract in people with high vitamin C levels of ca. 80 μmol/L (maximum plasma vitamin C saturation) as compared to those with 40 μmol/L, but very high doses of vitamin C (> 1000 mg/day), which are rather pharmacological than nutritional and then may be harmful, are required to reach those levels (74-76). Further studies are required to ascertain whether, as some studies appear to suggest, dietary changes, i.e. diets rich in antioxidant systems, may prevent occurrence or progression of cataract, which is a significant cause of morbidity and disability, and also involves significant healthcare costs.

4.3 Changes in the oral cavity

Virtually 70% of elderly subjects have xerostomy or a dry mucosa, which clearly affects food intake. Periodontal disease, tooth loss, and presence of dentures, so common in the elderly, impair adequate food salivation and subsequent swallowing (77).

A study conducted by Posner et al. (78) found a close relationship between dental disease and malnutrition in the elderly population. This is an easily treatable cause of malnutrition.

4.4 Gastrointestinal changes

Changes affecting the ability to digest and absorb food occur during normal aging. Hypochlorhydria, that may promote bacterial overgrowth and also impair vitamin B_{12} absorption, causing pernicious anemia and mental changes, is not uncommon from 60 years of age (39).

At about 50-60 years of age, calcium transport mechanism is impaired, and calcium absorption is therefore decreased.(40). Use of excessively fiber-rich diets, which are in fashion today, may contribute to a lower absorption of this mineral, albeit fiber rich foods are usually also rich in dietary minerals, and the effect of fiber to increase mineral absorption in the large intestine is still not completely understood. A lactase deficiency is not uncommon in elderly people.(79). When faced with this situation, rather than removing dairy products from the diet, which would decrease calcium intake, it is essential to recommend the use of dairy products treated with lactase or fermented products (yogurt, junket).

Adult celiac disease is a cause of malnutrition in this population group more common than usually thought, and is therefore a diagnosis to be considered (80). Adult celiac disease may cause few symptoms, but is associated with discomfort and/or pain related to food intake, so that this is voluntarily decreased, which may cause weight loss and micronutrient deficiencies. Gluten-free diets prevent these symptoms.

Constipation is very common in the elderly population. It is related to an inadequate food intake distributed in few meals, to inadequate fluids, and to the presence of depression (81) Use of laxatives in the elderly has been independently associated with the occurrence of hypoalbuminemia.

4.5 Cardiovascular changes

Cardiovascular disease currently accounts for 70% of deaths in people over 75 years of age, who have a greater prevalence of high blood pressure and hypercholesterolemia as compared to young adults. Some studies have shown therapeutic diets in patients with hypercholesterolemia and HBP to cause weight loss, hypoalbuminemia, and orthostatic hypotension (which may in turn cause falls in an already susceptible population) (82,83).

4.6 Renal changes

Kidney function may decrease 50% between 30 and 80 years of age. Elderly subjects have a poorer management of acid-base changes and protein and electrolyte overload. Some authors suggest that geriatric nephropathy could be the result of excess protein consumption throughout life (84). In fact, mean protein consumption in the United States is 166% of the RDAs, and consumption is too high in the whole Western society. Thus, additional intake over the recommended 1-1.25 g/kg intake of proteins per day in elderly subjects may not be encouraged.

4.7 Musculoskeletal changes

As already discussed, fat mass increases and lean body mass is lost during the normal aging process, so that healthy elderly people have a 30%-40% lower proportion of body protein (including muscle mass). The loss involves both the somatic and visceral compartments. Fat is mainly deposited in the trunk. Height loss due to vertebral fractures induced by osteoporosis is common in the elderly. Elderly women tend to restrict calorie intake to a lesser extent than men. Waist-hip ratio tends to be higher in males of a similar age and BMI.

Muscle changes associated with old age may be responsible for the decreased energy expenditure seen in the elderly. These changes are not reversed with dietary measures, but may be reversed with regular exercise. Physical activity helps maintaining integrity of skeletal muscle and bone, and also reduce the risk of falls and fractures. An adequate and balanced diet is indispensable to maintain an adequate muscle function. An inverse relationship has been seen between muscle strength and circulating vitamin D levels, and the presence of vitamin D receptors in muscle and the positive action of vitamin D on muscle metabolism are known (85). Studies have shown that calcium and vitamin D supplementation reduces the risk of falls in the elderly, probably as the result of increased muscle strength and maintenance of neuromuscular coordination. The association of calcium and vitamin D is more effective than vitamin D alone for reducing the risk of falls in women over 70 years of age (86). A group of elderly women were administered calcium 1200 mg/day plus vitamin D 800 IU, and another group received calcium alone for 12 weeks. The calcium and vitamin D group experienced a 49% decrease in the risk of fall as compared to the group on calcium alone, in relation with the improvement of muscle function (87).

4.8 Neurological changes

Neurological changes are very common in the elderly and may have a great impact on their nutritional status. For instance, tremor from Parkinson's disease, motor deficits induced by stroke, Alzheimer's disease, or other type of dementia may seriously impair intake. Experimental use of substances that may act as precursors for altered neurotransmitters in conditions such as Parkinson's or Alzheimer's disease is of great interest, although the role of these nutrients, such as tyrosine, tryptophan, and choline still remains to be fully elucidated.

A potential interesting relationship is the established one between carbohydrate-rich, low protein diets and attention deficit and a decreased alert status in elderly people. This could be due to a decreased serotonin synthesis.(88)

4.9 Immune system changes

Healthy elderly people experience, rather than a decreased immune response, a dysregulation inducing changes in both humoral and cellular immunity (89). Malnutrition is a factor that may induce a decreased immune response. Lymphocyte count in peripheral blood usually decreases by 10%-15% with age (90). Not only the number, but also the type is modified, so that there is a predominance of immature lymphocytes, natural killer (NK) and T helper 2 (TH2) cells, with a decrease in T helper 1 (TH1) cells. In vitro studies show a lower reactivity of lymphocytes from elderly people upon exposure to lectin or antigen stimulation. Thus, peripheral blood lymphocytes from elderly subjects have a lower capacity to respond to different stimuli as compared to those from young people.

These changes are probably responsible for the decreased cell-mediated immunity seen in the elderly. Energy and protein malnutrition also induces changes in cell immunity in the elderly that makes them more susceptible to lung infections (91).

In the elderly, malnutrition also induces changes in the function of polymorphonuclears and monocytes, and may therefore modify clinical symptoms of inflammation. Thus, an infected and malnourished elderly subject may not experience fever due to the poor release of interleukin-1. Lack of cytokine release results in a decreased mobilization of body reserves in the malnourished elderly, which causes an inadequate nutrient supply to lymphocytes, which further impairs the defense mechanisms. Some studies have shown that the intake of dietary supplements is able to reverse the majority of changes seen in cell-mediated immunity (92).

Aging also modifies humoral immunity. An increased antibody response appears to occur in the elderly, as shown by the fact that the levels of some immunoglobulins and monoclonal antibodies are increased. However, elderly people have a lower, shorter, and slower response to primary immunization as compared to young people (93).

In addition, antibodies from elderly people have less affinity for antigens and a reduced action spectrum as compared to young people. Long-term exposure to antigens during the life may be responsible for the impaired immune response seen in old age. Malnutrition further impairs this response, so that the absence of response to immunization in hospitalized elderly individuals has been attributed to malnutrition in 75% of cases (94).

As in humoral immunity studies, use of dietary supplements in malnourished elderly patients improves antibody response (92). They may therefore be indicated in such patients before they are vaccinated. A study where yogurt (95) was used as a supplement to the usual diet of subjects enrolled showed an improvement in immunological indices while supplementation was maintained, but energy intake continued to be low. We may therefore speculate about the role of calcium in such improvement.

4.10 Anorexia in the elderly

Elderly people appear to have an impaired appetite regulation and response to food intake that, among other things, causes a greater difficulty as compared to young adults to recover their initial weight after a weight loss or gain. The elderly tend to satiate earlier than young people after a meal; this early satiety is thought to be the consequence of signals emitted from the stomach. Most studies agree that a slower gastric emptying and a faster passage from the gastric fundus to the antrum occur with age (96). Food therefore remains in the antrum for a longer time, inducing a greater gastric distention.

Cholecystokinin, the major satiating hormone par excellence, may influence decreased intake in the elderly, particularly when malnourished, in whom it is secreted in greater amounts after food intake as compared to young subjects. However, further studies are required to elucidate the role of this hormone in age-related decrease in food intake.

Stimulation of κ opioid receptors by dynorphin plays a significant role in fat intake both in animals and humans. The number of opioid receptors decreases with age (97), which may explain the decreased fat consumption with age, much in the same way as hypodipsia in the elderly has been related to an age-related decrease in the number of μ opioid receptors.

Sex hormones also influence food intake, which is increased by testosterone and decreased by estrogens (98). Age-related decrease in intake is lower in women as compared to men, which may be related to the decrease in estrogen levels occurring in menopause. In men, testosterone levels decrease with age, which may lead not only to a decreased intake, but also to the loss of lean mass and the gain of fat mass predominately in the trunk reported in elderly males.

Leptin is a recently reported hormone, the product of the ob gene and synthesized by adipose tissue (99). Leptin serum levels are therefore related to the amount of body fat. Leptin concentrations tend to decrease with age, which suggests that the leptin signaling system may be impaired with time. This decrease may be the result of a decreased production and/or an increased clearance (100). Thus, leptin may also play a role in age-related anorexia.

4.11 Polymedication and treatment compliance

Elderly patients have a high incidence of comorbidities requiring long-term concomitant use of several drugs. This influences bone health in two aspects, one of them directly derived from the harmful side effect on bone metabolism of some drugs such as corticosteroids, glitazones, heparins, antiepileptics, etc. On the other hand, compliance with treatment against bone mass loss is difficult, with frequent discontinuation of calcium and vitamin D supplements, but also of specific treatments such as bisphosphonates, strontium salts, teriparatide, etc. (101).

4.12 Psychosocial changes

Unfortunately, frequent causes of malnutrition in the elderly population include short financial resources at the end of active working life and the social isolation faced by many elderly individuals in Western society. In elderly patients, depression is a much more frequent cause of malnutrition than in young people. Depression may cause malnutrition due not only to the anorexia usually associated with it, but also because of the loss of the capacity to enjoy social events, which very often occur around a table (102).

Furthermore, elderly people may have significant difficulties to buy food as a consequence of concomitant diseases affecting their mobility and/or strength.

5. Conclusions

Bone mass, one of the main components that determine bone strength, the reduction of which characterizes osteoporosis, experiences a gradual physiological decrease with age, and reach lower levels in the elderly. To this physiological loss should be added the negative effect caused by the decreased plasma levels of gonadal hormones and other factors, including nutritional deficiencies or changes conditioned by functional status and the presence of other concomitant diseases, such as endocrine and rheumatic diseases, the use of some drugs, etc.

Nutrition should be balanced during the elderly period of life, and adequate calcium intake, with or without vitamin D supplements, would attenuate bone mass loss. Other recommendations to be considered include avoidance of a high protein content in the diet, due to its high phosphate content; avoidance of strict vegetarian diets, especially vegan diets, because of their high phytate and oxalate contents, and also due to their relatively low content of some vitamins.

6. References

[1] Turner CH: Biomechanics of bone: determinants of skeletal fragility and bone quality. Osteopor Int. 2002; 13:97-104

[2] Cooper C, Cawley M, Bhalla A et al. Childhood growth, physical activity, and peak bone mass in women. J Bone Mineral Res. 1995; 10: 940-947

[3] Johnston CC Jr, Slemenda CW: Peak bone mass, bone loss and risk of fracture. Osteoporos Int. 1994; 4 suppl 1: 43-5.

[4] Díaz Curiel M. Fisiopatología de la osteoporosis en el anciano. Geriatrika. 1992; 8:61-65.

[5] Khosla S, Riggs BL. Pathophysiology of age-related bone loss and osteoporosis. Endocrinol Metab Clin N Am 2005; 34: 1015-1030.

[6] Hannan MT, Felson DT, Dawson-Hughes B et al. Risk factors for longitudinal bone loss in elderly men and women: the Framingham Osteoporosis Study. J Bone Miner Res 2000; 15: 710-720

[7] Chapurlat R. Epidemiology of osteoporosis. J Soc Biol. 2008;202:251-5

[8] Heaney RP, Recker RR, saville PD. Menopausal changes in bone remodeling. J Lab Clin Med 1978; 92: 864-970.

[9] Slemenda CW, Longcope C, Zhou L, Hui SL, Peacock M, Johnston CC. Sex steroids an bone mass in older men. Positive association with serum estrogens and negative associations with androgens. J Clin Invest. 1997; 100:1755-9.
[10] Murphy S, Khaw K, Cassidy A and Compston JE. Sex hormones and bone mineral density in elderly men. Bone and Mineral 1993; 20:133-140.
[11] Jankowski CM, Gozansky WS, Schwartz RS, Dahl DJ, Kittelson JM, Scott SM, Van Pelt RE, Kohrt WM. Effects of dehydroepiandrosterone replacement therapy on bone mineral density in older adults: a randomized, controlled trial.J Clin Endocrinol Metab. 2006;91:2986-93
[12] Paccou J, Dewailly J, Cortet B. Reduced levels of serum Reduced levels of serum IGF-1 is related to the presence of osteoporotic fractures in male idiopathic osteoporosis. Joint Bone Spine. 2011 Jul 29. [Epub ahead of print]
[13] Degli Uberti EC, Ambrosio MR, Cella SG, Margutti AR, Trasforini G, Rigamonti AE, Petrone E, Müler EF. Defective hypothalemic growth hormone (GH)-relasing hormone activity may contribute to declining GH secrection with age in man.J Clin Endocrinol Metab. 1997; 82:2885-8.
[14] Brixen K, Kassem M, Eriksen EF, Nielsen HK, Flyvbjerg A, Mosekilde L. Growth hormone (GH) and adult bone remodeling: the potential use of GH in treatment of osteoporosis. J Pediatr Endocrinol. 1993 Jan-Mar;6(1):65-71.
[15] Feldman D, Malloy PJ, Gross C. Vitamin D: Metabolism and action. En: Osteoporosis. Marcus R, Feldman D, Kelsey J, ed. San Diego, Academia Press 1996; 205-235.
[16] Chapuy MC, Arlot M, Delmas PD, Meunier PJ. Effect of calcium and cholecalciferol treatment for three years on hip fractures in elderly women, Br Med J. 1994; 308:1081-1082.
[17] Boonen S, Vanderschueren D, Cheng XG, Verdeke G, Dequeker J, Ceusens P, Broos P, Bouillon R. Age-related (type II) femoral neck osteoporosis in men: biochemical evidence for both hypovitaminosis D and androgen deficiency induced bone resorption. J Bone Miner Res. 1997; 12:2119-2126.
[18] Dawson Hughes B, Dalla GE, Krall EA, et al. A controlled trial of the effect of calcium on bone density in postmenopausal women. N Engl J Med 1990;323:878-83
[19] Dawson Hughes B, Harris SS, Krall EA, Dalla GE. Effect of calcium and vitamin D supplementation on bone density in men and women 65 years of age or older. N Engl J Med 1997;337:670-6
[20] Hollis BW. Circulating 25-hydroxyvitamin D levels indicative of vitamin D sufficiency: implications for establishing a new effective dietary intake recommendation for vitamin D. J Nutr 2005; 135:317.
[21] http://www.iom.edu/Reports/2010/Dietary-Reference-Intakes-for-Calcium-and-Vitamin-D/Report-Brief.aspx (Accessed on December 01, 2010)
[22] MacLaughlin J, Holick MF. Aging decreases the capacity of human skin to produce vitamin D3. J Clin Invest 1985; 76:1536.
[23] Quesada JM. "Metabolismo de la vitamina D en ancianos". Rev. Esp. Geriatr. y Gerontol. 1988; 23,2: 71-79
[24] Castillo Suárez, M; Sosa Henríquez, M . Modificación de las hormonas calciotropas y los marcadores bioquímicos de remodelamiento óseo, en función de la edad y el sexo, en una población anciana institucionalizada. Rev Esp Geriatr Gerontol. 1998;33:349-56

[25] Bouxsein ML, Marcus R. Overview of exercise and bone mass.Rheum Dis Clin North Am 1994; 20: 787-99
[26] 26) Bhutto A, Morley JE. The clinical significance of gastrointestinal changes with aging. Curr Opin Clin Nutr Metab Care. 2008 Sep;11(5):651-60.
[27] Guizoz Y. The Mini Nutricional Assesment (MNA)). Review of the literatura-what does it tell us?. J Nutr Health Aging. 2006; 10: 466-487
[28] Rubenstein LZ, HarkerJO, Salva A, Guipoz Y, Vellas B. Screening for undernutrition in geriatric practice: developing the short-form mini-nutricional assesment (MNS-SF). J Gerontol A Biol Sci Med Sci 2001; 56: 66-72
[29] Grieger JA, Nowson CA: Nutrient intake and plate waste from an Australian residential care facility. Eur J Clin Nutr. 2007;61:655-63
[30] Dumartheray EW, Krieg MA, Cornuz J, Whittamore DR, Lanham-New SA, Burckhardt P. Energy and nutrient intake of Swiss women aged 75-87 years. J Hum Nutr Diet. 2006;19:431-5
[31] Morley JE. Decreased food intake with aging. J Gerontol A Biol Sci Med Sci. 2001; 2: 81-8
[32] Looker AC, Harris TB, Madans JH, Sempos CT. Dietary calcium and hip fractures risks: The NHANES I Epidemiology follow-up study. Osteoporosis Int. 1993; 3:177-184.
[33] Asuman L.M. Nutrition in the elderly. En: Shils M.E. de.Modern Nutrition in health and disease. (8ª edic.). London: Lea Fegifer. 1994; 770-780..
[34] Marshall TA, Stumbo PJ, Warren JJ, Xie XJ. Inadequate nutrient intakes are common and are associated with low diet variety in rural, community-dwellingelderly. J Nutr. 2001 Aug;131(8):2192-6.
[35] Walrand S, Boirie Y. Optimizing protein intake in aging. Curr Opin Clin Nutr Metab Care. 2005 Jan;8(1):89-94.
[36] Munrohn H.N. Protein nutriture of a group of free-living elderly. Am. J. Clin. Nutr. 1987; (5):215.
[37] Cashman KD. Diet, nutrition, and bone health. J Nutr. 2007,137(11 Suppl):2507S-2512S
[38] Martínez-Ramírez MJ, Palma S, Martínez-González MA, Delgado-Martínez AD, de la Fuente C, Delgado-Rodríguez M. Dietary fat intake and the risk of osteoporotic fractures in the elderly. Eur J Clin Nutr. 2007;61:1114-20
[39] Hurwitz A, Brady DA, Schaal SE, et al. Gastric acidity in older adults. JAMA 1997, 278 (8): 659-62
[40] Tucker KL. Dietary intake and bone status with aging. Curr Pharm Des. 2003, 9(32):2687-704
[41] Rodríguez-Rodríguez E, Navia Lombán B, López-Sobaler AM, Ortega Anta RM; Grupo de investigaci6n: 920030. Review and future perspectives on recommended calcium intake. Nutr Hosp. 2010 May-Jun;25(3):366-74.
[42] Cashman KD. Calcium intake, calcium bioavailability and bone health. Br J Nutr. 2002 May;87 Suppl 2:S169-77.
[43] Armbrecht HJ. Effect of age and the milk sugar lactose on calcium absorptionby the small intestine. Adv Exp Med Biol. 1989;249:185-92.
[44] Griessen M, Ammann P, Selz L y cols. Comparison of the effect of medium-chain triacylglycerols on calcium absorption in health subjects. *Am J Clin Nutr* 1999; 69:1237-1242.

[45] Drueke TB. Mechanisms of action of calcium absorption factors that influence bioavailability. Nutr Factors Hypert 1990;12:155-73.
[46] Standing Comité on the Scientific Evaluation of Dietary reference Intakes. Dietary reference intakes of calcium, phosphorus, magnesium, vitamin D and fluoride. Washington D.C.: Nacional Academy Press;1997
[47] Nacional Osteoporosis Foundation. The clinician's guide to prevention and treatment of Osteoporosis. 2008
[48] Calvo MS: The effect of high phosphorus intake on calcium homeostasis. Adv Nutr Res. 1994; 9:183-207
[49] Calvo MS: Dietary phosphorus calcium metabolism and bone J Nutrition. 1993; 123: 1627-1633.
[50] McGartland C, Robson PJ, Murray L, Cran G, Savage MJ, Watkins D, Rooney M, Boreham C. Carbonated soft drink consumption and bone mineral density in adolescence: the Northern Ireland Young Hearts project. J Bone Miner Res. 2003 Sep;18(9):1563-9.
[51] Rude RK: Magnesium metabolism and deficiency. Endocrinol Metabol Clin N America. 1993; 22:377-395
[52] Vaquero MP. Magnesium and trace elements in the elderly: intake, status and recommendations. J Nutr Health Aging. 2002;6(2):147-53
[53] Andrès E, Loukili NH, Noel E, Kaltenbach G, Abdelgheni MB, Perrin AE, Noblet-Dick M, Maloisel F, Schlienger JL, Blicklé JF. Vitamin B12 (cobalamin) deficiency in elderly patients. CMAJ. 2004 Aug 3;171(3):251-9
[54] Cagnacci A, Bagni B, Zini A, Cannoletta M, Generali M, Volpe A. Relation of folates, vitamin B12 and homocysteine to vertebral bone mineral density change in postmenopausal women. A five-year longitudinal evaluation. Bone 2008; 42: 314-320
[55] Morris MS, Jacques PF, Selhub J. Relation between homocysteine and B-vitamin status indicators and bone mineral density in older americans. Bone 2005; 37: 234-42
[56] Stone KL, Bauer DC, Sellmeyer D, Cummings SR. Low serum vitamin B12 levels are associated with increased hip bone loss in older women: a prospective study. J Clin Endocrinol Metab 2004; 89: 1217-21
[57] Pérez-Llamas F, López-Contreras MJ, Blanco MJ, López-Azorín F, Zamora S,Moreiras O. Seemingly paradoxical seasonal influences on vitamin D status in nursing-home elderly people from a Mediterranean area. Nutrition. 2008; 24:414-20
[58] Linnebur SA, Vondracek SF, Vande Griend JP, Ruscin JM, McDermott MT. Prevalence of vitamin D insufficiency in elderly ambulatory outpatients in Denver, Colorado. Am J Geriatr Pharmacother. 2007;5:1-8.
[59] Van Dam RM, Snijder MB, Dekker JM, Stehouwer CD, Bouter LM, Heine RJ, Lips P. Potentially modifiable determinants of vitamin D status in an older population in the Netherlands: the Hoorn Study. Am J Clin Nutr. 2007;85:755-61
[60] Need AG, O'Loughlin PD, Morris HA, Coates PS, Horowitz M, Nordin BE. Vitamin D metabolites and calcium absorption in severe vitamin D deficiency. J Bone Miner Res. 2008;23:1859-63.
[61] Rusell RM y Suter PM: Vitamin requirements of elderly people: an update. Am J Clin Nutr. 1993; 58: 4-14.

[62] Ooms ME, Roos JC, Bezemer PD et al. Prevention of bone loss by vitamin D supplementation in elderly women: a randomized double-blind trial. J Clin Endocrinol Metab. 1995; 80:1052-1058.
[63] Chapuy MC, Arlot ME, Duboeuf F, et al.Vitamin D3 and calcium to prevent hip fractures in the elderly women. N Engl J Med 1992;327:1637-42
[64] Dawson Hughes B, Harris SS, Krall EA, Dalla GE. Effect of calcium and vitamin D supplementation on bone density in men and women 65 years of age or older. N Engl J Med 1997;337:670-6
[65] Moro-Álvarez MJ, Díaz-Curiel M. Pharmacological treatment of osteoporosis for people over 70. Aging Clin Exp Res 2007;19:246-255
[66] Calvo MS, Whiting SJ. Public health strategies to overcome barriers to optimal vitamin D status in populations with special needs. J Nutr. 2006;136:1135-9.
[67] Mundy GR. Nutritional modulators of bone remodeling during aging. Am J Clin Nutr. 2006;83:427S-430S
[68] Goodpaster B, Park SW, Harris T. The loss of skeletal muscle strength,mass, and quality in older adults: the health, aging and body composition study. J Gerontol Med Sci 2006; 61: 1059-1064
[69] Medina Mesa R, Dapcich V. Fisiología del envejecimiento. En libro blanco de la alimentación de los mayores. Ed. Muñoz M, Aranceta J, Guijarro JL, Ed. Panamericana 2004:15-22.
[70] Salvá A, Pera G. Sreening for malnutrition in dwelling elderly. Public Health Nutr 2001; 4: 1375-1378.
[71] Schiffman SS. Changes in taste and smell: drug interactions and food preferences. Nutr Rev. 1994; 52:115.
[72] Sperduto RD, Ferris FL, Kurinij N. Do we have a nutritional treatment for age related cataract or macular degeneration? Arch Ophthalmol. 1990; 108:1403-1405.
[73] Rautiainen S, Lindblad BE, Morgenstern R, Wolk A. Vitamin C supplements and the risk of age-related cataract: a population-based prospective cohort study in women. Am J Clin Nutr. 2010 Feb;91(2):487-93
[74] Levine, M. Conry-Cantilena, C. Wang, Y. Welch, R.W. Washko, P.W. Dhariwal, K.R. Park, J. B. Lazarev, A. Graumlich, J.F. King, J. Cantilena, L. R. (1996).
[75] Vitamin C pharmacokinetics in healthy volunteers: evidence for a recommended dietary allowance. Proc Natl Acad Sci USA, 93, 3704–3709.
[76] Podmore, I.D. Griffiths, H.R. Herbert, K.E. Mistry, N. Mistry, P. Lunec, J. (1998). Vitamin C exhibits pro-oxidant properties. Nature, 392, 559
[77] Turner MD, Ship JA. Dry mouth and its effects on the oral health of elderly people. J Am Dent Assoc. 2007 Sep;138 Suppl:15S-20S.
[78] Posner BM, Jette A, Smiegelski C, Miller D, Mitchell P. Nutritional risk in New England elders. J Gerontol. 1994; 49:123-132.
[79] Di Stefano M, Veneto G, Malservisi S, Strocchi A, Corazza GR. Lactose Malabsorption and intolerance in the elderly. Scand J Gastroenterol. 2001 Dec;36(12):1274-8.
[80] Riestra S, Fernandez E, Rodrígo L, Garcia S, Ocio G. Prevalence of coeliac disease in the general population of northern Spain. Strategies of serologic screening. Scand J Gastroenterol 2000; 35; 398-402.
[81] Bouras EP, Tangalos EG. Chronic constipation in the elderly. Gastroenterol in North Am. 2009 Sep;38(3):463-80.

[82] Buckler DA, Kelber ST, Goodwin JS. The use of dietary restrictions in malnourished nursing home patients. J Am Geriatr Soc. 1994; 42:1100-1102
[83] Galizia G, Abete P, Mussi C, Noro G, Morrione A, Langellotto A, et al. Role of early symptoms in assessment of syncope in elderly people: results from the Italian group for the study of syncope in the elderly. J Am Geriatr Soc. 2009;57:18-23.
[84] Rudman D. The kidney senescence: A model for aging. Nutr Rev. 1998; 46:209-214.
[85] Kaloostian CL, Shil AB. Effects of vitamin D on muscle strength and mobility in older women. J Am Geriatr Soc. 2011 Apr;59(4):771
[86] Murad MH, Elamin KB, Abu Elnour NO, Elamin MB, Alkatib AA, Fatourechi M, Almandoz JP, Mullan RJ, Lane MA, Liu H, Erwin PJ, Hensrud DD, Montori VM. The effect of vitamin D on falls: A Systematic Review and Meta-Analysis. J Clin Endocrinol Metab. 2011 Jul 27.
[87] Bischoff HA, Stahelin HB, Dick W et al. Effects of vitamin D and calcium supplementation on falls: a randomized controlled trial J Bone Miner Res 2003;18:343-51
[88] Fischer K, Colombani PC, Langhans W, Wenk C. Carbohydrate to protein ratio in food and cognitive performance in the morning. Physiol Behav. 2002 Mar;75(3):411-23.
[89] Ginaldi L, De Martinis M, D'Ostilio A, Marini L, Loreto MF, Martorelli V,Quaglino D. The immune system in the elderly: II.Specific cellular immunity.Immunol Res. 1999; 20:109-15.
[90] Lesourd BM. Nutrition and immunity in the elderly: modification of immune responses with nutritional treatments. Am J Clin Nutr. 1997 Aug;66: 478S-484S.
[91] Pawelec G, Barnett Y, Forsey R, Frasca D, Globerson A, McLeod J, Caruso C, Franceschi C, Fülöp T, Gupta S, Mariani E, Mocchegiani E, Solana R. T cells and aging, January 2002 update. Front Biosci. 2002 May 1;7:d1056-183
[92] Chandra RK, Joshi PAB, Woodford G, Chandra S. Effect of short term nutritional supplementation on cell mediated immunity and lymphocyte subsets. Nutr Res. 1982; 2:223-232.
[93] Ginaldi L, De Martinis M, D'Ostilio A, Marini L, Loreto MF, Corsi MP, Quaglino D. The immune system in the elderly: I. Specific humoral immunity. Immunol Res.1999;20:101-8.
[94] Bellei NC, Carraro E, Castelo A, Granato CF. Risk factors for poor immune response to influenza vaccination in elderly people. Braz J Infect Dis. 2006 Aug;10(4):269-73.
[95] Lesourd B. Long term yogurt consumption by elderly patients: effect on the immune system. NI: Yogurt: myth versus reality. McLean, VA: Donna Curtis. 1996; 90-101.
[96] Gainsborough N, Maskrey VL, Nelson ML, Keating J, Sherwood RA, Jackson SH, Swift CG. The association of age with gastric emptying. Age Ageing. 1993 ;22(1):37-40.
[97] Silver AJ, Morley JE. Role of the opioid system in the hypodipsia associated with aging. J Am Geriatr Soc. 1992 Jun;40(6):556-60.
[98] Butera PC. Estradiol and the control of food intake. Physiol Behav. 2010; 99(2):175-80.
[99] Kelesidis T, Kelesidis I, Chou S, Mantzoros CS. Narrative review: the role of leptin in human physiology: emerging clinical applications. Ann Intern Med. 2010; 152(2):93-100.
[100] Ostlund RE, Yang JW, Klein S, Gingerich R. Relation between plasma leptin concentration and body fat, gender, diet, age and metabolic covariates. J Clin Endocrinol Metab. 1996; 81:3909-3913.

[101] Richards JB, Cherkas LF, Spector TD. An analysis of which anti-osteoporosis therapeutic regimen would improve compliance in a population of elderly adults.Curr Med Res Opin. 2007;23:293-9

[102] Riobo P, Gonzalo MA, Moreno I y cols: Nutrición y función mental en el anciano Rev Clin Esp. 1993, 193:252-254.

5

Potato Antioxidant Compounds: Impact of Cultivation Methods and Relevance for Diet and Health

Danièle Evers and Hannah Deußer
Environment and Agro-Biotechnologies Department, Centre de Recherche Public – Gabriel Lippmann, Belvaux Luxembourg

1. Introduction

Potato is currently the fourth most important food crop in the world after maize, wheat and rice, with a production of 329 millions tons (FAO, 2009). As for the harvested area, potato ranks 7th after wheat, rice, maize, barley, sorghum and rapeseed worldwide. In terms of consumption, potato ranks third after rice and wheat. Interestingly, the importance of potato in the diet is higher in developed as compared to developing countries, accounting for 130 kcal per person per day for the developed world and for 41 kcal per person per day in the developing world (Burlingame et al., 2009). In Europe, the per capita consumption reaches almost 90 kg year^{-1}, whereas in developing countries, its per capita consumption is smaller, reaching around 20 kg year^{-1} (André et al., 2007a, FAO, 2009)).

Potato is mainly known to supply dietary fibre, carbohydrates, high-quality proteins, vitamins and minerals. According to the USDA National Nutrient Database, contents would be 2.4 g for dietary fiber, 15.7 g for carbohydrates, 1.7 g for protein content, 19.7 g for vitamin C, each per 100 g of white, raw potato (flesh and skin) (Singh & Kaur, 2009). As for the minerals, iron and zinc contents are around 0.52 mg and 0.29 mg per 100 g (Singh & Kaur, 2009). Potatoes are also known as sources of antioxidant compounds, including polyphenols, carotenoids and vitamins, pointing to their relevance not only as a starchy food, but also as a vegetable.

Data on the antioxidant content of potatoes can be found in several national food composition tables. Table 1 summarizes some of the data available on contents of ascorbic acid, β-carotene, α-tocopherol, lutein and zeaxanthin. Phenolic compounds were not specified in these databases.

Potato's genetic diversity is huge; more than 4000 different wild varieties have been collected at the International Potato Center in Lima (Peru). They are diverse regarding tuber shape, flesh and skin colour, flavour, storage quality and cooking quality (André et al., 2007a).

The classification of cultivated potatoes has been reviewed several times and still is a matter of debate. Huaman and Spooner (2002) classified all landrace populations of cultivated

Compound	DK	F	D	CH	USA white	USA red
Ascorbic acid (mg/100 g)	26.4	11.1	18.8	17.0	19.7	8.6
β-Carotene (μg/100 g)	10	2	5	5	5	4
Lutein + Zeaxanthin (μg/100 g)	n.a.	n.a.	n.a.	n.a.	13	21
α-Tocopherol (mg/100 g)	0.10	0.05*	0.05	0.06*	0.01	0.01

Table 1. Antioxidant contents of potatoes. Data were obtained from national food composition tables. DK = potato, raw, Denmark (National Food Institute - Technical University of Denmark, 2009), F = potato, boiled, France (French Information Center on Food Quality, 2008); D = potato, peeled, raw, Germany (Max-Rubner-Institut - Karlsruhe, 2010); CH = potato, peeled, raw, Switzerland (ETH Zürich, 2009); USA white = potato, white, raw, U.S.A. (Nutrient Data Laboratory, 2010); USA red = potato, red, raw, U.S.A. (Nutrient Data Laboratory, 2010); n.a. = not analyzed; * = α-tocopherol equivalents.

potatoes in a single species *Solanum tuberosum* with 8 cultivar groups: Ajanhuiri Group (2x), Andigenum Group (4x), Caucha Group (3x), Chilotanum Group (4x), Curtilobum Group (5x), Juzepczukii Group (3x), Phureja Group (2x) and Stenotonum Group (2x). Modern cultivars were classified into a ninth cultivar group, the Tuberosum Group. However, it is worth mentioning that Spooner et al. (2007) reviewed the classification combining morphological data with molecular fingerprinting data.

In South America, the center of potato origin and diversity, potato constitutes the main staple crop and farmers cultivate up to 50 varieties in a field (FAO, 2008). This not only enables them to be protected from a complete loss in the case of a disease or an abiotic stress, it also allows them to get a more diversified diet. Indeed, the composition of the tubers varies according to the cultivar, agricultural practices, climate and soil (Rodriguez et al., 2010); moreover cooking and processing may have an effect on tuber composition (Xu et al., 2009).

Biofortification programs aiming at enriching nutrient contents of edible plants for health improvement and disease prevention are ongoing. They include biofortification through fertilization, breeding or biotechnology (White & Broadley, 2005). Nutrient biofortification of food crops may not only include elevated mineral and amino acid levels, but also enhanced antioxidant levels (Diretto et al., 2007, Rommens et al., 2008). Traditional agricultural approaches, such as breeding, can improve the nutritional value to some extent (Hirschi, 2009); especially if the corresponding trait is strongly dependent on the genotype, the selection of adequate progenitors allows to expect good progress in breeding (Burgos et al., 2009a).

The present chapter gives an overview on the potato antioxidants and on parameters impacting their contents in the tuber. Moreover, it will give an insight into potential health-promoting effects and bioavailability of antioxidants.

2. Antioxidants in potato

The concept of potato as a source of antioxidants is not widely spread. However, recent studies have placed potato into the perspective of an antioxidant-rich crop. More precisely, potatoes contain phenolic compounds including hydroxycinnamic acids, the predominant being chlorogenic acid (André et al., 2007b, Brown, 2005) and flavonoids, for example catechin, epicatechin and anthocyanins. Potato contains low amounts of carotenoids, such as β-carotene (Brown, 2005), indicating that potato is not a good source of pro-vitamin A carotenes; more important are the oxygenated carotenoids, the xantophylls, such as neoxanthin, violaxanthin, antheraxanthin, lutein and zeaxanthin (Griffiths et al., 2007). As for the vitamins, potato contains on average 20 mg per 100 g FW of vitamin C (Brown, 2005) and concerning vitamin E, mainly α-tocopherol is present at concentrations between 55 and 416 μg per 100 g FW (André et al., 2007b). Hereafter, the types and contents of potato antioxidants will be discussed in more detail.

2.1 Polyphenols

Polyphenols are secondary plant metabolites. In potatoes, mainly hydroxycinnamic acid derivatives and flavonoids occur. As examples, the chemical structures of chlorogenic acid, a hydroxycinnamic acid ester with quinic acid, and kaempferol 3-*O*-rutinoside, a flavonol glycoside, are shown in Figure 1.

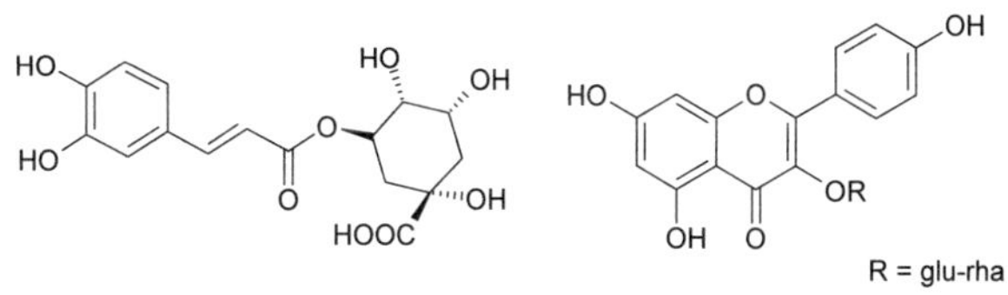

Fig. 1. Chemical structures of chlorogenic acid (left) and kaempferol 3-*O*-rutinoside (right).

Plant material	Total phenolic content (mg/g DW)	Sample preparation	Reference
S. tuberosum	2.5 - 4.8	crude, freeze-dried	(Lachman et al., 2008)
unspecified	4.5 - 6.8	crude, freeze-dried	(Stratil et al., 2006)
Solanum sp.	1.8 - 11	crude, freeze-dried	(Navarre et al., 2011)
unspecified	0.9 - 3.0	crude, freeze-dried	(Xu et al., 2009)
S. tuberosum	1.1 - 27.4[a]	crude, freeze-dried	(André et al., 2007a, André et al., 2009a)
unspecified	0.32[b]	crude	(Chiou et al., 2007)
S. tuberosum	0.3 - 0.5	crude, freeze-dried	(Rumbaoa et al., 2009)
S. tuberosum	15.1	crude	(Natella et al., 2010)
unspecified	0.6 - 2.3[a]	crude	(Blessington et al., 2010)
Solanum sp.	3 - 16 [a]	crude	(Campos et al., 2006)
S. tuberosum	1.5 - 3.2	cooked, peeled, mashed	(Kaspar et al., 2011)
S. tuberosum	2.1 - 7.8 [a]	crude	(Reddivari et al., 2007)
S. tuberosum	1.0 - 2.9	crude, freeze-dried	(Leo et al., 2008)
S. tuberosum	1.3 - 13.4	crude	(Madiwale et al., 2011)
S. tuberosum	1.8	crude, freeze-dried	(Mäder et al., 2009)
S. tuberosum	8 - 78	crude, freeze-dried	(Stushnoff et al., 2010)

Table 2. Total phenolic content of potatoes determined with the Folin-Ciocalteau assay. As far as not otherwise indicated, the results are expressed as gallic acid equivalents. [a] As chlorogenic acid equivalents, [b] as caffeic acid equivalents. Contents reported in another unit than mg/g dry weight (DW) were recalculated utilizing an average moisture content of 80%.

Here, we present an overview on current available literature on total phenolic contents (Table 2), anthocyanins (Table 3) and individual polyphenols (Table 4). Total phenolic contents have been determined by using a spectrophotometric assay, the Folin-Ciocalteau assay. This assay estimates the phenolic content in plants, but other plant compounds (for example thiol derivatives, vitamins, and amino acids) may also be detected (Everette et al., 2010). Total anthocyanins, which impart for colour in red and purple potatoes, have been determined using the pH differential method (Giusti & Wrolstad, 2001). This method gives the sum of all anthocyanins present in the sample. Individual polyphenols have been quantified by HPLC with UV-detection.

Plant material	Total anthocyanin content (mg/g DW)	Sample preparation	Reference
S. tuberosum	0 - 21.4 [a]	crude, freeze-dried	(André et al., 2009a, André et al., 2009b)
S. tuberosum	0.3 - 1.8 [b]	crude, freeze-dried	(Brown, 2008)
Solanum sp.	0 - 4 [b]	crude	(Campos et al., 2006)
S. tuberosum	0 - 6.2	cooked, peeled, mashed	(Kaspar et al., 2011)
Solanum sp.	0 - 1.2 [b]	crude	(Brown et al., 2007)
S. tuberosum	1.0 -5.5	crude	(Madiwale et al., 2011)
S. tuberosum	2 - 45	crude, freeze-dried	(Stushnoff et al., 2010)

Table 3. Total anthocyanin content of potatoes determined with the pH differential method. The results are expressed as [a]petanin or [b]cyanidin 3-O-glucoside. In some references the standard compound was not stated. Contents reported in another unit than mg/g dry weight (DW) were recalculated utilizing an average moisture content of 80%.

Plant material	Phenolic content (μg/g DW)	Sample preparation	Reference
Chlorogenic acid			
S. tuberosum cvs. Siikli and Timo	455 – 600	cooked, freeze-dried	(Mattila & Hellström, 2007)
S. tuberosum cv. Désirée	1500	crude, freeze-dried	(Lukaszewicz et al., 2004)
S. tuberosum cv. Sava	361 – 520	crude, peeled, freeze-dried	(Soltoft et al., 2010)
S. tuberosum	421 -2185	crude, freeze-dried	(Zhu et al., 2010)
Solanum sp.	220 -4730	crude, freeze-dried	(Navarre et al., 2011)
Solanum sp.	55 - 1340	crude	(del Mar Verde Mendez et al., 2004)
unspecified	420.5 – 3183.4	crude, freeze-dried	(Xu et al., 2009)
S. tuberosum	216 – 12746	crude, freeze-dried	(André et al., 2007b, André et al., 2009a, André et al., 2009b)
unspecified	6.5	crude	(Chiou et al., 2007)
unspecified	33 - 6370	crude, freeze-dried	(Im et al., 2008)
S. tuberosum	600 - 2100	crude, freeze-dried	(Navarre et al., 2010)

Plant material	Phenolic content (µg/g DW)	Sample preparation	Reference
S. tuberosum cv. Norkotah and Ranger	2200 and 1000	crude, freeze-dried	(Shakya & Navarre, 2006)
S. tuberosum	343 - 1249	crude	(Hajslova et al., 2005)
unspecified	39.5 - 65	crude	(Blessington et al., 2010)
S. tuberosum	470 - 920	crude, freeze-dried	(Leo et al., 2008)
S. tuberosum	718	crude, freeze-dried	(Mäder et al., 2009)
Caffeic acid			
S. tuberosum cv. Désirée	800	crude, freeze-dried	(Lukaszewicz et al., 2004)
Solanum	5 - 476	crude, freeze-dried	(Navarre et al., 2011)
Solanum sp.	5 - 135	crude	(del Mar Verde Mendez et al., 2004)
unspecified	0 – 93.8	crude, freeze-dried	(Xu et al., 2009)
S. tuberosum	7 – 143	crude, freeze-dried	(André et al., 2007b, André et al., 2009a, André et al., 2009b)
unspecified	5 - 293	crude, freeze-dried	(Im et al., 2008)
S. tuberosum cv. Norkotah and Ranger	100 and 200	crude, freeze-dried	(Shakya & Navarre, 2006)
S. tuberosum	50 - 120	crude, freeze-dried	(Leo et al., 2008)
S. tuberosum	203	crude, freeze-dried	(Mäder et al., 2009)
Quercetin 3-*O*-rutinoside (= Rutin)			
S. tuberosum cv. Monalisa	180	crude	(Tudela et al., 2002)
Solanum sp.	0 - 141	crude, freeze-dried	(Navarre et al., 2011)
S. tuberosum	0 – 256	crude, freeze-dried	(André et al., 2007b, André et al., 2009a, André et al., 2009b)
S. tuberosum	6.6 – 7.4	crude, freeze-dried	(Navarre et al., 2010)
unspecified	26.5 - 49	crude	(Blessington et al., 2010)
Kaempferol 3-*O*-rutinoside			
Solanum sp.	0 - 433	crude, freeze-dried	(Navarre et al., 2011)
S. tuberosum	0 – 227	crude, freeze-dried	(André et al., 2007b, André et al., 2009a, André et al., 2009b)
S. tuberosum	5.5 – 7.0	crude, freeze-dried	(Navarre et al., 2010)
(+)-Catechin			
Solanum sp.	90 - 1305	crude	(del Mar Verde Mendez et al., 2004)
S. tuberosum	430 - 1570	crude, freeze-dried	(Leo et al., 2008)
S. tuberosum	13	crude, freeze-dried	(Mäder et al., 2009)

Table 4. Content of individual polyphenols in potatoes determined by HPLC-UV. Contents reported in another unit than µg/g dry weight (DW) were recalculated utilizing an average moisture content of 80%.

As can be seen in tables 2, 3 and 4, potatoes contain between 0.3 and 78 mg/g DW of phenolic compounds, whereas the total anthocyanin content was shown to be 0 in white or yellow fleshed potatoes and up to 45 mg/g DW in some red and purple fleshed cultivars. Regarding the individual polyphenols, chlorogenic acid is the predominant compound in potatoes with contents ranging from 6.5 to 12746 µg/g DW. The contents of caffeic acid, quercetin 3-*O*-rutinoside, kaempferol 3-*O*-rutinoside and (+)-catechin ranged from 0 to 800 µg/g DW, 0 to 256 µg/g DW, 0 to 433 µg/g DW, and 13 to 1570 µg/g DW, respectively. Generally, purple and red fleshed cultivars contained higher amounts of polyphenols than cultivars with a cream or white flesh.

Additionally to the phenolic compounds listed in Table 4, other compounds have been described, namely neochlorogenic acid, cryptochlorogenic acid, caffeoyl putrescine, *p*-coumaric acid, ferulic acid, cinnamic acid, syringic acid, sinapic acid, gallic acid, vanillin, vanillic acid, *p*-hydroxybenzoic acid, *p*-hydroxyphenylacetic acid, myricetin, (-)-epicatechin, p-coumaroylhydroxyagmatine, and petanin (André et al., 2007b, André et al., 2009b, Blessington et al., 2010, Chiou et al., 2007, del Mar Verde Mendez et al., 2004, Mäder et al., 2009, Navarre et al., 2011).

2.2 Carotenoids

The carotenoid levels in potatoes determine whether the tuber flesh is white (low carotenoid content), yellow (moderate content), or orange (high content). Carotenoids in potatoes belong to the groups of the bicyclic carotenes and the xanthophylls. Carotenes (e.g. β-carotene) are pure polyen hydrocarbons whereas xanthophylls (e.g. lutein) contain oxygen groups (hydroxyl-, epoxy-, or carbonyl-groups). The chemical structures of two carotenoids from potatoes are shown in Figure 2.

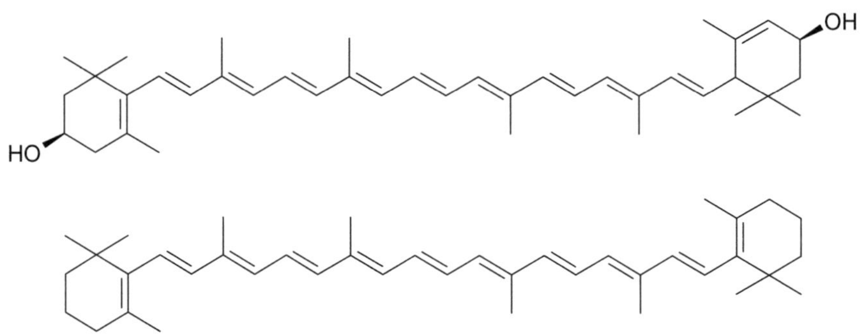

Fig. 2. Chemical structures of the carotenoids lutein (above) and β-carotene (below).

The total carotenoid contents and concentrations of individual carotenoids reported in literature are summarized in Table 5 and Table 6. To achieve a better comparability, values that were not expressed in µg/g DW were recalculated to this basis. The major carotenoids in potatoes are violaxanthin, lutein, zeaxanthin, neoxanthin, and antheraxanthin, but according to the publications cited in Table 6 their contribution varies among the cultivars. The composition of carotenoids in the skin and the flesh of the potato tubers is similar (Burmeister et al., 2011).

The determination of the total carotenoid content was mostly conducted in saponificated carotenoid extracts using a spectrophotometric assay, whereas individual carotenoids were identified and quantified by HPLC-UV analysis. Near-infrared reflectance spectroscopy (NIRS) - an analytical method that is less expensive and more rapid than HPLC - was also applied for the determination of individual carotenoids (Bonierbale et al., 2009).

Plant material	Carotenoid content (µg/g DW)	Sample preparation	Reference
S. tuberosum	2.5 - 17.5	crude	(Brown, 2008)
S. tuberosum	3.9 - 7	crude	(Blessington et al., 2010)
S. tuberosum	10 - 25	crude	(Campos et al., 2006)
S. tuberosum cv. Pentland Javelin	1.6	crude, freeze-dried, white fleshed	(Morris et al., 2004)
S. tuberosum cv. Desiree	4.9	crude, freeze-dried, cream/yellow-fleshed	(Morris et al., 2004)
S. phureja cv. DB375\1	36.3	crude, freeze-dried	(Morris et al., 2004)
S. tuberosum	1.9 - 43.9	crude	(Nesterenko & Sink, 2003)
S. tuberosum cv. Yukon Gold and Superior	3.2 and 5.6	crude, without skin	(Lu & Haynes, 2001)
Hybrid from S. phureja and S. stenotomum	6.8 and 71.8	crude, without skin	(Lu & Haynes, 2001)
S. tuberosum	7.5 - 233	crude	(Kotikova et al., 2007)
Cross between S. tuberosum and S. phureja	17.7 - 34.9		(Kobayashi et al., 2008)
S. phureja	0.6 - 42.7	crude, freeze-dried	(Griffiths et al., 2007)
S. tuberosum	1.3 - 58	cooked, without skin, mashed	(Kaspar et al., 2011)
Solanum sp.	1.7 - 84.2	crude	(Brown et al., 2007)
S. tuberosum	2.6 - 14.8	crude, freeze-dried	(Burmeister et al., 2011)
S. phureja	8.9 - 9.9	crude, freeze-dried	(Burmeister et al., 2011)
S. phureja	4.9 - 92	crude	(Burgos et al., 2009b)
S. tuberosum	4.8 - 46.5	crude	(Reddivari et al., 2007)
S. phureja	1.0 - 21.4	crude, freeze-dried	(Bonierbale et al., 2009)
S. tuberosum	1 - 3	crude, freeze-dried	(Leo et al., 2008)
S. tuberosum	2.8 - 36.2	crude, freeze-dried	(André et al., 2007a)

Table 5. Total carotenoid content in potatoes reported in literature. Contents reported in another unit than µg/g dry weight (DW) were recalculated utilizing an average moisture content of 80%.

Plant material	Carotenoid content (µg/g DW)	Sample preparation	Reference
Lutein			
S. tuberosum	0.5	crude	(Blessington et al., 2010)
S. tuberosum	1.0 - 6.0	crude	(Nesterenko & Sink, 2003)
S. tuberosum cv. Yukon Gold and Superior	1.2 and 0.8	crude, without skin	(Lu & Haynes, 2001)
Hybrid from S. phureja and S. stenotomum	2.8 - 26.6	crude, without skin	(Lu & Haynes, 2001)
Cross between S. tuberosum and S. phureja	0.8 - 11.6		(Kobayashi et al., 2008)
S. tuberosum	n.d. - 0.03	crude	(Zhou et al., 2011)
S. tuberosum cv. Baltica	3.3	cooked and mashed	(Bub et al., 2008)
S. tuberosum cv. Red Laura and Shetland Black	3.2 and 0.5	crude, freeze-dried	(Burmeister et al., 2011)
S. phureja cv. Mayan Twilight and Mayan Gold	2.1 and 2.7	crude, freeze-dried	(Burmeister et al., 2011)
S. phureja	2.8 - 10.6	crude	(Burgos et al., 2009b)
S. tuberosum cv. Désirée	0.7	crude, freeze-dried	(Ducreux et al., 2005)
S. tuberosum	1.1 - 17.7	crude, freeze-dried	(André et al., 2007b)
S. phureja	0.6 - 1.9	crude, freeze-dried	(Bonierbale et al., 2009)
S. tuberosum	5.0 - 11.4	crude, freeze-dried	(André et al., 2009b)
Violaxanthin			
S. tuberosum	0.2 - 6.2	crude	(Nesterenko & Sink, 2003)
S. tuberosum cv. Yukon Gold and Superior	2.6 and 1.0	crude, without skin	(Lu & Haynes, 2001)
Hybrid from S. phureja and S. stenotomum	1.2 - 22.0	crude, without skin	(Lu & Haynes, 2001)
S. tuberosum	0.2 - 3.3	crude, without skin	(Breithaupt & Bamedi, 2002)
S. tuberosum	n.d. - 0.03	crude	(Zhou et al., 2011)
S. tuberosum cv. Red Laura and Shetland	6.1 and 1.3 (9-cis)	crude, freeze-dried	(Burmeister et al., 2011)

Plant material	Carotenoid content (µg/g DW)	Sample preparation	Reference
Black			
S. phureja cv. Mayan Twilight and Mayan Gold	2.9 and 3.1 (9-cis)	crude, freeze-dried	(Burmeister et al., 2011)
S. phureja	traces - 20.5	crude	(Burgos et al., 2009b)
S. tuberosum cv. Désirée	2.2	crude, freeze-dried	(Ducreux et al., 2005)
S. phureja cv. Mayan Gold	11.7	crude, freeze-dried	(Ducreux et al., 2005)
Zeaxanthin			
Cross between S. tuberosum and S. phureja	7.7 - 24.6		(Kobayashi et al., 2008)
S. tuberosum	n.d. - 0.5	crude	(Zhou et al., 2011)
S. tuberosum cv. Baltica	0.6	cooked and mashed	(Bub et al., 2008)
S. phureja	traces - 64.5	crude	(Burgos et al., 2009b)
S. tuberosum	0 - 17.7	crude, freeze-dried	(André et al., 2007b)
S. tuberosum	2.7 - 4.3	crude, freeze-dried	(André et al., 2009b)
Antheraxanthin			
S. tuberosum	0.4 - 2.4	crude, without skin	(Breithaupt & Bamedi, 2002)
S. phureja	0.3 - 18.8	crude	(Burgos et al., 2009b)
S. phureja cv. Mayan Gold	4.2	crude, freeze-dried	(Ducreux et al., 2005)
S. phureja	0.03 - 3.54	crude, freeze-dried	(Bonierbale et al., 2009)
Lutein-5,6-epoxide (Taraxanthin)			
S. tuberosum cv. Yukon Gold and Superior	0.9 and 0.5	crude, without skin	(Lu & Haynes, 2001)
Hybrid from S. phureja and S. stenotomum	1.1 - 27.4	crude, without skin	(Lu & Haynes, 2001)
β-Carotene			
S. tuberosum cv. Baltica	0.7	cooked and mashed	(Bub et al., 2008)
S. phureja	0 - 1.4	crude	(Burgos et al., 2009b)
S. tuberosum	0 - 2.2	crude, freeze-dried	(André et al., 2007b)

Table 6. Individual carotenoid content of potatoes reported in literature. Contents reported in another unit than µg/g dry weight (DW) were recalculated utilizing an average moisture content of 80%.

Total carotenoid contents in potatoes were reported to be between 0.6 and 233 µg/g DW. None of the carotenoids were quantifiable in all potatoes under study and a huge range of the contents of the individual compounds was observed. In potatoes where lutein was quantifiable, the contents ranged between 0.5 and 26.6 µg/g DW. The contents of violaxanthin, zeaxanthin, antheraxanthin, lutein-5,6-epoxide, and β-carotene ranged between 0.2 and 22.0 µg/g DW, 0.6 and 64.5 µg/g DW, 0.03 and 18.8 µg/g DW, 0.5 and 27.4 µg/g DW, and 0.7 and 2.2 µg/g DW, respectively.

2.3 Ascorbic acid

Regarding the content of ascorbic acid (vitamin C, chemical structure is shown in Figure 3) in potatoes, a huge number of publications are available. Quantification was done either by HPLC-UV, a spectrophotometric assay (after addition of 2,6-dichloroindophenol or 2,4-dinitrophenylhydrazine) or by fluorescence measurement (after addition of sodium acetate and O-phenylene diamine, for references see Table 7). In most cases, the total ascorbic acid content including dihydroascorbic acid was determined. Table 7 shows the contents of ascorbic acid that have been described in literature. The amounts ranged from 0.2 to 5.6 mg/g DW.

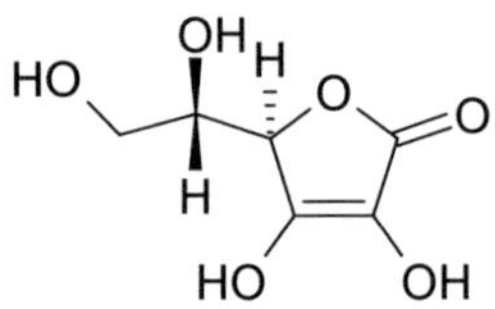

Fig. 3. Chemical structure of ascorbic acid (vitamin C).

2.4 Vitamin E

In addition to the antioxidants described above, potatoes also contain vitamin E. Vitamin E includes four tocopherols (α, β, γ, and δ) and four tocotrienols (α, β, γ, and δ), among which α -tocopherol has the highest biological activity in humans. In potatoes, α-tocopherol is the predominant vitamin E representative. γ-Tocopherol and α-tocotrienol were detectable in minor amounts (André et al., 2007b, Chun et al., 2006, Crowell et al., 2008). The amounts of α-tocopherol found in potatoes are summarized in Table 8. When recalculated to the DW of potatoes, the amounts ranged between 0.8 and 34.8 µg/g DW.

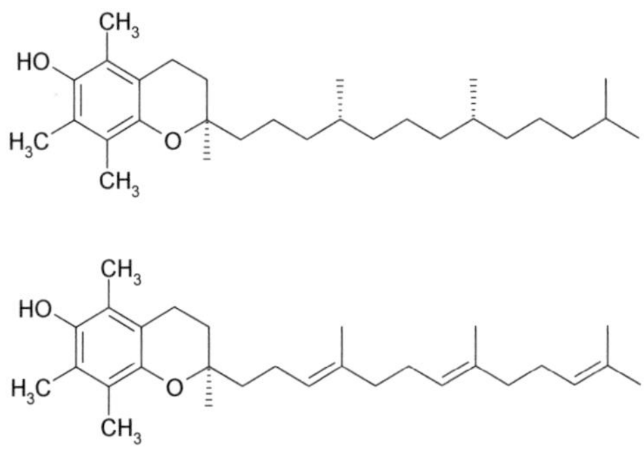

Fig. 4. Chemical structures of the vitamin E representatives α-tocopherol (above) and α-tocotrienol (below).

Plant material	Ascorbic acid content (mg/g DW)	Sample preparation	Reference
S. tuberosum	1.4	Skin, crude	(Singh et al., 2011)
S. tuberosum	0.5 – 1.7	crude, freeze-dried	(Dale et al., 2003)
S. tuberosum L. cv. Favorite	4.2	crude	(Qin et al., 2011)
S. tuberosum cvs. Bintje, Piccolo and Purple Majesty	0.4 – 0.6	crude, freeze-dried	(Navarre et al., 2010)
S. tuberosum ssp. Andigenum	0.4 – 0.7	crude	(Jimenez et al., 2009)
S. tuberosum cv. Spunta	1.4	crude	(Jimenez et al., 2009)
S. tuberosum	0.6 – 1.5	crude, freeze-dried	(Love & Salaiz, 2004)
S. tuberosum cvs. Russet Burbank and Shepody	0.5 – 1.2	crude, freeze-dried	(Rogan et al., 2000)
S. tuberosum cv. Norkotah and Ranger	0.4 and 0.8	crude, freeze-dried	(Shakya & Navarre, 2006)
S. tuberosum	0.3 – 0.7	crude	(Hajslova et al., 2005)
S. tuberosum and S. x chaucha	2.0 – 5.6	crude	(Rodriguez et al., 2010)
S. tuberosum	0.5 – 1.0	crude, freeze-dried	(Leo et al., 2008)
S. tuberosum	0.8 – 2.1	crude	(Han et al., 2004)
S. tuberosum	1.9	crude	(Natella et al., 2010)
S. tuberosum	0.2 – 1.2	crude	(Burgos et al., 2009a)
S. tuberosum	0.2 – 1.8	crude, freeze-dried	(André et al., 2007a, André et al., 2009b)

Table 7. Ascorbic acid content in potatoes reported in literature. Contents reported in another unit than mg/g dry weight (DW) were recalculated utilizing an average moisture content of 80%.

Plant material	α-Tocopherol content (µg/g DW)	Sample preparation	Reference
S. tuberosum cv. Spunta	2.1	crude	(Crowell et al., 2008)
S. tuberosum cv. Spunta	1.4	crude	(Crowell et al., 2008)
S. tuberosum	2.7 –34.8	Andean cultivars, crude, freeze-dried	(André et al., 2007b, André et al., 2009b)
S. tuberosum cv. Nicola	0.8	crude, freeze-dried	(André et al., 2007b)
S. tuberosum cv. Vitelotte	2.3	crude, freeze-dried	(André et al., 2007b)
S. tuberosum	3.5	crude	(Chun et al., 2006)
S. tuberosum	3.0	boiled	(Chun et al., 2006)
S. tuberosum	0.7 – 1.4	stored, crude	(Spychalla & Desborough, 1990)
S. tuberosum cv. Désirée	22.5	crude, freeze-dried	(Ducreux et al., 2005)

Table 8. Vitamin E (α-tocopherol) content in potatoes reported in literature. Contents reported in another unit than µg/g dry weight (DW) were recalculated utilizing an average moisture content of 80%.

3. Impact of cultivation, storage and processing on antioxidant compounds in potato

3.1 Impact of cultivation conditions

Environmental or abiotic stresses may have an impact on antioxidant contents, as these can generate reactive oxygen species in plants, thereby causing oxidative stress and impacting the antioxidant responses. Many plant secondary metabolites are determinants of both plant stress tolerance and nutritional quality (Jansen et al., 2008). In a study on potato exposed to drought stress (André et al., 2009c), highly cultivar-specific responses were observed. Thus, in yellow tubers, changes in the contents of antioxidants were weak, whereas in pigmented (red- and purple fleshed) tubers, variations (reduction or increase) were high. Hamouz et al. (2010) reported higher total polyphenol contents in potatoes grown under drought in a warm lowland location and under low temperatures during the vegetation period, both reflecting extreme climatic conditions. Carotenoid contents increased in most of the cases, whereas vitamin C was not affected by drought, except in one of the investigated cultivars. In a review on plant stress and human health (Jansen et al., 2008), polyphenol and vitamin C changes upon exposure of a range of plants to UV-B radiation were variable and dependent on the experimental system under investigation. It can be said that the effect of a stress on the nutritional composition of plants is still poorly understood and needs further investigations.

Higher anthocyanins and total phenolics were observed at locations with longer days and cooler temperatures (Reyes et al., 2004); according to the authors, light and day length may have an influence on the phenylpropanoid metabolism thereby affecting/favouring the synthesis of some compounds. Lachman et al. (2008) compared total phenolic contents from potato tubers grown in 4 different localities and attributed higher total phenolic contents (a difference of approximately 12% between highest and lowest contents) at harvest mainly to lower temperatures in the end of the vegetation period.

Concerning cultivation methods (conventional versus organic), the nutrient composition of plants is affected by the differences in organic and conventional farming, mainly due to fertilization management (Soltoft et al., 2010). However, it cannot clearly be said that significant differences in health-promoting compounds exist between the two farming systems. Potatoes produced in organic cultivation when compared to conventional cultivation contain higher levels of chlorogenic acid (Hajslova et al., 2005, Soltoft et al., 2010). Vitamin C content in the tubers from organically grown potatoes were not significantly higher than those from conventional farming (Hajslova et al., 2005, Rembialkowska, 1998). More generally, Faller and Fialho (2010) found that polyphenol content and antioxidant capacity in plant foods (several vegetables and fruits) were mostly similar or slightly higher in organic agriculture. Further studies are however necessary to clarify the impact of organic versus conventional agriculture on the contents of antioxidants in potato.

3.2 Post-harvest storage

Potatoes are mainly grown as single crop, with subsequent storage for three to ten months, a period in which they may undergo high metabolic activities (Pinhero et al., 2009).

It is known that ascorbic acid in fresh foods, including potato, is not stable postharvest (Pinhero et al., 2009) with levels decreasing to 40 to 60% of the initial value (Woolfe, 1987) or even to about 25% of the original one (Dale et al., 2003). Similarly, according to Burgos et al. (2009a), ascorbic acid concentrations decreased as the storage time increased.

As for carotenoids, Griffiths et al. (2007) found that postharvest storage reduces carotenoid contents, whereas Morris et al. (2004) describe only slight or no total carotenoid content changes at all during storage at cold temperatures. However, according to Morris et al. (2004), individual carotenoid levels do change during storage. Blessington et al. (2010) describe higher carotenoid contents (contents increase between 1 and 100%) in stored (4°C and 20°C) potatoes as compared to non-stored potatoes.

Storage at 4°C and 20°C also increased the total phenolic contents compared to non-stored potatoes (Blessington et al., 2010).

According to Dale et al. (2003), storage would be the major effect impacting contents of antioxidants, whereas location, year and genotype also would play a role. Interestingly for vitamin C, there is a high level of consistency in the ranking of genotypes across years, indicating heritability, a trait to be exploited during breeding. Burgos et al. (2009a) corroborate this information and found that the genotype effect is higher than the environment and than the genotype x environment interaction.

3.3 Cooking

Vitamin C changes during cooking have been evaluated by Burgos et al. (2009a) by boiling, microwaving and baking tubers. Losses of vitamin C were most important in baked and microwaved tubers as compared to boiled tubers. The percentage of retention ranged from 53 to 97%, from 6 to 66% and from 6 to 39% in boiled, baked and microwaved potatoes. In new potatoes, harvested at a young developmental stage, contents of total phenolics and vitamin C did not decrease after cooking by any method, presumably because of the small size of these tubers and the short cooking time required (Navarre et al., 2010).

Similarly, according to Xu et al. (2009), antioxidant capacity, mainly determined by the potato variety, was slightly influenced by cooking conditions depending on the cultivar. According to Blessington et al. (2010), when comparing different cooking methods of potatoes prepared with skin (baking, boiling, frying, microwaving), boiled samples were lower in total carotenoid contents, whereas for total phenolics, increased contents were observed after baking, frying and microwaving as compared to uncooked samples. Mulinacci et al. (2008) describe unaltered phenolic acid contents after cooking and microwaving unpeeled potatoes; however, in coloured potatoes, a decrease in the total anthocyanin content was observed (decrease in the range 16-29%). It has been shown by Dao and Friedman (1992) that peeling influences the contents of health-promoting compounds; they showed that potato peels contain high levels of phenolics, making them a promising material for the generation of functional foods. Mattila and Hellström (2007) observed a decrease in phenolics when comparing a peeled and cooked potato with an uncooked potato. It was also suggested that during the cooking process (cooking with skin), phenolics might migrate from the peel into the cortex and the internal tissues suggesting an improved extractability from cooked samples.

A more specific processing is the so-called chuno production. Chuno corresponds to a traditional Andean freeze and sun-dried potato, in which water is completely removed by mechanical pressing facilitated through freeze and thaw cycles implicating destruction of cellular structure (Penarrieta et al., 2011). During this process, a loss of antioxidants and phenolic compounds was shown, though some compounds seemed to be transferred from the peel to the flesh during the process, as also reflected by the dark to black colour of the chuno.

4. Nutritional relevance and health-beneficial properties of potato antioxidants

Oxidative stress is a disturbance of the equilibrium between pro-oxidants and antioxidants, in favour of the former (Kaspar et al., 2011). This imbalance may lead to cellular damage, as oxidation of cellular lipids, proteins and DNA imparts cellular function and increases susceptibility to a number of chronic diseases. Antioxidant molecules may scavenge reactive oxygen species, thereby limiting oxidative stress (Robert et al., 2006).

4.1 Vitamins

Potato is known to be a good source of vitamin C in the human diet. In the human body, ascorbate plays a role as a water-soluble antioxidant and as cofactor in reactions catalyzed by a number of metal-dependent oxygenases (André et al., 2010). Besides its positive effect on human health, it is also important to mention that vitamin C plays an important role as enhancer of iron bioavailability from potato (Yun et al., 2004).

As for vitamin E, its major biological role is to prevent lipid peroxidation and to protect poly unsaturated fatty acids and low density lipoproteins from oxidation by free radicals (André et al., 2010).

4.2 Carotenoids and polyphenols

As previously said, potatoes are good sources of antioxidants, such as the hydrophilic polyphenols and only moderate sources of the lipophilic carotenoids. Dietary carotenoids are associated with health benefits. On one hand, the provitamin A activity of carotenoids, such as beta-carotene, alpha-carotene and beta-cryptoxanthin is well-known; on the other hand, non pro-vitamin A carotenoids, such as lutein and zeaxanthin for example, have important antioxidant activity and are known to provide protection against age-related macular degeneration (Griffiths et al., 2007).

A correlation between polyphenol intake and reduced incidence of cancers, cardiovascular and neurodegenerative diseases was shown by Arts and Hollman (2005); however these positive effects could not only be attributed to their antioxidant properties. Polyphenols also exert their health beneficial effects through modulation of cellular signalling processes: as inflammation modulatory agents, as regulators of cell proliferation and differentiation, angiogenesis and apoptosis and as modulators of signalling cascades and apoptotic processes (reviewed in (Stevenson & Hurst, 2007)).

Few studies are available in the literature on health-promoting effects of potato antioxidants and will be presented hereafter. The impact of the consumption of pigmented potatoes on oxidative stress and inflammatory damage in man has been studied by Kaspar et al. (2011). Men consumed either white, yellow (high concentrations in phenolic acis and carotenoids) or purple fleshed (high concentrations of phenolic acids and anthocyanins) potato once per day in a randomized 6-week study with good compliance. The consumption of pigmented potato resulted in elevated antioxidant status and reduced inflammation and DNA damage, as reflected by e.g. decreased inflammatory cytokine and C-reactive protein concentrations. Another study on the lipid-lowering effect of potato in rats showed that feeding rats a potato-enriched diet led to a decrease in cholesterol and triglyceride levels in plasma, decreased cholesterol level in the liver and improved antioxidant status (Robert et al., 2006, Robert et al., 2008). These results suggest that potato consumption may enhance antioxidant defense and improve the lipid metabolism. Similarly in a study performed by Han et al. (2006) on rats fed with anthocyanin-rich purple potato flake extracts, it was shown that these extracts have antioxidant capacity with regard to radical scavenging activity and inhibition of linoleic acid oxidation; moreover, they would enhance hepatic Mn-SOD, Cu/Zn-SOD and GSH-Px mRNA expression suggesting a reduced hepatic lipid peroxidation and an improved antioxidant potential in the rats.

In a study performed by Thompson et al. (2009) on induced breast cancer in rats, a greater inhibition of carcinogenesis was shown when the rats were fed with a red pigmented cultivar as compared to a White Russet Burbank. The red cultivar had high levels of anthocyanin and chlorogenic acid derivatives, previously reported to inhibit the growth of human breast cancer cells grown in monolayer culture (Hakimuddin et al., 2004).

5. Bioavailability

Studies on the bioavailability of antioxidants and/or clinical outcomes following the administration of antioxidants from potato are scarce. Concerning nutrition, the term bioavailability describes the quantity of an ingested nutrient that is used by the body in its original or metabolized form. Many problems are to be faced, such as those related to cultivar variability, sample preparation and metabolism in the organism as well as interaction with gut microflora.

A few studies on the bioavailability of β-carotene from sweet potato are available (Bengtsson et al., 2009, Failla et al., 2009). It is worth mentioning that the bioavailability of carotenoids is dependent on a number of parameters including the physicochemical state, cooking style, other components of the meal (e.g. fat content), to name only a few. Failla et al. (2009) pointed to a relatively poor bioaccessibility (i.e. the release from the food matrix and the solubility in the gastro-intestinal fluids) of β-carotene from sweet potato together with a poor micellarization; this latter process corresponds to the transfer of the carotenoids from the food matrix into mixed bile salt micelles, during the small intestinal phase of digestion, a process that was however improved by the addition of oil. In a study performed by Bub et al. (2008) on genetically modified, zeaxanthin-enriched potato (270 μg/100g), the concentration of zeaxanthin was significantly increased in chylomicrons, a group of lipoproteins reflecting newly absorbed carotenoids, after the consumption of genetically modified potatoes (by three men, randomized, controlled double-blinded) as compared to no increase after consumption of control potatoes (12.9 μg/100g).

There are a number of studies on the bioavailability of polyphenols, such as e.g. D'Archivio et al. (2010), Manach et al. (2004), Scalbert and Williamson (2000), and Williamson and Manach (2005), though none specifically on potatoes. Within polyphenols, the bioavailability greatly differs and decreases from isoflavones, to flavonols, to flavan—3-ols, to anthocyanins and proanthocyanidins. The difficulty in realizing polyphenol biovailability studies lies in the fact that they may undergo substantial modifications following ingestion. Some compounds may be absorbed in the small intestine. Others reach the colon where they undergo modifications such as hydrolization of glycosides into aglycones by the colonic microflora. Prior to the passage into the blood stream, other modifications might occur including methylation, sulfation, glucuronidation (D'Archivio et al., 2010). Since chlorogenic acid is the major phenolic acid present in potato, it is worth mentioning a study on the bioavailability of chlorogenic acid from coffee (Stalmach et al., 2010). This study implicating human ileostomy volunteers indicates a potential absorption of 29% of the intake by the small intestine; thus approximately one third of ingested chlorogenic acid in foods would be absorbed and enter the bloodstream; in healthy subjects with a functioning colon, the remaining part would reach the large intestine. A recent publication on the bioavailability of quercetin 3-*O*-rutinoside included healthy volunteers and those with an ileostomy. After consumption of tomato juice fortified with quercetin 3-*O*-rutinoside, no metabolites were detected in the plasma and urine of the ileostomists and 86% of the ingested quercetin 3-*O*-rutinoside was detected in the ileostomy bag. In healthy subjects, this amount reaches the colon. In the colon, the quercetin 3-*O*-rutinoside is converted to phenolic acids by the microflora. Amounts of these phenolic acids corresponding to 22% of the quercetin 3-*O*-rutinoside intake have been detected in the urine of the healthy volunteers (Jaganath et al., 2006).

Informations on the bioavailability of α-tocopherol are scarce. An *in vitro* study with broccoli showed that 20% of the applied α-tocopherol has been incorporated into the aqueous phase by micellarization in a digestion model (Granado et al., 2006). The bioavailability of α-tocopherol in humans was assessed after the consumption of d_6- α-tocopherol spiked apples. When the apples were consumed together with a breakfast containing no fat, 10% of the 22 mg d_6- α-tocopherol were detected in the plasma of the probands. Increasing the breakfast fat content to 6% and 21%, 20% and 33% of the d_6- α-tocopherol has been detected, respectively (Bruno et al., 2006).

The determination of the bioavailability of ascorbic acid in doses present in food is difficult. The plasma levels of healthy persons after oral ingestion of vitamin C are saturated at 70 to 80 µM (Duconge et al., 2008) and the ingestion of additional ascorbic acid does not lead to a simple increase of the plasma levels. A study with seven healthy vitamin C depleted volunteers showed that the bioavailability was 100% at doses of 200 mg. When using higher doses, the bioavailability was lower than 50% (Levine et al., 1996). When stable, isotopically labelled ascorbic acid was administered to four healthy probands, the unspecific measurement of ascorbic acid in their plasma revealed a 12% increase of the plasma level. When the samples were analysed for their isotope content, an increase of 3 to 6% was observed. The authors of the study stated that the ingested vitamin C dose enters an existing pool in the body, but this pool is continuously fluxed with vitamin C present in the body before the beginning of the study (Bluck et al., 2005). Ascorbic acid bioavailability after ingestion of a fortified beverage and orange juice was approximately 65%, but the standard deviations in this study have been quite high (Carter et al., 2010).

6. References

André, C., Larondelle, Y., & Evers, D. (2010). Dietary antioxidants and oxidative stress from a human and plant perspective: a review. *Curr Nutr Food Sci.* 6, 2-12

André, C. M., Ghislain, M., Bertin, P., Oufir, M., Herrera, M. R., Hoffmann, L., Hausman, J. F., Larondelle, Y., & Evers, D. (2007a). Andean potato cultivars (*Solanum tuberosum* L.) as a source of antioxidant and mineral micronutrients. *J Agric Food Chem.* 55, 2, 366-378

André, C. M., Oufir, M., Guignard, C., Hoffmann, L., Hausman, J. F., Evers, D., & Larondelle, Y. (2007b). Antioxidant profiling of native Andean potato tubers (*Solanum tuberosum* L.) reveals cultivars with high levels of beta-carotene, alpha-tocopherol, chlorogenic acid, and petanin. *J Agric Food Chem.* 55, 26, 10839-10849

André, C. M., Oufir, M., Hoffmann, L., Hausman, J. F., Rogez, H., Larondelle, Y., & Evers, D. (2009a). Influence of environment and genotype on polyphenol compounds and in vitro antioxidant capacity of native Andean potatoes (*Solanum tuberosum* L.). *J Food Compos Anal.* 22, 517-524

André, C. M., Schafleitner, R., Guignard, C., Oufir, M., Aliaga, C. A., Nomberto, G., Hoffmann, L., Hausman, J. F., Evers, D., & Larondelle, Y. (2009b). Modification of the health-promoting value of potato tubers field grown under drought stress: emphasis on dietary antioxidant and glycoalkaloid contents in five native andean cultivars (*Solanum tuberosum* L.). *J Agric Food Chem.* 57, 2, 599-609

André, C. M., Schafleitner, R., Legay, S., Lefevre, I., Aliaga, C. A., Nomberto, G., Hoffmann, L., Hausman, J. F., Larondelle, Y., & Evers, D. (2009c). Gene expression changes related to the production of phenolic compounds in potato tubers grown under drought stress. *Phytochemistry.* 70, 9, 1107-1116

Arts, I. C. & Hollman, P. C. (2005). Polyphenols and disease risk in epidemiologic studies. *Am J Clin Nutr.* 81, 1 Suppl, 317S-325S

Bengtsson, A., Larsson, A. M., & Svanberg, U. (2009). In vitro bioaccessibility of beta-carotene from heat-processed orange-fleshed sweet potato. *J Agric Food Chem.* 57, 20, 9693-9698

Blessington, T., Nzaramba, M. N., Scheuring, D. C., Hale, A. L., Reddivari, L., & Miller, J. C., Jr. (2010). Cooking methods and storage treatments of potato: Effects on carotenoids, antioxidant activity, and phenolics. *Am J Pot Res.* 87, 479-491

Bluck, L. J. C., Jones, K. S., Coward, W. A., & Bates, C. J. (2005). The "anomalous" absorption of labelled and unlabelled vitamin c in men. *Br J Nutr.* 93, 627-632

Bonierbale, M., Grüneberg, W., Amoros, W., Burgos, G., Salas, E., Porras, E., & zum Felde, T. (2009). Total and individual carotenoid profiles in Solanum phureja cultivated potatoes: II. Development and application of near-infrared reflectance spectroscopy (NIRS) calibrations for germplasm characterization. *J Food Compos Anal.* 22, 509-516

Breithaupt, D. E. & Bamedi, A. (2002). Carotenoids and carotenoid esters in potatoes (*Solanum tuberosum* L.): New insights into an ancient vegetable. *J Agric Food Chem.* 50, 7175-7181

Brown, C. R. (2005). Antioxidants in potato. *Am J Potato Re.* 82, 163-172

Brown, C. R. (2008). Breeding for phytonutrient enhancement of potato. *Am J Pot Res.* 85, 298-307

Brown, C. R., Culley, D., Bonierbale, M., & Amoros, W. (2007). Anthocyanin, carotenoid content, and antioxidant values in native south american potato cultivars. *Hortscience.* 42, 7, 1733-1736

Bruno, R. S., Leonard, S. W., Park, S., Zhao, Y., & Traber, M. G. (2006). Human vitamin E requirements assessed with the use of apples fortified with deuterium-labeled α-tocopheryl acetate. *Am J Clin Nutr.* 83, 299-304

Bub, A., Mösender, J., Wenzel, G., Rechkemmer, G., & Briviba, K. (2008). Zeaxanthin is bioavailable from genetically modified zeaxanthin-rich potatoes. *Eur J Nutr.* 47, 99-103

Burgos, G., Auqui, S., Amoros, W., Salas, E., & Bonierbale, M. (2009a). Ascorbic acid concentration of native Andean potato varieties as affected by environment, cooking and storage. *J Food Compos Anal.* 22, 533-538

Burgos, G., Salas, E., Amoros, W., Auqui, M., Munoa, L., Kimura, M., & Bonbierale, M. (2009b). Total and individual carotenoid profiles in *Solanum phureja* of cultivated potatoes: I. Concentrations and relationships as determined by spectrophotometry and HPLC. *J Food Compos Anal.* 22, 503-508

Burlingame, B., Mouillé, B., & Charrondière, R. (2009). Nutrients, bioactive non-nutrients and anti-nutrients in potatoes. *J Food Compos Anal.* 22, 494-502

Burmeister, A., Bondiek, S., Apel, L., Kühne, C., Hillebrand, S., & Fleischmann, P. (2011). Comparison of carotenoid and anthocyanin profiles of raw and boiled *Solanum tuberosum* and *Solanum phureja* tubers. *J Food Compos Anal.* doi:10.1016/j.jfca.2011.03.006,

Campos, D., Noratto, G., Chirinos, R., Arbizu, C., Roca, W., & Cisneros-Zevallos, L. (2006). Antioxidant capacity and secondary metabolites in four species of Andean tuber crops: native potato (*Solanum* sp.), mashua (*Tropaeolum tuberosum* Ruiz & Pavón), Oca (*Oxalis tuberosa* Molina) and ulluco (*Ullucus tuberosus* Caldas). *J Sci Food Agric.* 86, 1481-1488

Carter, B., Monsivais, P., & Drewnowski, A. (2010). Absorption of folic acid and ascorbic acid from nutrient comparable beverages. *J Food Sci.* 75, 9, H289-H293

Chiou, A., Salta, F. N., Kalogeropoulos, N., Mylona, A., Ntalla, I., & Andrikopoulos, N. K. (2007). Retention and distribution of polyphenols after pan-frying of French fries in oils enriched with olive leaf extract. *J Food Sci.* 72, 8, S574-S584

Chun, J., Lee, J., Ye, L., Exler, J., & Eitenmiller, R. R. (2006). Tocopherol and tocotrienol contents of raw and processed fruits and vegetables in the United States diet. *J Food Compos Anal.* 19, 196-204

Crowell, E. F., McGrath, J. M., & Douches, D. S. (2008). Accumulation of vitamin E in potato (*Solanum tuberosum*) tubers. *Transgenic Res.* 17, 2, 205-217

D'Archivio, M., Filesi, C., Vari, R., Scazzocchio, B., & Masella, R. (2010). Bioavailability of the polyphenols: status and controversies. *Int J Mol Sci.* 11, 4, 1321-1342

Dale, M. F. B., Griffiths, D. W., & Todd, D. T. (2003). Effects of genotype, environment, and postharvest storage on the total ascorbate content of potato (*Solanum tuberosum*) tubers. *J Agric Food Chem.* 51, 244-248

Dao, L. & Friedman, M. (1992). Chlorogenic acid content of fresh and processed potatoes determined by ultraviolet spectrophotometry. *J Agric Food Chem.* 40, 2152-2156

del Mar Verde Mendez, C., Rodriguez Delgado, M. A., Rodriguez Rodriguez, E. M., & Diaz, R. C. (2004). Content of Free Phenolic Compounds in Cultivars of Potatoes Harvested in Tenerife (Canary Islands). *J Agric Food Chem.* 52, 1323-1327

Diretto, G., Al-Babili, S., Tavazza, R., Papacchioli, V., Beyer, P., & Giuliano, G. (2007). Metabolic engineering of potato carotenoid content through tuber-specific overexpression of a bacterial mini-pathway. *PLoS One.* 2, 4, e350-1-e350-8

Duconge, J., Miranda-Massari, J. R., Gonzalez, M. J., Jackson, J. A., Warnock, W., & Riordan, N. H. (2008). Pharmacokinetics of vitamin C: insights into the oral and intravenous administration of ascorbate. *PRHSJ.* 27, 1, 7-19

Ducreux, L. J. M., Morris, W. L., Hedley, P. E., Shepherd, T., Davies, H. V., Milliam, S., & Taylor, M. A. (2005). Metabolic engineering of high carotenoid potato tubers containing enhanced levels of β-carotene and lutein. *J Exp Bot.* 56, 81-89

ETH Zürich (30-4-2009), Schweizer Nährwertdatenbank, In: *SwissFIR*, 18-7-2011, Available from http://nwdb.ethz.ch:8080/nwdb/request?xml=MessageData&xml=MetaData&xsl=SearchField&lan=de&pageKey=Start

Everette, J. D., Bryant, Q. M., Green, A. M., Abbey, Y. A., Wangila, G. W., & Walker, R. B. (2010). Thorough study of reactivity of various compound classes toward the Folin-Ciocalteu reagent. *J Agric Food Chem.* 58, 14, 8139-8144

Failla, M. L., Thakkar, S. K., & Kim, J. Y. (2009). In vitro bioaccessibility of beta-carotene in orange fleshed sweet potato (Ipomoea batatas, Lam.). *J Agric Food Chem.* 57, 22, 10922-10927

Faller, A. L. K. & Fialho, E. (2010). Polyphenol content and antioxidant capacity in organic and conventional plant foods. *J Food Compos Anal.* 23, 561-568

FAO (2009), In: *FAOSTAT*, 10-5-2011, Available from http://faostat.fao.org

French Information Center on Food Quality (2008), French Food Composition Table, In: *French Agency for food, environmental and occupational health safety*, 18-7-2011, Available from http://www.afssa.fr/TableCIQUAL/

Giusti, M. M. and Wrolstad, R. E. (2001). Characterization and measurement of anthocyanins by UV-visible spectroscopy, In: *Current Protocols in Food Analytical Chemistry*, Wrolstad, R. E., Acree, T. E., Decker, E. A., Penner, M. H., Reid, D. S., Schwartz, S. J., Shoemaker, C. F., Smith, D. M., and Sporns, P., pp.(F1.2.1-F1.2.13), John Wiley & Sons, Inc., 9780471142911, New York

Granado, F., Olmedilla, B., Herrero, C., Perez-Sacristan, B., Blanco, I., & Blazquez, S. (2006). Bioavailability of carotenoids and tocopherols from broccoli: *In vivo* and *in vitro* assessment. *Exp Biol Med.* 231, 1733-1738

Griffiths, D. W., Dale, M. F. B., Morris, W. L., & Ramsay, G. (2007). Effects of season and postharvest storage on the carotenoid content of *Solanum phureja* potato tubers. *J Agric Food Chem.* 55, 379-385

Hajslova, J., Schulzova, V., Slanina, P., Janne, K., Hellenäs, K. E., & Andersson, C. (2005). Quality of organically and conventionally grown potatoes: Four-year study of micronutrients, metals, secondary metabolites, enzymic browning and organoleptic properties. *Food Addit Contam.* 22, 6, 514-534

Hakimuddin, F., Paliyath, G., & Meckling, K. (2004). Selective cytotoxicity of a red grape wine flavonoid fraction against MCF-7 cells. *Breast Cancer Res Treat.* 85, 1, 65-79

Hamouz, K., Lachman, J., Hejtmankova, A., Pazderu, K., Cizek, M., & Dvorak, O. (2010). Effect of natural and growing conditions on the content of phenolics in potatoes with different flesh colour. *Plant Soil Environ.* 56, 8, 368-374

Han, J.-S., Kozukue, N., Young, K.-S., Lee, K.-R., & Friedman, M. (2004). Distribution of ascorbic acid in potato tubers and in home-processed and commercial potato foods. *J Agric Food Chem.* 52, 6516-6521

Han, K. H., Sekikawa, M., Shimada, K., Hashimoto, M., Hashimoto, N., Noda, T., Tanaka, H., & Fukushima, M. (2006). Anthocyanin-rich purple potato flake extract has antioxidant capacity and improves antioxidant potential in rats. *Br J Nutr.* 96, 6, 1125-1133

Hirschi, K. D. (2009). Nutrient Biofortification of food crops. *Annu Rev Nutr.* 29, 401-421

Huaman, Z. & Spooner, D. M. (2002). Reclassification of landrace populations of cultivated potatoes (*Solanum* sect. *Petota*). *Am J Bot.* 89, 6, 947-965

Im, H. W., Suh, B. S., Lee, S. U., Kozukue, N., Ohnisi-Kameyama, M., Levin, C. E., & Friedman, M. (2008). Analysis of phenolic compounds by high-performance liquid chromatography and liquid chromatography/mass spectrometry in potato plant flowers, leaves, stems, and tubers and in home-processed potatoes. *J Agric Food Chem.* 56, 9, 3341-3349

Jaganath, I. B., Mullen, W., Edwards, C. A., & Crozier, A. (2006). The relative contribution of the small and large intestine to the absorption and metabolism of rutin in man. *Free Radical Res.* 40, 10, 1035-1046

Jansen, M. A. K., hectors, K., O-Brien, N. M., Guisez, Y., & Potters, G. (2008). Plant stress and human health: do human consumers benefit from UV-B acclimated crops? *Plant Science.* 175, 449-458

Jimenez, M. E., Rossi, A. M., & Samman, N. C. (2009). Phenotypic, agronomic and nutritional characteristics of seven varieties of Andean potatoes. *J Food Compos Anal.* 22, 613-616

Kaspar, K. L., Park, J. S., Brown, C. R., Mathison, B. D., Navarre, D. A., & Chew, B. P. (2011). Pigmented potato consumption alters oxidative stress and inflammatory damage in men. *J Nutr.* 141, 1, 108-111

Kobayashi, A., Ohara-Takada, A., Tsuda, S., Matsuura-Endo, C., Takada, N., Umemura, Y., Nakao, T., Yoshida, T., Hayashi, K., & Mori, M. (2008). Breeding of potato variety "Inca-no-hitomi" with a very high carotenoid content. *Breeding Sci.* 58, 77-82

Kotikova, Z., Hejtmankova, A., Lachman, J., Hamouz, K., Trnkova, E., & Dvorak, P. (2007). Effect of selected factors on total carotenoid content in potato tubers (*Solanum tuberosum* L.). *Plant Soil Environ.* 53, 8, 355-360

Lachman, J., Hamouz, K., Orsak, M., Pivec, V., & Dvorak, P. (2008). The influence of flesh colour and growing locality on polyphenolic content and antioxidant activity in potatoes. *Sci Hortic.* 117, 109-114

Leo, L., Leone, A., Longo, C., Lombardi, D. A., Raimo, F., & Zacheo, G. (2008). Antioxidant compounds and antioxidant activity in "early potatoes". *J Agric Food Chem.* 56, 11, 4154-4163

Levine, M., Conry-Cantilena, C., Wang, Y., Welch, R. W., Washko, P. W., Dhariwal, K. R., Park, J. B., lazarev, A., Graumlich, J. F., King, J., & Cantilena, L. R. (1996). Vitamin C pharmacokinetics in healthy volunteers: Evidence for a recommended dietary allowance. *Proc Natl Acad Sci USA* 93, 3704-3709

Love, S. L. & Salaiz, T. (2004). Stability of expression and concentration of ascorbic acid in North American potato germplasm. *Hortscience*. 39, 1, 156-160

Lu, W. & Haynes, K. (2001). Carotenoid content and color in diploid potatoes. *J Amer Soc Hort Sci*. 126, 6, 722-726

Lukaszewicz, M., Matysiak-Kata, I., Skala, J., Fecka, I., Cisowski, W., & Szopa, J. (2004). Antioxidant capacity manipulation in transgenic potato tuber by changes in phenolic compounds content. *J Agric Food Chem*. 52, 1526-1533

Mäder, J., Rawel, H., & Kroh, L. W. (2009). Composition of phenolic compounds and glycoalkaloids α-solanine and α-chaconine during commercial potato processing. *J Agric Food Chem*. 57, 14, 6292-6297

Madiwale, G. P., Reddivari, L., Holm, D. G., & Vanamala, J. (2011). Storage elevates phenolic content and antioxidant activity but suppresses antiproliferative and pro-apoptotic properties of colored-flesh potatoes against human colon cancer cell lines. *J Agric Food Chem*. doi:10.1021/jf201073g,

Manach, C., Scalbert, A., Morand, C., Remesy, C., & Jimenez, L. (2004). Polyphenols: food sources and bioavailability. *Am J Clin Nutr*. 79, 5, 727-747

Mattila, P. & Hellström, J. (2007). Phenolic acids in potatoes, vegetables, and some of their products. *J Food Compos Anal*. 20, 152-160

Max-Rubner-Institut - Karlsruhe (2010), Bundeslebensmittelschlüssel, In: *Bundesministerium für Ernährung, Landwirtschaft und Verbraucherschutz*, 18-7-2011, Available from http://www.bls.nvs2.de/index.php?id=39

Morris, W. L., Ducreux, L., Griffiths, D. W., Stewart, D., Davies, H. V., & Taylor, M. A. (2004). Carotenogenesis during tuber development and storage in potato. *J Exp Bot*. 55, 975-982

Mulinacci, N., Ieri, F., Giaccherini, C., Innocenti, M., Andrenelli, L., Canova, G., Saracchi, M., & Casiraghi, M. C. (2008). Effect of cooking on the anthocyanins, phenolic acids, glycoalkaloids, and resistant starch content in two pigmented cultivars of *Solanum tuberosum* L. *J Agric Food Chem*. 56, 24, 11830-11837

Natella, F., Belelli, F., Ramberti, A., & Scaccini, C. (2010). Microwave and traditional cooking methods: Effect of cooking on antioxidant capacity and phenolic compounds content of seven vegetables. *J Food Biochem*. 34, 796-810

National Food Institute - Technical University of Denmark (16-3-2009), Nutrition Information and facts, In: *Danish Food Composition Databank - ed.7.01*, 18-7-2011, Available from http://www.foodcomp.dk/v7/fcdb_default.asp

Navarre, D. A., Pillai, S. S., Shakya, R., & Holden, M. J. (2011). HPLC profiling of phenolics in diverse potato genotypes. *Food Chem*. 127, 34-41

Navarre, D. A., Shakya, R., Holden, J., & Kumar, S. (2010). The effect of different cooking methods on phenolics and vitamin C in developmentally young potato tubers. *Am J Pot Res*. 87, 350-359

Nesterenko, S. & Sink, K. C. (2003). Carotenoid profiles of potato breeding lines and selected cultivars. *Hortscience*. 38, 6, 1173-1177

Nutrient Data Laboratory (12-2-2010), USDA National Nutrient Database for Standard Reference, In: *United States Department of Agriculture*, 18-6-2011, Available from http://www.nal.usda.gov/fnic/foodcomp/search/

Penarrieta, J. M., Salluca, T., Tejeda, L., Alvarado, J. A., & Bergenstahl, B. (2011). Changes in phenolic antioxidants during chuno production (traditional Andean freeze and sun-dried potato). *J Food Compos Anal*. 24, 580-587

Pinhero, R. G., Coffin, R., and Yada, R. Y. (2009). Post-harvest storage of potatoes, In: *Advances in Potato Chemistry and Technology*, Singh, J. and Kaur, L., pp.(339-370), Elsevier Inc., 978-0-12-374349-7, Amsterdam

Qin, A., Shi, Q., & Yu, X. (2011). Ascorbic acid contents in transgenic potato plants overexpressing two dehydroascorbate reductase genes. *Mol Biol Rep*. 38, 1557-1566

Reddivari, L., Hale, A. L., & Miller, J. C., Jr. (2007). Genotype, location, and year influence antioxidant activity, carotenoid content, phenolic content, and composition in specialty potatoes. *J Agric Food Chem*. 55, 8073-8079

Rembialkowska, E. (1998). [A comparison of selected parameters of potatoes health quality from ecologically oriented and conventional farms]. *Rocz Panstw Zakl Hig*. 49, 2, 159-167

Reyes, L. F., Miller, J. C., Jr., & Cisneros-Zevallos, L. (2004). Environmental conditions influence the content and yield of anthocyanins and total phenolics in pruple- and red-flesh potatoes during tuber development. *Am J Potato Res*. 81, 187-193

Robert, L., Narcy, A., Rayssiguier, Y., Mazur, A., & Remesy, C. (2008). Lipid metabolism and antioxidant status in sucrose vs. potato-fed rats. *J Am Coll Nutr*. 27, 1, 109-116

Robert, L., Narcy, A., Rock, E., Demigne, C., Mazur, A., & Remesy, C. (2006). Entire potato consumption improves lipid metabolism and antioxidant status in cholesterol-fed rat. *Eur J Nutr*. 45, 5, 267-274

Rodriguez, G. B., Rios, M. D., Rodriguez Rodriguez, E. M., & Diaz, R. C. (2010). Influence of the cultivar on the organic acid and sugar composition of potatoes. *J Sci Food Agric*. 90, 13, 2301-2309

Rogan, G. J., Bookout, J. T., Duncan, D. R., Fuchs, R. L., Lavrik, P. B., Love, S. L., Mueth, M., Olson, T., Owens, E. D., Raymond, P. J., & Zalewski, J. (2000). Compositional analysis of tubers from insect and virus resistant potato plants. *J Agric Food Chem*. 48, 5936-5945

Rommens, C. M., Richael, C. M., Yan, H., Navarre, D. A., Ye, J., Krucker, M., & Swords, K. (2008). Engineered native pathways for high kaempferol and caffeoylquinate production in potato. *Plant Biotechnol J*. 6, 9, 870-886

Rumbaoa, R. G. O., Cornago, D. F., & Geronimo, I. M. (2009). Phenolic content and antioxidant capacity of Philippine potato (*Solanum tuberosum*) tubers. *J Food Compos Anal*. 22, 546-550

Scalbert, A. & Williamson, G. (2000). Dietary intake and bioavailability of polyphenols. *J Nutr*. Supplement, 2073S-2085S

Shakya, R. & Navarre, D. A. (2006). Rapid screening of ascorbic acid, glycoalkaloids, and phenolics in potato using high-performance liquid chromatography. *J Agric Food Chem*. 54, 5253-5260

Singh, A., Sabally, K., Kubow, S., Donnely, D. J., Gariepy, Y., Orsat, V., & Raghavan, G. S. V. (2011). Microwave-assisted extraction of phenolic antioxidants from potato peels. *Molecules*. 16, 2218-2232

Singh, J. and Kaur, L. (2009). Advances in potato chemistry and technology, In: Elsevier, Burlington, USA

Soltoft, M., Nielsen, J., Holst, L. K., Husted, S., Halekoh, U., & Knuthsen, P. (2010). Effects of organic and conventional growth systems on the content of flavonoids in onions and phenolic acids in carrots and potatoes. *J Agric Food Chem.* 58, 19, 10323-10329

Spooner, D. M., Nunez, J., Trujillo, G., Herrera, M. R., Guzman, F., & Ghislain, M. (2007). Extensive simple sequence repeat genotyping of potato landraces supports a major reevaluation of their gene pool structure and classification. *Proc Natl Acad Sci USA* 104, 49, 19398-19403

Spychalla, J. P. & Desborough, S. L. (1990). Superoxide dismutase, catalase, and alpha-tocopherol content of store potato tubers. *Plant Physiol.* 94, 1214-1218

Stalmach, A., Steiling, H., Williamson, G., & Crozier, A. (2010). Bioavailability of chlorogenic acids following acute ingestion of coffee by humans with an ileostomy. *Arch Biochem Biophys.* 501, 1, 98-105

Stevenson, D. E. & Hurst, R. D. (2007). Polyphenolic phytochemicals-just antioxidants or much more? *Cell Mol Life Sci.* 64, 2900-2916

Stratil, P., Klejdus, B., & Kuban, V. (2006). Determination of total content of phenolic compounds and their antioxidant activity in vegetables--evaluation of spectrophotometric methods. *J Agric Food Chem.* 54, 3, 607-616

Stushnoff, C., Ducreux, L. J. M., Hancock, R. D., Hedley, P. E., Holm, D. G., McDougall, G. J., McNicol, J. W., Morris, J., Morris, W. L., Sungurtas, J. A., Verrall, S. R., Zuber, T., & Taylor, M. A. (2010). Flavonoid profiling and transcriptome analysis reveals new gene-metabolite correlations in tubers of *Solanum tuberosum* L. *J Exp Bot.* 61, 4, 1225-1238

Thompson, M. D., Thompson, H. J., mcGinley, J. N., Neil, E. S., Rush, D. K., Holm, D. G., & Stushnoff, C. (2009). Functional food characteristics of potato cultivars (*Solanum tuberosum* L.): phytochemical composition and inhibition of 1 methyl 1nitrosurea induced breast cancer in rats. *J Food Compos Anal.* 22, 571-576

Tudela, J. A., Cantos, E., Espín, J. C., Tomás-Barberán, F. A., & Gil, M. A. (2002). Induction of antioxidant flavonol biosynthesis in fresh-cut potatoes. Effect of domestic cooking. *J Agric Food Chem.* 50, 5925-5931

White, P. J. & Broadley, M. R. (2005). Biofortifying crops with essential mineral elements. *Trends Plant Sci.* 10, 12, 586-593

Williamson, G. & Manach, C. (2005). Bioavailability and bioefficacy of polyphenols in humans. II. review of 93 intervention studies. *Am J Clin Nutr.* 81, 243S-255S

Woolfe, J. A. (1987). Effects of storage, cooking and processing on the nutritive value of potatoes, In: Woolfe, J. A., pp. (83-161), Cambridge University Press, Cambridge

Xu, X., Li, W., Lu, Z., Beta, T., & Hydamaka, A. W. (2009). Phenolic content, composition, antioxidant activity, and their changes during domestic cooking of potatoes. *J Agric Food Chem.* 57, 21, 10231-10238

Yun, S., Habicht, J.-P., Miller, D. D., & Glahn, R. P. (2004). An in vitro digestion/Caco-2 cell culture system accurately predicts the effects of ascorbic acid and polyphenolic compounds on iron bioavailability in humans. *J Nutr.* 134, 2717-2721

Zhou, X., McQuinn, R., Fei, Z., Wolters, A.-M. A., van Eck, J., Brown, C., Giovannoni, J. J., & Li, L. (2011). Regulatory control of high levels of carotenoid accumulation in potato tubers. *Plant Cell Environ.* 34, 1020-1030

Zhu, F., Cai, Y. Z., Ke, J., & Corke, H. (2010). Compositions of phenolic compounds, amino acids and reducing sugars in commercial potato varieties and their effects on acrylamide formation. *J Sci Food Agric*. 90, 13, 2254-2262

6

Beneficial Effects of Fragrances in Beverages on Human Health

Hitoshi Aoshima
Yamaguchi University
Japan

1. Introduction

Foods contain proteins, carbohydrates, lipids and vitamins, among others, and are essential to human life. Beverages such as tea, coffee and liquor however are not essential to life, but are consumed in most countries around the world for mental and physical arousal, or rest and relaxation. Beverages usually contain a physiologically active component such as caffeine or ethanol, which acts mainly on neurotransmitter receptors in the brain, affects mental state, emotion and consciousness, and activates the reword system. It is thought in general that foods and beverages are appreciated for their taste, but the olfactory and somatic sensory systems are also important. The aromas in coffee, teas and liquors induce a desire to drink them. Aromas in beverages are especially important for mental relaxation.

Scent is not essential to human life, though pheromones are used by insects such as ants and bees to maintain social systems, and by some moths and animals to attract partners for copulation. However, fragrances in essential oils have been used as antibacterial compounds and perfumes since very early times. Various leaves have been used to rap foods in order to protect them from bacteria and to add flavor. Recently, aromatherapy has become popular for mental relaxation, to combat the mental and physical stress due to busy modern lifestyles. Most people feel some emotional effect of aroma. The effects of aromatic compounds have not been clarified well from a scientific perspective, since they are quite weak compared with those of medical drugs.

Smells are thought to stimulate the olfactory system and produce signals that project to the olfactory bulb, where smell images are produced, analyzed and recognized by the brain (Buck, 2000). The olfactory bulb is part of the limbic system, along with the hippocampus, amygdala and hypothalamus. Olfactory stimulation is likely to have some effect on these organs. The hippocampus is important for memory establishment and recollection, while the amygdala is related to fear and stress responses. The hypothalamus controls the autonomic nervous, endocrine and immune system. Thus, fragrances have some effect on our mental state, mood or consciousness, through stimulation of the olfactory system (Shepherd, 2006). Reportedly, some fragrances enhanced sympathetic nervous activity and suppressed parasympathetic activity, while others have the opposite effect on autonomic nervous systems in rats.

Most fragrant substances are lipophilic, absorbed into blood through the skin, lungs, stomach and intestines, and enter the brain through the blood-brain barrier non-selectively.

Many fragrant compounds in essential oils and beverages potentiate the response of ionotropic γ-aminobutyric acid receptors ($GABA_A$ receptors) caused by γ-aminobutyric acid (GABA), though they do not act as agonists. Since drugs such as benzodiazepine tranquilizers, barbiturate central nervous depressants, neuro-steroids, general anesthetics and ethanol, also potentiate the response of $GABA_A$ receptors and induce their activity, fragrant compounds may also affect mental state, mood or consciousness, when incorporated into the brain (Aoshima & Hamamoto, 1999). The GABAergic nervous system projects to the hypothalamus, which controls the autonomic nervous, endocrine and immune systems and has some effect on the functions of these systems. For example, potentiation of the GABAergic nervous system suppresses the release of corticotropin-releasing hormone (CRH) from hypothalamus.

Fragrant compounds have psychological effects, stimulating the limbic system and triggering memories, a phenomenon known as the "Proust effect". Proust vividly described how a tea-soaked madeleine brought back powerful childhood memories (Chu & Downes, 2000). Smell is usually perceived together with visual, auditory or tactile stimulation. These sensory systems work synergistically to affect the mental and physical state of humans. For physical and mental health, it is essential to balance the sympathetic and parasympathetic nervous systems in the autonomic nervous system, since this balance is closely related to the endocrine and immune systems. Beverages can be used to balance these systems. Some fragrances in beverages play an important role in enhancing the parasympathetic nervous system and inducing physical and mental relaxation, while others enhance the sympathetic nervous system and induce mental arousal. Moreover, most fragrant compounds in beverages potentiate the response of $GABA_A$ receptors, which induces a tranquilizing effect on the human mind. Thus, the fragrant compounds in beverages affect the homeostasis of mental and physical conditions together with active components such as caffeine and ethanol. Recent studies on the olfactory system are summarized in detail, and the physiological activities of aromatic (fragrant) compounds in beverages are discussed in this review.

2. Production of fragrant compounds

More than twenty thousand compounds, which have a molecular weight of less than about 400 and stimulate the olfactory system are estimated to be present in the world. Most fragrant compounds are produced by plants through two major pathways as described below. Beverages contain many fragrant compounds derived from raw materials. Liquors are produced from fruits, grains and sweet potatoes, teas such as green tea, oolong tea and black tea are produced from tea (*Camellia sinensis*) leaves, and coffee is produced from coffee (*Coffea arabica*) beans.

2.1 Fragrances from plants

Terpene compounds in essential oils are produced from isoprene (2-methyl-1,3-butadiene) derivatives in plants. Condensation of isopentenyl diphosphate (IPP) and dimethylallyl diphosphate (DMAPP) produces geranyl diphosphate. IPP is produced from mevalonate or deoxyxylulose phosphate, while DMAPP is produced from IPP by an isomerase. Isomerization, oxidation, cyclization and dephosphorization produce various monoterpenes. Then, sesquiterpene, diterpene, and sesterterpene are produced by the addition of an isoprene residue from IPP (Dewick, 2002). Triterpene and tetraterpene are produced from the dimerization of sesquiterpenes and diterpenes, respectively.

The other pathway involves oxylipin. Leaf aldehyde ((*E*)-2-hexanal), leaf alcohol ((*Z*)-3-hexanol), and (*Z*,*Z*)-3,6-nonadienal are produced by the oxygenation of unsaturated fatty acids by lipoxygenase and their scission by lyase (Hatanaka, 1993). Jasmonic acid is produced from lipid hydroperoxide by enzymes such as allene oxide synthase (Matsui, 20066). The structural formulae of popular fragrant compounds produced through these two pathways are shown in Fig. **1**.

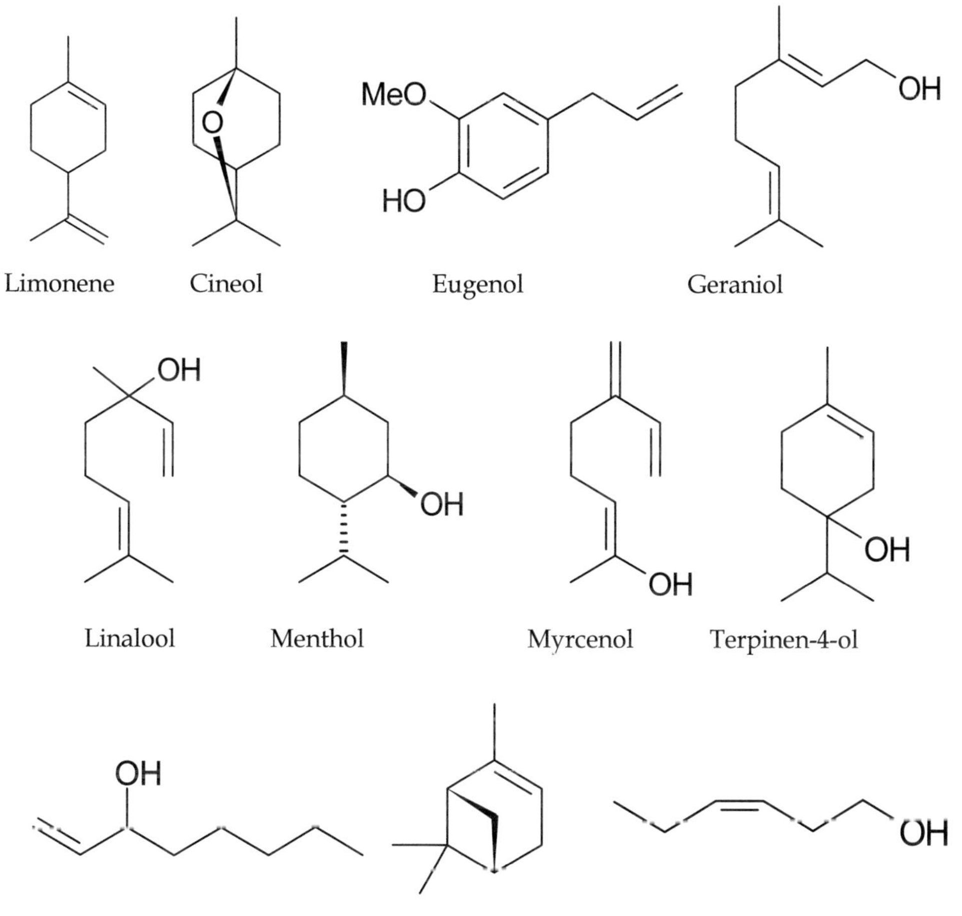

Fig. 1. Chemical structure of several popular fragrant compounds.

2.2 Fragrances in beverages

Many fragrant compounds in beverages origin from raw materials such as tea leaves, coffee beans, fruits or grains (Maase, 1991). Many fragrant compounds are also produced during the processing of beverages (Maase, 1991). Not only ethanol, but also other alcohols such as *n*-propanol, *iso*-buthanol and *iso*-amyl alcohol are produced during fermentation by yeast when liquors are made from fruits and grains. During the fermentation process, fusel aldehydes and carboxylic acids are also produced by the oxidation of fusel alcohols, and esters such as ethyl acetate and *iso*-amyl acetate are produced by condensation between the alcohols and carboxylic acids. Whiskey and red wines are stored in oak barrels for many years for aging, during which aromas (Fig, **2**) and pigments move into the liquors from oak wood. Hops are added to the beer to give it a characteristic bitter taste and many fragrant compounds. Hops have floral fragrances such as linalool, geraniol and 1-octen-3-ol. Higher alcohols such as myrcenol and humulenol from myrcene and humulene in hops are produced during the boiling of sweet wort, the extract of the mixture of malt and hops. Liqueurs are produced from liquors by addition of various herbs, fruits or nuts, which contain many fragrant compounds.

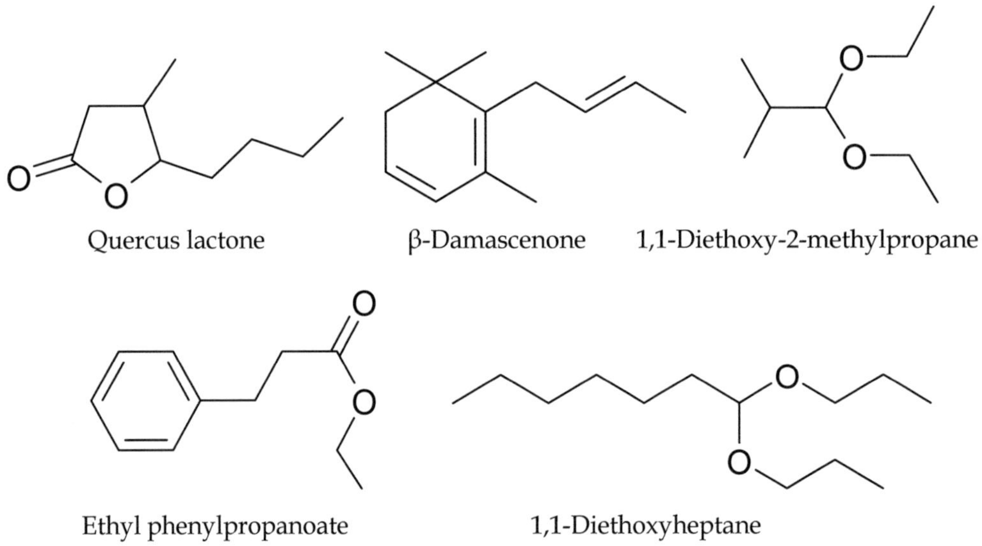

Fig. 2. Some fragrant compounds in whiskey (or whisky).

Green, oolong and black teas are produced from tea (*Camellia sinensis*) leaves, which contain caffeine as an active component, by heating, crumpling and drying (Maase, 1991). To produce green tea, tea leaves are heated first to stop the enzymatic reaction in tea leaves by applying hot steam or heating in a caldron, then crumpled and dried. Tea leaves are crumpled without destroying the cells in the production of oolong tea, while they are crumpled to destroy the cells in the production of black tea. During the crumpling, enzymes such as oxidases or plolymerases of catechins and glucosidases of glycosides of fragrant compounds catalyze various reactions. The composition of fragrant substances, polyphenols and pigments in different teas changes and is dependent on the processing. Then these teas are heated and dried to produce oolong tea or black tea. Typical fragrant compounds in green tea and black tea are summarized in Table **1**. Herbal teas are produced from herbs by drying and contain many fragrant components (Nagashima, 2010). Herbal teas do not usually contain bio-active compounds such as caffeine, but their aromatic substances may act on $GABA_A$ receptors. It is thought that herbal teas induce mental relaxation through both olfactory stimulation and potentiation of the response of $GABA_A$ receptors. It has been also found that green tea polyphenol (-)-epigallocatechinn gallate (EGCG) has anxilolyric-like effects by interacting with the $GABA_A$ receptor (Vignes, et al., 2006).

Coffee is produced from coffee beans, which contain caffeine as an active component, by roasting and crushing. During roasting, phenol derivatives, which are smoky, spicy or burnt, are produced from thermal degradation of lignin. The following compounds are also produced during roasting and play an important role in determining the flavors of coffee (Table **2**) (Maase, 1991). Furfural derivatives are produced by a caramelization of sugars and

Compound	Green tea	Black tea	Characteristic
cis-3-Hexen-1-ol	++	++	Green odor
Nerolidol	+++	+++	Woody, milky and deep odor
α-Cadinol	+++	NM	Woody, milky and deep odor
Benzyl alcohol	+++	++	Jasmine and ylang-ylang like odor
cis Jasmone	+++	+++	Jasmine like odor
β-Ionone	+++	+++	Fragrant orange-colored olive like odor
Linalool	++	+++	Lavender and daphne like odor
Phenylethyl alcohol	+	++	Roselike odor
Geraniol	+	+++	Roselike odor
Indole	+++	NM	Floral odor at low concentration
Pyrrole	++	NM	Floral odor at low concentration

+: small amount, ++: middle amount, +++: large amount, NM: not determined. Amount of fragrant compounds in teas are expressed qualitatively, since they are very variable and depend on tea leaves and their processing.

Table 1. Fragrances in green and black tea. (Yamanishi, 1992)

Compound	Character
Limonene	Weak good odor
β-Myrcene	Pleasant odor
1-Octen-3-ol	Mushroom like odor
(E)-2-Nonenal	Green cucumber like odor
2-Methylbenzaldehyde	Plum like odor
Methyl phenyl acetate	Honey or jasmine like odor
4-Butanolide	Weak sweet odor
Maltol	Sweat caramel like odor
Phenol derivatives	Smoky, spicy and burnt odor
Furane, thiophen derivatives	Toasted, caramel-like and nutty burnt odor
Pyrrole derivatives	Caramel like odor
Oxazole derivatives	Natty, sweet and green odor
Pyridine derivatives	Green, bitter, roasted and burnt odor
Pyrazine derivatives	Sweet and toasted odor

Table 2. Fragrances in coffee beverages (Maase, 1991)

have a toasted penetrating odor and (or) caramel-like and nutty burnt flavor. Sulfur-containing furfural compounds are important to the flavor of roasted coffee. Thiophene, pyrrole, oxazole, thazole, pyridine and pyrazine derivatives also contribute to the complex and attractive flavors of coffee. Arabica coffees have better, milder and sweeter fragrances than Robusta ones.

3. Olfactory system

3.1 Olfactory system and smell images

The olfactory system has been studied extensively ever since Buck and Axel first reported the olfactory receptors (Buck & Axel, 1991), G protein-coupled receptors whose genes have seven trans-membrane domains. These receptors are thought to activate adenylate cyclase, increasing the cyclic adenosine monophosphate (cAMP) concentration. The family of olfactory receptors has about 1000 members in rodents, and 380 members in humans. The olfactory cell has only one type of olfactory receptor. Fragrant compounds bind to several olfactory receptors differing in affinity. The axons of about ten thousand sensory neurons in olfactory cells with the same species of receptors project to the same glomerulus in the olfactory bulb. The glomeruli are excited dependent on the concentration of the fragrant compounds and the affinity for the receptors. Thus, information on aromas received in the olfactory epithelium is converted to topological maps of activated glomeruli, *i.e.* smell images, which are analyzed and perceived by the brain (Fig. **3**) (Buck, 2000;Shepherd, 2006). The existence of smell images was confirmed directly using high-resolution functional magnetic resonance imaging (fMRI) in mice (*Mus musculus*). A homologous chemical series such as aldehydes with different chain lengths elicits patterns that overlap but have different spatial patterns of activity in the glomerular layer of the olfactory bulb (Xu et al., 2003).

The smell images in the olfactory bulb are subjected to processing by the olfactory cortex and relayed to the primary olfactory cortex in the orbitofrontal cortex, a part of the prefrontal lobe, through mitral cells and the piriform (Fig. **3**). The signals in the piriform are also relayed to the entorhinal cortex, hippocampus, amygdala, and hypothalamus in the limbic system. Since the olfactory perceptual system is closely linked to systems for learning, memory, emotion and reward, aromas have various effects on mental state, *i.e.* consciousness, emotion and instinct. Aromas also influence the autonomic nervous system, endocrine system and immune system, which are controlled by the hypothalamus (Julius & Katz, 2004).

It is important to clarify how the smell images in the olfactory bulb are interpreted in the brain. However, it is reported that stimulation of a specific glomerulus by an odor induces a specific behavior in fruitflies or mice, suggesting the presence of specific circuits from the olfactory cells, *i.e.* olfactory receptors to the brain.

3.2 Innate and learning pathways

The fruitfly (*Drosophila melanogaster*) exhibits robust and innate olfactory-based avoidance behavior in response to CO_2. Specialized neurons with Gr21a/Gr63a CO_2 receptors in the antenna and a dedicated neuronal circuit in the higher olfactory system mediate CO_2 detection and avoidance. Both 1-hexanol and 2,3-butanedione which are often emitted from bananas and yeasts inhibit the response of the CO_2 receptor, which allows fruitflies to find and eat their foods (Turner & Ray, 2009). Fruitflies are markedly attracted to food odors.

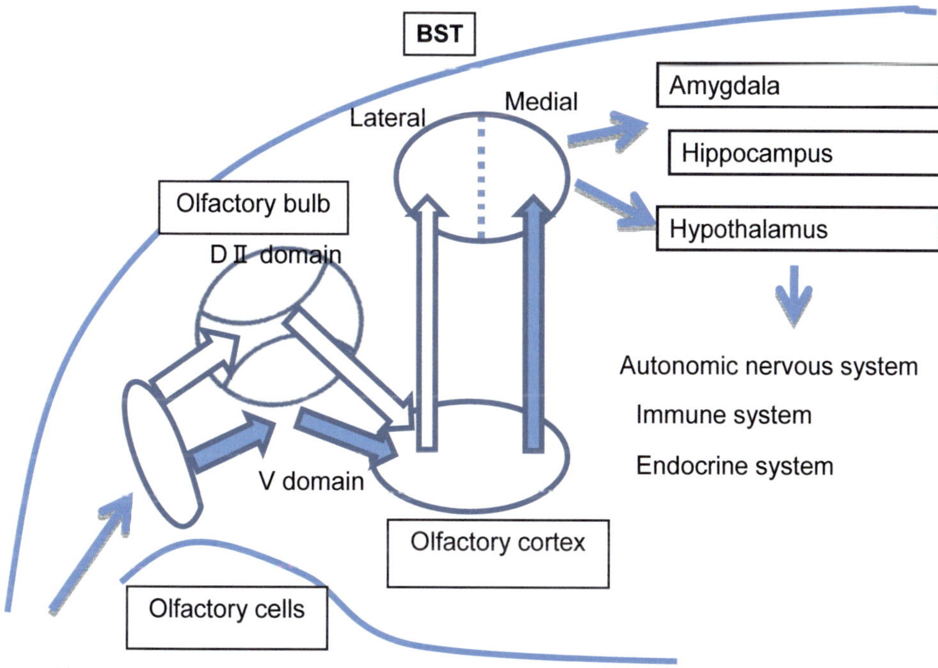

Fig. 3. Olfactory system and innate (gray arrays) and learning (white arrays) pathways. (Kobayakawa et al., 2007).

Semmelhack and Wang used genetic tools to dissect the contribution of each of six glomeruli activated by apple cider vinegar and found that an absence of activity in two glomeruli markedly reduced the attractive effect (Semmelhack & Wang, 2009). When each of these two glomeruli was selectively activated, the flies showed as robust an attraction to the vinegar as wild-type flies. A higher concentration of the vinegar excited an additional glomerulus and was less attractive to the flies. Activation of the extra glomerulus was necessary and sufficient to mediate the behavioral switch. These results indicate that individual glomeruli, rather than the overall pattern of activation (smell images), mediate the innate behavioral output of fruitflies.

Carey *et al.* expressed 72 odorant receptors of the mosquito *Anopheles gambiae* in mutant neurons of *Drosophila melanogaster* lacking endogenous odorant receptors, and characterized electrical responses to 110 chemically diverse odorants (Carey et al., 2010). The receptors of *A. gambiae* responded strongly to components of human odor and might aid in the process of human recognition. The odorants were differentially encoded by the two species in ways consistent with their ecological needs to find their foods.

Koyabakawa et al. (2007) generated mutant mice in which olfactory sensory neurons in a specific area of the olfactory epithelium are ablated by targeted expression of the diphtheria toxin gene. The mutant mice lacked innate responses to aversive odorants, even though they were capable of detecting them and could be conditioned for aversion with the remaining glomeruli. In mice, aversive information caused by trimethyl-thiazoline secreted from the anal gland of foxes was received in the olfactory bulb by separate sets of glomeruli, those dedicated to innate responses and those for learned responses. The aversive signals are transferred through a dorsal domain for class II odorant receptors (D_{II} domain) in the olfactory bulb, the olfactory cortex, the medial aspect in the bed nucleus of the stria terminalis and hypothalamus (gray arrays), while the learned signals are transferred through a ventral domain for class II odorant receptors (V domain) in the olfactory bulb, the lateral division in the bed nucleus of the stria terminalis (white arrays) (Fig. **3**). It is thought that humans have no pheromone, which affects the growth or behavior of creatures, since humans have no vomeronasal organ, which detects specifically pheromones and induces specific effects (Buck, 2000). However, humans may have innate olfactory circuits for dangerous compounds such as H_2S and NH_3, which are present around active volcanoes or produced from rotten foods.

3.3 Orthonasal and retronasal stimulation

Smell is unique in having a dual nature, that is, it can sense signals originating outside (orthonasal) and inside (retronasal) the body (Shepherd, 2006). Orthonasal stimulation refers to sniffing in through the external nares of the nose to activate sensory cells in the olfactory epithelium. Good flavors attract people to beverages. Retronasal stimulation occurs during the ingestion of food and beverage, when volatile molecules released in the mouth are pumped, by movements of the mouth, from the back of the oral cavity up through the nasopharynx to the olfactory epithelium. This stimulation is especially important when foods are taken together with beverages such as tea and whiskey. The re-emergence of powerful childhood memories in response to a tea-soaked madeleine, so vividly described by Marcel Proust (Chu & Downes, 2000), would have occurred primarily through the retronasal pathway. It is likely that fragrant compounds in beverages are detected by the odor-reward association learning system, since humans feel better as they drink in part because of an addiction to caffeine or ethanol.

3.4 Adaptation and masking

Cilia in olfactory receptor cells produce electrical signals on the binding of aromatic compounds to olfactory receptors as shown in Fig. **4**. It is thought that these receptors activate adenylate cyclase (○), increasing the cAMP concentration. The cAMP opens cyclic nucleotide-gated Ca^{2+} channels, causing the influx of Ca^{2+} into the cells and depolarization of the cell membrane. Then, Ca^{2+} opens Ca^{2+}–activated Cl^- channels and causes further membrane depolarization, since the olfactory cells have abnormal Cl^- concentrations between the inside and outside of the membrane, *i.e.*, almost equal Cl^- concentrations between the inside and outside. Thus the signals induced by aromas are amplified and produce action potentials in the cells. These action potentials are transferred to the glomeruli in the olfactory bulb.

Adaptation to odorants is thought to begin at the level of olfactory receptor cells, presumably through modulation of their transduction machinery. Kurahashi & Menini (1997) studied the adaptational mechanism in intact olfactory cells of newts by using a combination of odorants and caged cAMP photolysis which produces current responses. Odorant- and cAMP-induced responses showed the same adaptations in a Ca^{2+}–dependent manner, indicating that the adaptation occurs entirely downstream of the adenylate cyclase. The Ca^{2+}–activated Cl^- channels did not show adaptations when Ca^{2+} was applied to the cells by caged Ca^{2+} photolysis. Thus, the principal mechanism underlying odorant adaptation is actually modulation of the cAMP-gated channel by Ca^{2+} feedback, *i.e.*, a change in affinity of the channel for the ligand.

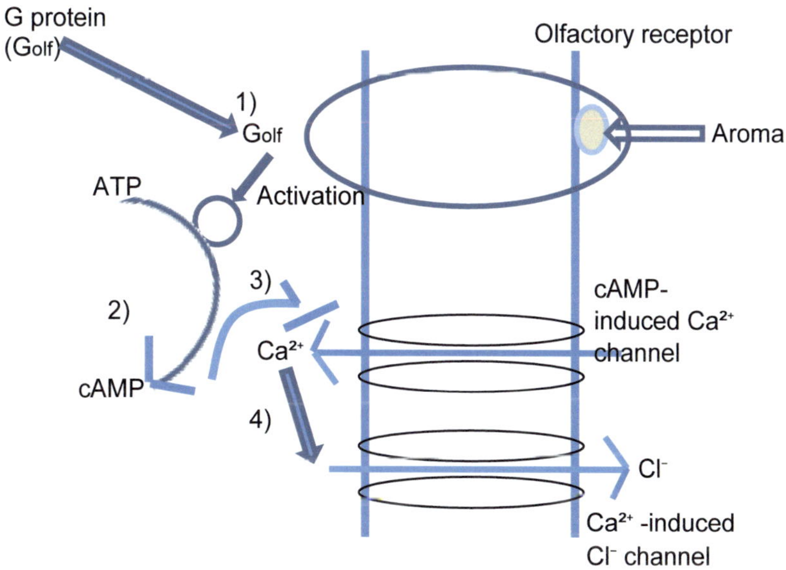

Fig. 4. Production of electrical signals in olfactory receptor cells (Takeuchi et al., 2009).

Other mechanisms of adaptation in the central nervous system were proposed by Linster et al. (2009), who showed that odor-specific adaptations in piriform neurons, mediated at least partially by synaptic adaptations between the olfactory bulb outputs and piriform cortex pyramidal cells, are highly odor specific, while those observed at the synaptic level are specific only to certain odors.

Olfactory masking has been used to erase unpleasant sensations by bad odorants. Takeuchi et al. (2009) measured the currents of both cAMP-gated Ca^{2+} channels and Ca^{2+}-activated channels in the cilia of olfactory receptor cells by the photolysis of caged compounds, using the whole cell patch clamp method. They found that 16 odorants suppressed the response of the cAMP-gated Ca^{2+} channels with different sensitivities, but showed no effect on the response of the Ca^{2+}-activated channels. Ringer's solution pre-exposed to odorant-containing air affected the cAMP-induced current of single cells. Using the same odorants, in parallel, they measured human olfactory masking with 6-rate scoring tests and obtained a correlation coefficient of 0.8 with the channel block. Thus olfactory masking in the sensory cilia is induced by the inhibition of cAMP-gated Ca^{2+} channels by odorants. Tripral, geraniol, linalool dihydromyrcenol and benzaldehyde among 16 chemicals showed strong adaptation in terms of both cAMP-gated Ca^{2+} channels and human olfactory masking tests. Since these compounds differ significantly in structure, adaptation of the olfactory system seems to occur for many types of odorants. These findings may facilitate the design of masking chemicals, targeting olfactory manipulation at the ciliary membrane. Geraniol, linalool or myrcenol in beverages such as tea and beer may induce masking of unpleasant odors such as amines or mercaptans produced in foods such as meat and fish.

Other mechanisms for olfactory masking are also proposed. Takahashi et al. (2004) mapped alkylamine-responsive glomeruli to a subregion of the aliphatic acid-responsive and aldehyde-responsive cluster in odor maps of the rat olfactory bulb and found that fennel and clove, species known to add flavor and mask fatty, fish odors, activated glomeruli in the surrounding clusters and suppressed the alkylamine-induced and acid-aldehyde-induced responses of mitral cells, suggesting that the masking is mediated, in part, by lateral inhibitory connections in the maps.

4. Effects through the olfactory system

4.1 Positron emission tomography

Odors are believed to be identified based on patterns of glomerular activity from information encoded in the olfactory bulb. Shepherd's group studied the spatial patterns of activity elicited in the rat olfactory bulb by odors such as amyl acetate, camphor, cage air, dimethyl disulfide, and pure air, using the 2-deoxyglucose autoradiography technique (positron emission tomography: PET), and found them to be different but overlapping (Stewart et al., 1979). The regions of activity were greatest in extent and density with the highest odor concentrations. These results define the regions within which more restricted and isolated foci appear at lower concentrations. The results provide evidence for the specific role of spatial factors in the neural processing of the quality and concentration of an odor. Further, Johnson et al. (2005) studied whether interactions between fragrant functional groups and the hydrocarbon structure influence activity in glomerular response modules, and the effects of the positions of functional groups on spatial representations of aliphatic odorants and the chemotopic representations of aromatic odorants in the rat olfactory bulb. They concluded that different odors are represented by different patterns of spatial activity in the olfactory bulb.

4.2 Electroencephalography

The effects of fragrances on human mood have been studied by electroencephalography (EEG). It is known that α waves increase empirically when people close their eyes and are relaxed. Field et. al. (2005) measured EEG patterns and heart rate for 11 healthy adults who sniffed a cosmetic cleansing gel with a lavender fragrance. The lavender smell had a significant transient effect of improving mood, allowing the subjects to feel more relaxed and perform math computations faster. A specific cosmetic fragrance can have a significant role in enhancing relaxation. Reportedly, fragrances of wine, whiskey, beer and coffee also increased α waves of human subjects, suggesting relaxation (Yokogoshi, 2006). Sniffing of six red or white wines increased α waves from 12 to 13 Hz compared to a 12% ethanol aqueous solution. The increase in α wave caused by the wines had a close correlation with mental state. With the increase of wine aroma concentrations, different kinds of wines showed some difference in their effect on α waves. Spectral information on the frequency fluctuation of α waves calculated for each individual is related to psychologically evaluated values of positive-negative moods and feelings of arousal to identify correspondence between the values of fluctuation characteristics and psychological conditions. A mixture of ethyl acetate and isoamyl acetate significantly exhibited a relaxing (lowering arousal) effect on humans, suggesting that ester flavors contribute to the increase in relaxed feeling one experiences while drinking beer (Yokogoshi, 2006).

Event-related potential has been studied by electroencephalography as fragrances of whiskey or coffee are administered by inhalation. These fragrances increased the amplitude of the late component at 300 msec (P300) of evoked potential change and decreased the latency, suggesting that they induced mental relaxation and improved the ability to manage information (Chuyen & Ishikawa, 2008; Yokogoshi, 2006).

4.3 Green grass odor

The equivalent mixture of (*Z*)-3-hexenol and (*E*)-2-hexenal (leaf alcohol and leaf aldehyde), "green odor", is present in teas and known to have a healing effect on the psychological damage caused by stress. Behavioral studies in humans and monkeys have revealed that green odor prevents the prolongation of reaction time caused by fatigue. Nakashima et. al. (2004) investigated the effect of the green odor on elevations in plasma adrenocorticotropic hormone (ACTH) levels induced by restriction stress in male rats. Rats that inhaled the odor while under stress showed a significant reduction in plasma ACTH levels in comparison with the vehicle-treated group. Sasabe *et. al.* investigated the regions of the brain activated by green odor using positron emission tomography (PET) with alert monkeys and found that not only the prepyriform area (the primary olfactory cortex) and the orbitofrontal cortex (the secondary olfactory cortex), but also the anterior cingulate gyrus were activated (Sasabe et al., 2003).

4.4 Effect of fragrances on the autonomic nervous system

Nagai's group measured the effect of olfactory stimulation with scents of essential oils and their main components on the autonomic nerves, lipolysis and appetite in urethane-anesthetized rats. They measured the effect on both renal sympathetic nervous activity (RSNA) and gastric vagal (parasympathetic) nervous activity (GVNA). They observed that olfactory stimulation with the scent of grapefruit oil and its major component, limonene, for ten minutes enhanced RSNA and suppressed GVNA, increased the plasma glycerol concentration, blood pressure and body temperature, and decreased appetite (Shen et al., 2005a), while stimulation with the scent of lavender oil and its major component, linalool, had the opposite effects on the autonomic nerves, the plasma glycerol concentration and appetite (Shen et al., 2005b). The effects of essential oils and their components on the autonomic nerves were induced through the olfactory system, since local anesthesia of the nasal mucosa with xylocaine or anosmic treatment using $ZnSO_4$ eliminated the autonomic changes. Intracerebral administration of diphenhydramine, a histaminergic H1-antagonist, abolished the effect of olfactory stimulation with the grapefruit essential oil and limonene on RSNA, GVNA and blood pressure. Bilateral lesions of the hypothalamic suprachiasmatic nucleus (SCN) eliminated the aroma-mediated increases in RSNA and blood pressure and decrease in GVNA. These results suggest that the smell of grapefruit essential oil affects autonomic neurotransmission and blood pressure through central histaminergic nerves and the SCN (Tanida et al., 2005). The effects of essential oils of grapefruit and lavender, and their components, are summarized in Table **3**. In further studies, essential oils of rosemary, lemon, fennel, estragon, ylang ylang, peppermint, geranium (Egypt), lemon grass and coriander, and cineol enhanced RSNA and suppressed GVNA as grapefruit oil did, while essential oil of chamomile suppressed RSNA and enhanced GVNA as lavender oil did (Niijima, 2008). Though aromatic compounds are usually thought to have tranquillizing effects, these results suggested that most essential oils acted on the sympathetic nervous activity in the autonomic nervous system. Herbal teas such as chamomile or lavender are expected to exhibit tranquillizing activity, while herbal teas containing citrus peel, peppermint, lemon grass or coriander induce mental arousal.

Aroma	Sym. NS	Parasym. NS	BP	PGC	Temperature	Weight
Grapefruit essential oil (rich in limonene)	↑	↓	↑	↑	↑	↓
Lavender essential oil (rich in linalool)	↓	↑	↓	↓	↓	↑

Sym. NS: Sympathetic nervous system, Parasym. NS: Parasympathetic nervous system,
BP: Blood pressure, PGC: Plasma glycerol concentration
↑: Increase, ↓:Decrease

Table 3. Effect of grapefruit and lavender essential oil on the autonomic nervous system in rats (Shen et al., 2005a,b).

The effect of fragrance of Scotch-type whiskey (Hibiki produced by Suntory Ltd., in Japan, and stored in an oak barrel for 17 years) on the autonomic nervous system was examined by a similar method (Niijima et al., 2009). Olfactory stimulation with fragrance of whiskey increased GVNA and inhibited RSNA. These observations suggest that fragrance of whiskey may activate gastric movement and secretion of gastric juice through the vagus nerve, and decrease energy expenditure through suppression of the sympathetic nerve activity. The effect of the administration of whiskey fragrance for ten minutes on the autonomic nerve system continued for 2 hours after the administration.

Fushiki's group studied the effect of inhaling the aroma of jasmine tea and lavender on the autonomic nervous system by conducting a power spectral analysis of heart rate variability in human subjects (Inoue et al., 2003;Kuroda et al., 2005). It is thought that parasympathetic nervous activity increases in relation to spectral integrated values for high-frequency components. The jasmine tea and lavender caused significant decreases in heart rate and significant increases in spectral integrated values for high-frequency components in comparison with the control, suggesting the activation of the parasympathetic nervous system which induces sedative effects. Dayawansa et. al. (2003) measured the effect of cedrol, the main component of the essential oil of cedar wood, on the autonomic function of healthy individuals. A spectral analysis of heart rate indicated an increase in high-frequency components (index of parasympathetic activity), and a decrease in the ratio of low-frequency to high-frequency components (index of sympathovagal balance) during inhalation, which is consistent with the idea of a relaxant effect of cedrol.

5. Effects of fragrances through incorporation into the body

It has been studied extensively how active components such as ethanol and caffeine act on neurotransmitter receptors and affect the mental state, inducing addiction. Ethanol potentiates the response of $GABA_A$ receptors composed of $\alpha_4\beta_3\delta$ and $\alpha_6\beta_3\delta$ subunits in outer synaptic regions under physiological conditions (Martin & Olsen, 2000), though it may also inhibit NMDA receptors and voltage-dependent channels. Caffeine inhibits adenosinA2a receptors noncompetitively under physiological conditions (Yoshimura, 2006), though it may also inhibit both $GABA_A$ receptors and phosphodiesterase. These activities are modulated further by the fragrant compounds in beverages.

5.1 Incorporation of fragrances into the body

Effects of aromatic compounds not only through stimulation of the olfactory system, but also through incorporation into the body are described in most technical books on aromatherapy. Incorporation into the body through the skin has been reported by some researchers. The rate at which compounds move through cell membranes generally increases with their hydrophobicity and a decrease in their molecular weight. Since one can feel the effects of ethanol on consciousness soon after drinking liquor, the idea that fragrants of higher alcohols are also incorporated into the brain makes sense. Reportedly, levels of fragrances such as linalool and its acetyl ester in blood reach a plateau 15-20 min after their administration to human skin by massage and decrease gradually for 2 h. Cal (2006) studied the percutaneous absorption of terpenes such as linalool, terpinen-4-ol, citronellol and α-pinene. Components of essential oils accumulated in the mouse brain following percutaneous absorption or exposure to vapor, since they are usually hydrophobic. Ylang ylang (*Cananga odorata*) oil caused a significant decrease in blood pressure and a significant increase in skin temperature when administered transdermally (Hongratanaworakit et al., 2006). At a behavioral level, subjects given ylang ylang oil dissolved in pure sweet almond oil rated themselves as calmer and more relaxed than subjects in the control group given only pure sweet almond oil. Linalool was applied percutaneously to 14 healthy subjects and induced deactivation with respect to physiology, that is, a decrease of systolic blood pressure and a smaller decrease of skin temperature, compared to the corresponding control group. However, the target of aromatic compounds after their incorporation into the body has not been clarified.

5.2 $GABA_A$ receptors and their potentiation

We have found that many fragrant compounds potentiated the electrical response of $GABA_A$ receptors expressed in *Xenopus* oocytes by injection of poly(A)$^+$RNA or cRNA of the receptors (Fig. **5**, Fig. **6 a** and **b**) (Aoshima & Tenpaku, 1997; Hossain & Aoshima, 2008). $GABA_A$ receptors are ligand-gated ion channels whose subunits have similar amino acid sequences to those of ionotropic nicotinic acetylcholine, serotonine (type 3), and glycine recptors (Chebib & Johnston, 2000; Johnston et al., 2006; Nicholls, 1994). They are thought to have heteropentamers made up of subunits likely derived from a common ancestor. The cDNAs of about 20 mamalian $GABA_A$ receptor proteins have been cloned, including α, β, γ, δ, ε, π, and ρ subunit cDNAs, together with a few splice variants. There may be 50-100 combinations expressed in the brain. The different combinations may result in different pharmacological characteristics, for example, affinity for GABA. $GABA_A$ receptors are mood-defining receptors and have a complex pharmacology, with binding sites for direct GABA agonists and antagonists, together with multiple allosteric sites for benzodiazepine tranquillizers, barbiturate central nervous system depressants, both synthetic and endogenous steroids, general anesthetics, and ethanol. Since many aromatic compounds are higher alcohols, it is not unusual that they also potentiate the response of $GABA_A$ receptors.

The effects of the functional groups of various six-carbon hydrocarbons on the response of $GABA_A$ receptors were studied by expressing the receptors in *Xenopus* oocytes (Aoshima et al., 2001). 1-Hexanol best potentiated the response. Hexanal, butyl acetate and hexylamine potentiated the response slightly, while hexanoic acid inhibited the response weakly in a competitive manner. The potentiating effect of alcohols increased along with chain length.

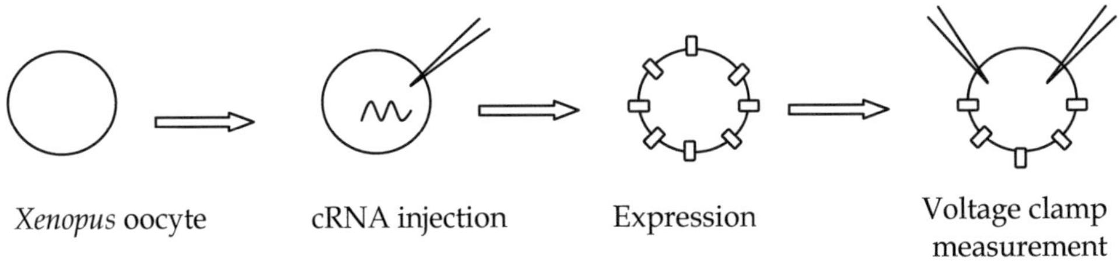

Fig. 5. The experimental procedures using the *Xenopus* oocyte expression system and electrophysiological measurements to examine the effects of fragrant compounds on the response of $GABA_A$ receptors.

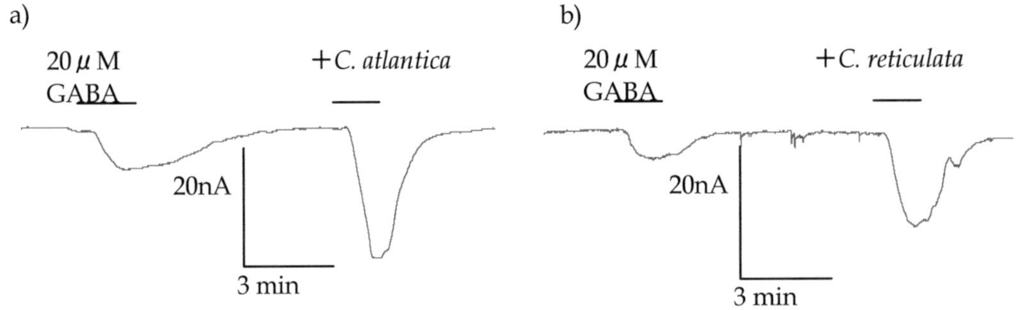

Fig. 6. Potentiation of the response of $GABA_A$ receptors expressed in *Xenopus* oocytes when essential oil of a) *C. Atlantica (Atlas cedar)* or b) *C. Reticulate (Mandarin orange)* was co-applied with GABA (Aoshima et al., 2009), taken with permission from Fragrance Journal Ltd, Japan.

The potentiation of the response depended on the GABA concentration, that is, less than 20 μM potentiated the response, while more than 100μM did not. The addition of aromatic compounds to GABA solutions shifted the dose-response curve of GABA to the lower concentrations, similar to anesthetics (Franks & Lieb, 1994).

Since ionotropic neurotransmitter receptors are thought to have evolved from a common ancestor, they have not only a similar structure but also a common functional mechanism. A minimum schematic model for nicotinic acetylcholine receptors (Hess et al., 1983) has been developed to explain the potentiation of the response of the $GABA_A$ receptors as shown in Fig. 7. The binding of two agonist molecules to the receptor (RL_2) with a dissociation constant of K_1 is necessary to open the channel. The receptors with two bound agonists are converted to an open channel (RL_2(open)), reaching equilibrium. The receptors occupied by the agonist(s) change their structure to a desensitized form (D) and close the channel, at a rate less than that of channel opening. The potentiator (P), *i.e.* aromatic compound, binds to the receptor with a dissociation constant of K_p. The receptor occupied by the potentiator (RP) binds the GABA molecule with a dissociation constant of K_{1p}. Since K_{1p} is less than K_1, the open form of the receptor with the potentiator increases and the response is potentiated in the presence of the potentiator, when the GABA concentration is low (Aoshima et al., 2001).

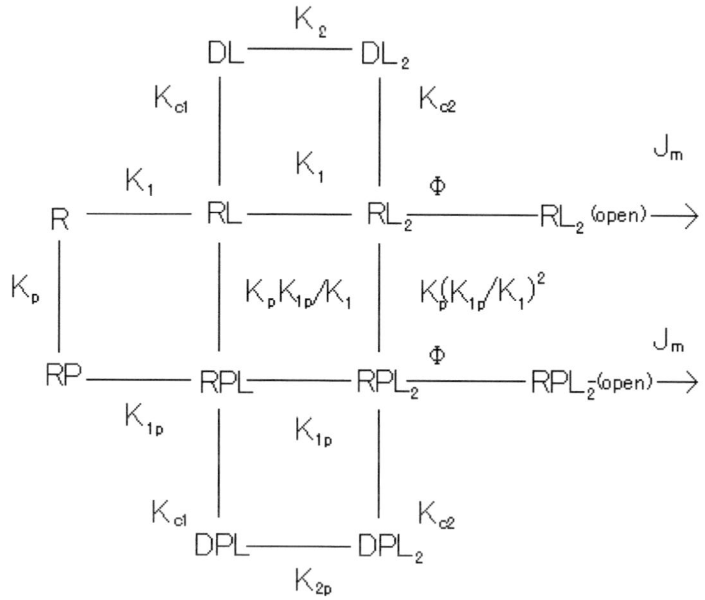

Fig. 7. The simplest schematic model accounting for the potentiation of the $GABA_A$ receptor-response (Aoshima et. al., 2001), taken with permission from the Japanese Biochemical Society. The affinity for GABA of the $GABA_A$ receptors increases ($K_{1p} < K_1$) when aromatic compounds bind the receptor with a dissociation constant of K_p, increasing the open-form of the receptor and potentiating the response.

5.3 Target of ethanol against $GABA_A$ receptors

It is reasonable to think that higher alcohols bind to the same site as ethanol. The potentiation site of ethanol is expected to be composed of α and β subunits, though the γ subunit is essential for potentiation of the response of the $GABA_A$ receptors by benzodiazepine. By using chimeric receptor constructs, Mihic et al. (1997) identified a region of 45 amino-acid residues that is both necessary and sufficient for the enhancement of receptor function. Within this region, two specific amino-acid residues in trans-membrane domains 2 and 3 (TM2 and TM3) are critical for allosteric modulation of $GABA_A$ receptors by alcohols and volatile anesthetics. The potentiation site of the receptors appears to have both hydrophobic and hydrophilic group-binding regions. The hydrophilic group-binding region, which recognizes the functional group, binds best to a hydroxyl group. Not only alcohols, but also phenol derivatives potentiated the response of the $GABA_A$ receptors strongly, though polyphenols inhibited the response (Aoshima et al., 2001). The hydrophobic group-binding region is large enough to bind a hydrocarbon of at least 10 carbon atoms (Aoshima et al., 2001).

Wallner et al. (2003) reported that ethanol at low concentrations enhanced the response of $\alpha_4\beta_3\delta$ and $\alpha_6\beta_3\delta$ $GABA_A$ receptors expressed in *Xenopus* oocytes. These receptors are usually present in extrasynaptic regions of neurons. Thus higher alcohols possibly bind to the same site and receptors as ethanol, and enhance the response of the $GABA_A$ receptors. However, effects of fragrant substances on the responses of $GABA_A$ receptors composed of various combinations of subtypes have not been examined yet.

5.4 Potentiation of the response of $GABA_A$ receptors by liquors

The potentiation of the $GABA_A$ receptors by various compounds is summarized in Table **4** (Aoshima et al., 2008). Flavors in beverages are very important and determine their quality. Effects of flavors in tea, coffee, whiskey and beer on $GABA_A$ receptor's responses have been studied using an *Xenopus* oocyte expression system and electrophysiological measurements (Hossain et al., 2007). Many components of the flavors in the beverages potentiated the response of the receptors. Since most of the aromatic compounds in essential oils and beverages are lipophilic, they are incorporated into the blood stream, cross the blood-brain barrier, and act on $GABA_A$ receptors in the brain, which may modulate mental state, mood, or consciousness.

Compound: formula	K_p (mM)	V_m (%)	K_{1p} (μM)
Cineol:$C_{10}H_{18}O$	0.11	255	34
Citral: $C_{10}H_{16}O$	0.21	181	41
Eugenol: $C_{10}H_{12}O_2$	0.18	284	32
Terpinen-4-ol: $C_{10}H_{18}O$	0.15	635	16
Linalool: $C_{10}H_{18}O$	0.32	332	31
Geraniol: $C_{10}H_{18}O$	0.78	258	34
1-Octen-3-ol: $C_{10}H_{16}O$	0.76	688	21
Myrcenol: $C_{10}H_{16}O$	0.35	353	31

K_p and V_m are the dissociation constant of the complex followed by the receptor and potentiator, and the maximum potentiation of the receptors in Figure **7** when all the potentiation sites were occupied by the potentiator (Aoshima et al., 2001, 2008). The chemical structures of the compounds are shown in Fig. **1**.

Table 4. Estimated constants, K_p, V_m and K_{1p}, of several fragrant compounds.

Most flavors specific to matured whiskey potentiated the response of the receptors. Though fragrant components are much less numerous than ethanol in whiskey, the dose-potentiation relationship showed that they potentiated the response of $GABA_A$ receptors a thousand times more than ethanol (Fig. **8**) (Hossain et al., 2002a). The potentiation of the $GABA_A$ receptor's response by whiskey can be measured by adding whiskey itself to the GABA solution, since whiskey is a distilled liquor and contained no GABA but various compounds are moved into whiskey from an oak barrel during aging. The whiskey, produced by Suntory Ltd. in Japan, also potentiated the response of $GABA_A$ receptors more than did ethanol. The potentiation increased with the aging period of the whiskey in oak barrels (Fig. **9**), with a very high correlation between the two (Koda et al., 2003). Various types of whiskies produced in different countries also potentiated the response, though the degree of the potentiation showed some variation. An aged whiskey with much fragrance may induce intoxication with less toxicity of ethanol than a less-aged whiskey, since whiskey amount consumed can be reduced.

Beer, brewed liquor, caused high GABA-like activity. An extract of beer was prepared using pentane (EXT) to examine the presence of modulators of $GABA_A$ receptors in beer. Though

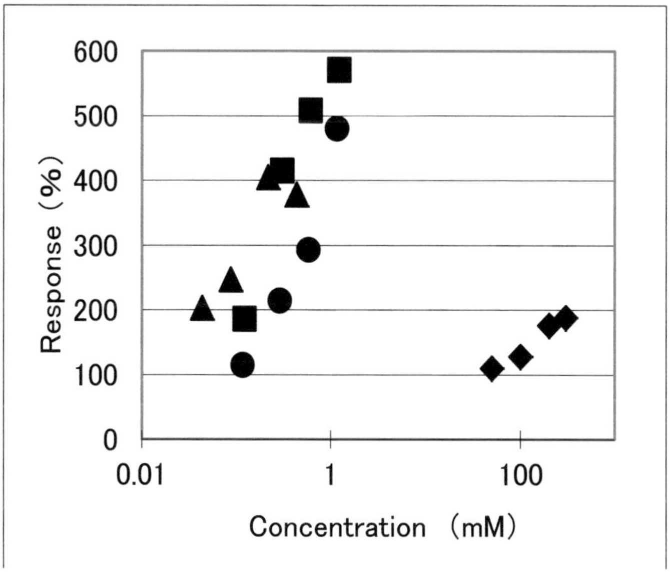

◆:ethanol, ●:quercus lactone, ▲:1,1-diethoxyheptane, ■:ethyl phenylpropanoate.

Fig. 8. Dose-potentiation relationship of some fragrances in whiskey and ethanol on the response of $GABA_A$ receptors expressed in *Xenopus* oocyte as shown in Fig. **5,** n = 4 (Hossain et al., 2002a), taken with permission from the American Chemical Society. The response caused by 0.25 μM GABA was taken as a control (100%).

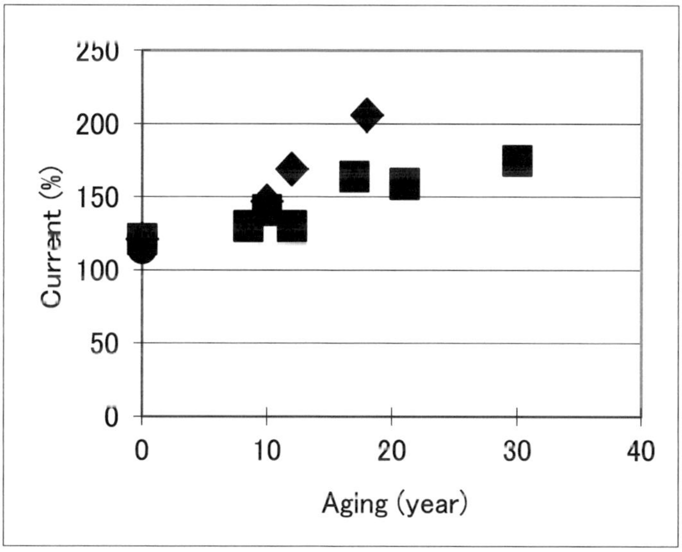

◆ : single malt whiskey, ■ : blended malt whiskey.

Fig. 9. Effect of aging period of whiskey on the response of the $GABA_A$ receptors caused by 0.25 μM GABA as shown in Fig. **5,** n = 4 (Koda et al., 2003), taken with permission from the American Chemical Society. The response caused by 0.25 μM GABA was taken as a control (100%).

beer itself contains GABA-like activity, EXT induces no electrical response in the $GABA_A$ receptor-expressing oocytes, indicating the absence of GABA-like activity in EXT. However, addition of this extract causes the potentiation of the $GABA_A$ receptor-response elicited by 0.25 μM GABA dose-dependently (Aoshima et al., 2006).

Beer is mostly produced from malted barley, water and hops, and is reported to contain fragrant compounds such as alcohols, esters, aldehydes, and hydrocarbons (Maarse, 1991). The fragrances in beer possibly come from two sources, fermentation and the addition of hops. Brewers' yeast produces not only ethanol, but also fusel alcohols such as various butanol and pentanol derivatives (Maarse, 1991), which potentiate the response of $GABA_A$ receptors (Aoshima et al., 2001). These alcohols are oxidized to form aldehydes and carboxylic acids. Then various esters are synthesized between alcohols and carboxylic acids. Aliphatic esters potentiated the $GABA_A$ receptor-response (Aoshima et al., 2006).

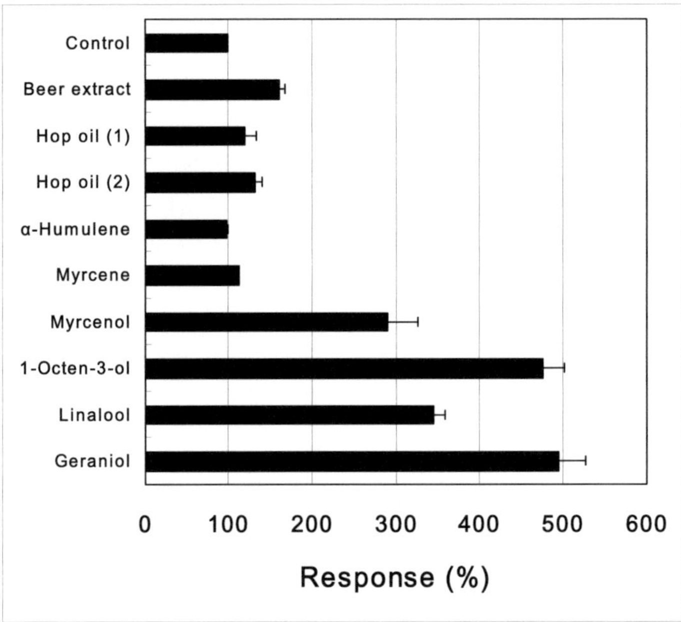

Fig. 10. Potentiation of the response of the $GABA_A$ receptors by fragrances in beer as shown in Fig. 5, n = 4 (Aoshima et al., 2006), taken with permission from the American Chemical Society. The response caused by 0.25 μM GABA was taken as a control (100%). Error baras represent the standard deviations.

The esters tended to potentiate the response less as their carbon chain length increased, though this may be attributed the solubility of the esters. Propyl acetate and ethyl propanoate showed similar dose-dependency in the potentiation of the $GABA_A$ receptor's response.

Sweet wort is boiled with hops to give its characteristic bitter taste. Hops includes various fragrances such as α-humulene, myrcene, linalool, geraniol and 1-octen-3-ol. α-Humulene and myrcene are typical hydrocarbons present in hops (Maarse, 1991) and humulenol and mycenol are produced from α-humulene and myrcene during the boiling of wort with hops. α-Humulene and myrcene have little effect on the $GABA_A$ receptor, but the alcohol myrcenol potentiates the response strongly. Linalool, geraniol, and 1-octen-3-ol also potentiate the receptor's response strongly (Fig. **10**). However these compounds do not induce the response of $GABA_A$ receptors, that is, they do not act as an agonist. Reportedly, a hops' (*Humulus lupulus L.*) CO_2 extract exhibited pentobarbital sleep-enhancing properties and antidepressant activity in rats (Zanoli et al., 2005). However, Schellenberg et al. (2004) reported that the fixed combination of valerian and hops acts via a central adenosine mechanism which is possibly the reason for its sleep-inducing and -maintaining activity. So further studies are necessary to clarify how hops affect our mood.

5.5 Effects of components in tea and coffee on the response of $GABA_A$ receptors

The effects of several components of brewed teas on the response of $GABA_A$ receptors were measured as shown in Fig. **11** (Hossain et al., 2002b, 2004). Most fragrant components potentiated the response, while methyl xanthines such as caffeine and polyphenols such as catechin inhibited it. An extract of green tea made with diethyl ether, which contains lipophilic components such as caffeine and catechin in green tea thought to be incorporated into the brain, inhibited the response elicited by GABA, possibly because the amounts of caffeine and catechin derivatives were much larger than those of fragrant components. The responses of $GABA_A$ receptors were measured as summarized in Fig. **5**, and the control was the response by by 1 μM GABA without the other components in teas.

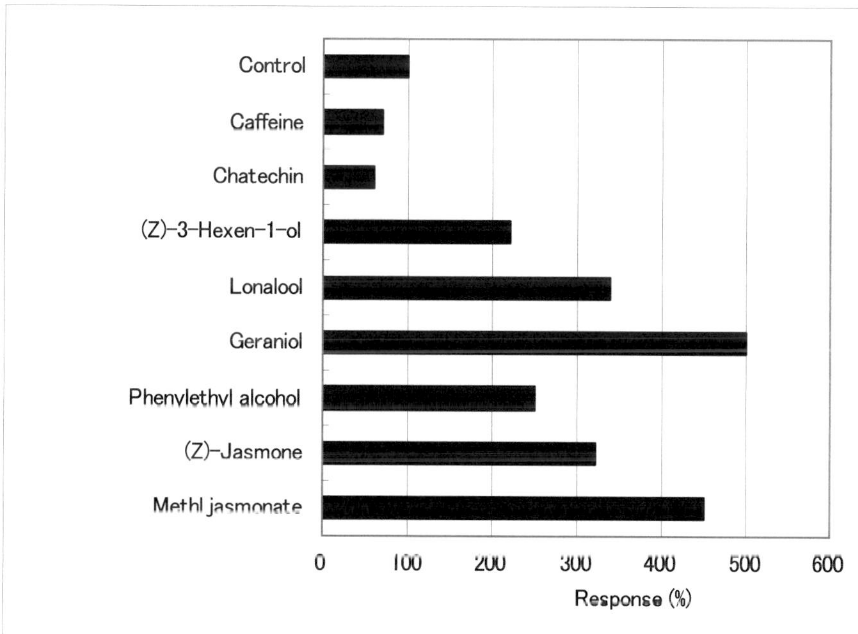

Fig. 11. Effects of compounds in teas on GABA-induced response of the $GABA_A$ receptors as shown in Fig. **5**, n = 4 (Hossain et al., 2002b, 2004) The response caused by 1 μM GABA was taken as a control (100%), taken with permission from the American Chemical Society.

Effects of many components of coffee on the response of $GABA_A$ receptors were measured as shown in Fig. **12** (Hossain et al., 2003). Most fragrant components potentiated the response, but methyl xanthines and chlorogenic acid inhibited it. The extract of coffee obtained with diethyl ether slightly potentiated the response at low concentrations, but inhibited it at high concentrations. This result suggests that the extract contains two types of components: fragrant components that potentiated the response with high affinity for the receptors and components such as caffeine and chlorogenic acid that inhibited the response with low affinity for the receptors.

Herbal teas contain various fragrant compounds with pleasant smells (Nagashima, 2010). It is believed that they have beneficial effects such as tranquillizing, anti-stress, anti-bacterial, anti-oxidative, anti-inflammatory and anti-fatigue activities, though some of these properties may come from polyphenols or pigments. Reportedly, some herbal teas increase appetite, the digestion of food, blood flow in vessels and urination, while others stimulate arousal.

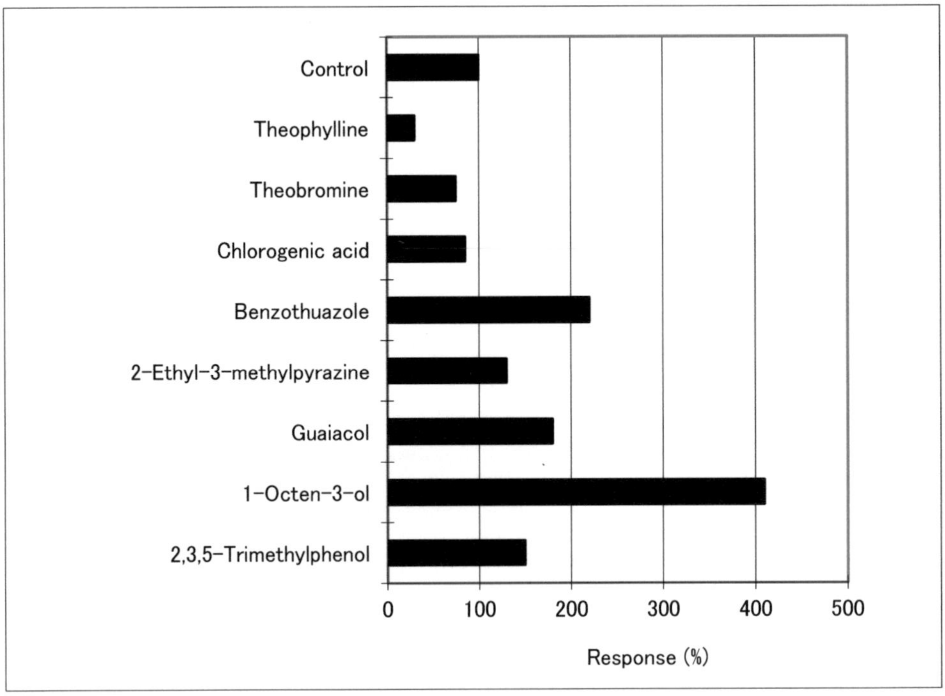

Fig. 12. Effects of compounds in coffee on GABA-induced response of the $GABA_A$ receptors as shown in Fig. 5, n = 4 (Hossain et al., 2003) The response caused by 0.25 μM GABA was taken as a control (100%), taken with permission from the American Chemical Society.

6. Pharmacological and animal behavioral studies

6.1 Extension of sleeping time and the block of convulsion in mice by fragrances

Pentobarbital potentiates the response of $GABA_A$ receptors and induces sleep in higher animals. The co-administration of aromas which potentiate the response of $GABA_A$ receptors in *Xenopus* oocytes, with pentobarbital is expected to extend sleeping time, since the pentobarbital and aromas have an additive effect on the receptors. A close relationship has been observed between the potentiation of the response of $GABA_A$ receptors expressed in *Xenopus* oocytes and the extension of sleeping time by essential oils of various trees (Aoshima et al., 2009). Extensions of sleeping time in mice by fragrances were induced by various methods such as intraperitoneal, inhalational and oral administration (Aoshima et al., 2009; Mubassara et al., 2008). Inhalation of whiskey by mice increased the sleeping time induced by pentobarbital more than did inhalation of ethanol (Fig. **13**) (Koda et al., 2003).

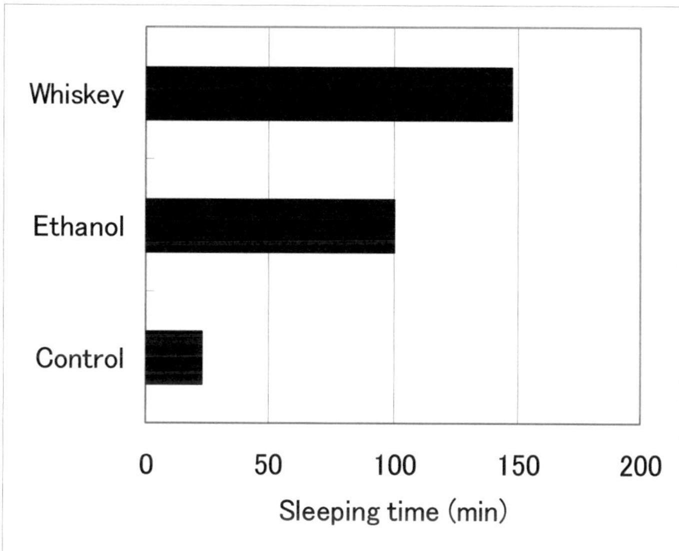

Fig. 13. Effect of ethanol (43%) and whiskey (Scotch-type whiskey made by Suntory Ltd. with 43% ethanol) on sleeping time induced by pentobarbital (sleeping drug) in mice, n = 5 (Koda et al., 2003), taken with permission from the American Chemical Society Control means that only sodium pentobarbital (50 mg/kg) was injected intraperitoneally into rats.

Aromatic components of beverages, *cis*-jasmone and methyl jasmonate in oolong tea, 1-octen-3-ol in coffee, and myrcenol in beer, potentiated the response of the $GABA_A$ receptors and extended sleeping time when they were co-administered with pentobarbital (Hossain et al., 2007). Therefore, aromatic compounds in beverages are possibly taken up by the brain, and thereby potentiate the response of the $GABA_A$ receptors additively, and extend sleeping time, though the possibility that aromatic compounds inhibit the decomposition of pentobarbital in the liver cannot be excluded. Inhalation or intraperitoneal administration of compounds such as terpinen-4-ol, which potentiate the response of $GABA_A$ receptors even without pentobarbital, could induce abnormal behavior similar to that caused by the administration of liquors or anesthetics (Aoshima et al., 2009).

Tsuchiya et al. (1992) reported that pentobarbital-induced sleeping time in mice was prolonged by terpinyl acetate and phenethyl alcohol, and shortened by lemon oil and jasmine oil, and that no effect of the aromatic compounds was observed on pentobarbital-induced sleep when using anosmic mice produced by intranasal zinc sulphate treatment, suggesting that the effect of odors on sleeping time is induced through the olfactory system. Thus, aromatic compounds at low concentrations may also affect sleep caused by pentobarbital through the olfactory system possibly via the autonomic nervous system.

Pentetrazole, an antagonist of $GABA_A$ receptors, induces convulsions through inhibition of $GABA_A$ receptor-elicited responses in the brain. Yamada et al. (1994) found that inhaling lavender oil vapor blocked pentetrazole-induced convulsions in mice, suggesting a potentiation of the $GABA_A$ receptor's response by lavender oil. The intraperitoneal administration of ethyl phenylpropanoate, which is one of the fragrances in whiskey and potentiates the response of $GABA_A$ receptors, delayed significantly convulsions caused by pentetrazole in mice, suggesting that ethyl phenylpropanoate potentiates the response of $GABA_A$ receptors against inhibition of the response by pentetrazole (Hossain et al., 2002a).

6.2 Anti-stress and anti-conflict effects of fragrances

Mental or physical stress induces the release of stress hormones such as corticotrophin-releasing hormone (CRH) and adrenocorticotropic hormone (ACTH) from the paraventricular nucleus or pituitary gland. Yamada et al. (1996) found that the intraperitoneal administration of diazepam, a benzodiazepine, suppressed plasma ACTH levels in ovariectomized rats under restriction-stress. Similar effects on the plasma ACTH level were observed in ovariectomized rats when chamomile, lemon and lavender oil, or their components were administered by inhalation. Inhalation of terpinen-4-ol and 1-octen-3-ol suppressed ACTH levels significantly (Fig. **14**) (Aoshima et al., 2009). Inhalation of linalool, cineol, α-terpineol and mastic tree oil also suppressed ACTH levels, but not significantly. The plasma ACTH levels of ovariectomized rats under restriction stress were greater than those of normal rats, since the ovariectomized rats, used as an experimental menopausal model, are likely to be affected by the stress. Inhalation of whiskey fragrances by rats decreased restriction-stress-induced increases in the plasma adrenocorticotropic hormone (ACTH) level, suggesting an anti-stress effect (Yokogoshi, 2006).

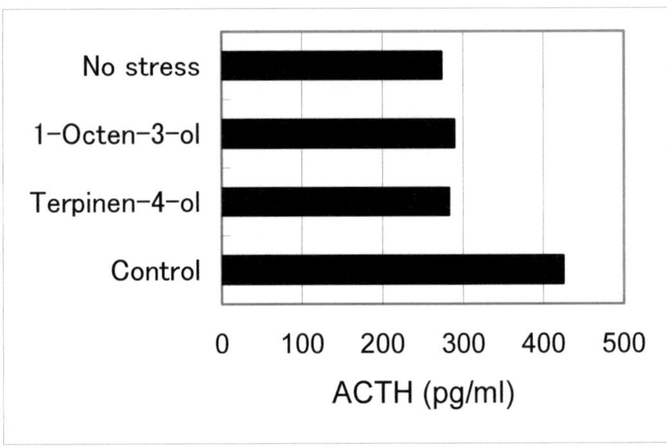

Fig. 14. Suppression of plasma ACTH levels in rats by inhalated aromas under restriction stress, n = 10 (Aoshima et al., 2009), taken with permission from Fragrance Journal Ltd., Japan. Rats were kept under restriction stress for 1 hour with inhalation of 1-octen-3-ol or terpinen-4-ol. Control rats were kept for 1 hour under restriction stress without inhalation of aromas. Plasma ACTH concentrations were measured by the immunoradiometric assay.

Anticonflict effects of rose oil and its components such as 2-phenethyl alcohol and citronellol were observed in Geller and Vogel conflict tests in mice administered intraperitoneally (Umezu, 1999, 2000). In mice, the intraperitoneal injection of fragrant compounds such as terpinen-4-ol significantly increased both the number of entries in open arms and time spent in the open arms in the elevated plus maze (Fig. **15**), indicating anti-anxiety effects as did muscimol and diazepam, an agonist and a potentiator of $GABA_A$ receptors, respectively. Effects of lavender and rose oils on rodents were also measured in the elevated plus maze and their anxiolytic effects were observed (Tsang & Ho., 2010).

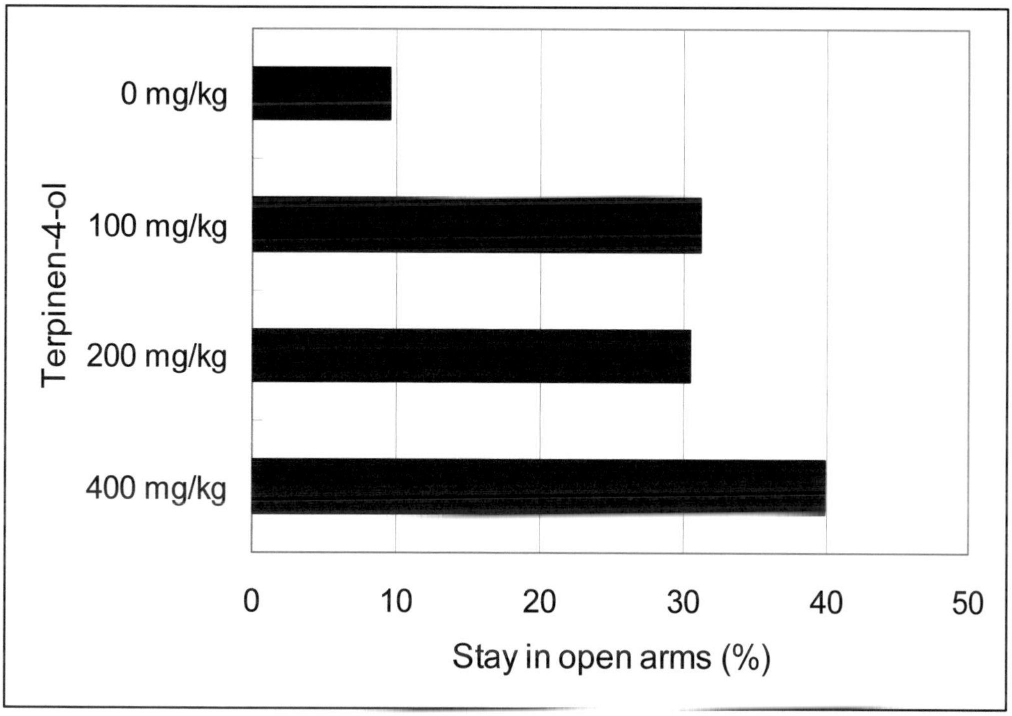

Fig. 15. Anti-anxiety-like effect of terpinen-4-ol in the elevated plus maze. The stay in the open arms of an elevated plus maze was measured for 15 min 30 min after the intraperitoneal administration of terpinen-4-ol dissolved in olive oil in mice.(n = 12). The control mice (0 mg/ kg) were intraperitoneally injected with only olive oil. As a positive control, increase of stay in open arms in mice was confirmed by injection of diazepam.

6.3 Ambulatory activity in mice

Peppermint oil is believed to be effective at treating mental fatigue. Umezu et. al. (2001) administered peppermint oil intraperitoneally to mice and found that their ambulatory (locomotor and rearing) activities, which were measured using a tilting-type ambulometer consisting of 10 bucket-like Plexiglas activity cages 20 cm in diameter (SAM-10, O'hara Co., Tokyo), increased dramatically. Intravenous administration of 1,8-cineol, menthone, isomenthone, menthol, (*R*)-(+)-pulegone, menthyl acetate or caryophyllene, which are components of peppermint oil, also induced a significant increase in ambulatory activity at much lower doses (20 mg/kg), Measurements of effects of a dopamine transporter-inhibitor, a tyrosine hydroxylase inhibitor, and antagonists of dopamine receptors on the ambulation of mice suggested that dopamine was involved in the ambulation-promoting effect of menthol (Umezu, 2009).

Fragrances in tea and coffee are expected to suppress the arousal effect of caffeine, since they potentiated the response of $GABA_A$ receptors (Hossain & Aoshima, 2008). However, ethanol enhanced the ambulation-increasing effect of caffeine in mice, though diazepam and pentobarbital, which potentiate the response of $GABA_A$ receptors specifically, reduce the effect of caffeine as expected (Kurihara, 1993). Terpinen-4-ol enhanced the ambulation-increasing effect of caffeine in mice similarly to ethanol (Fig, **16**) (unpublished result). Since a antagonist of N-methyl-D-aspartate (NMDA) receptor, MK-801, also enhanced the ambulation-increasing effect of caffeine in mice (Kurihara et al., 1992), the enhancement is possibly caused by the inhibition of NMDA receptors in the brain of mice.

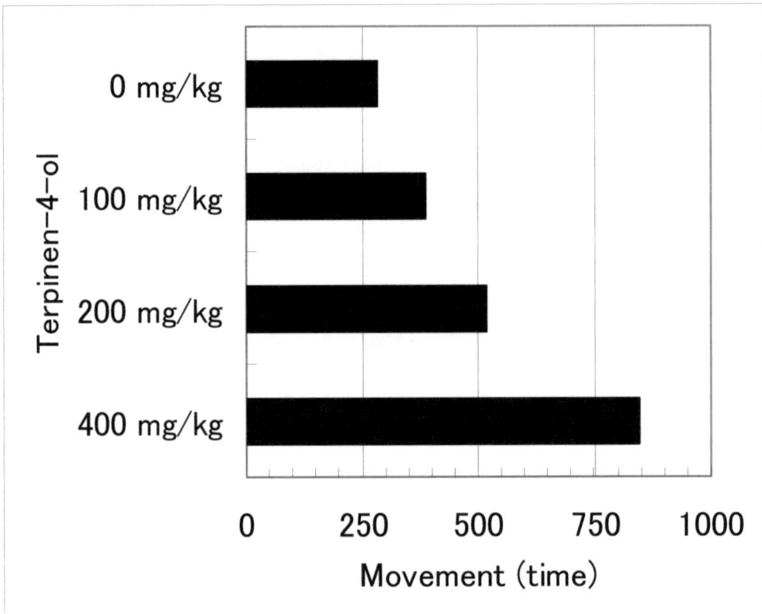

Fig. 16. Changes in ambulatory activity in mice after intraperitoneal administration of terpinen-4-ol and caffeine, measured by the tilting-type ambulometer for an hour, n = 10. Caffeine (10 mg/kg) was administered 30 min after administration of terpinen-4-ol dissolved in olive oil. As a control (0 mg/kg), caffeine was administered 30 min after administration of only olive oil.

7. Other activities

7.1 Anti-bacterial activity

Antibacterial activities of essential oils have been studied extensively. It is difficult to summarize these activities, since they vary in strength depending on the experimental conditions (Inoue, 2002). Terpenes derived from trees are named "phytoncids", since most of them have antibacterial activity. Until antibiotics were developed, essential oils have been used to disinfect medical equipment and to cure infective diseases, a practice known as medical aromatherapy. Some volatile compounds in herbs and spices have antibacterial activity and are used to preserve vegetables, fruits and foods (Yatagai, 2010).

One distinctive feature of essential oils or aromatic compounds is their very broad spectrum of activity, suppressing the growth of viruses, parasites, *mycoplasma,* bacteria and fungi, whereas antibiotics act on on a certain range of species. Essential oils are more active against bacteria, which are aerobic, lipophilic and producing spores. Essential oils suppress growth generally in the following order: molds > Gram-positive bacteria and acid-fast bacillus > yeast > Gram-negative bacteria (Inoue, 2002).

It is widely thought that aromatic compounds act on the cell membrane. At low concentrations, they stabilize the membrane structure similar to anesthetics and suppress the flux of ions such as Na^+ and Ca^{2+}. At high concentrations, they disturb the membrane structure, increase the flux of ions through cell membranes and cause cell death. A vapor usually has higher antibacterial activity than a liquid at similar concentrations. Aromatic compounds can kill bacteria resistant to antibiotics such as Methicillin-resistant *Staphylococus aureus* (Inoue, 2002).

Distilled liquors with high concentrations of ethanol were once used to sterilize of wounds. However, the development of antibiotics has reduced this use. Fragrant compounds should increase the sterilizing activity of liquors. Teas, especially green tea, have anti-microbial and anti-viral activities, but it is likely that these effects come from catechin derivatives such as epigallochatechin 3-*O*- gallate (Uesato et al., 2003).

7.2 Immunological activity

The immunological system is associated with the endocrine and autonomic nervous systems. These three systems are controlled by the hypothalamus and maintain homeostasis. Aromas restored immunological activity which decreased under strong physical and mental stress, through stimulation of the olfactory system. Reportedly, a forest bathing trip increased human natural killer (NK) activity and expression of anti-cancer proteins in female subjects (Li et al., 2008). Exposure of phytoncides (wood essential oils) containing pinenes to humans significantly increased NK activity and the percentages of NK, perfolin, granulysin, and granzymeA/B-expressing cells, and significantly decreased the percentage of Tcells, and the concentration of adrenaline and noradrenaline in urine, suggesting that phytocide exposure and decreased stresshormone levels partially contribute to increased NK activity (Li et al., 2006, 2009). In mice, a decrease in plaque-forming cells involving thymic involution induced by high-pressure stress (2.2 kg/cm^2) caused by compressed air in a chamber was restored by long term inhalation of tuberose, lemon, oak-moss and labdanum for 24 h following the stress (Fujiwara et al., 1998). Inhalational treatment of rats with a citrus fragrance normalized neuroendocrine hormone levels and immune function and was rather more effective than antidepressants, imipramine (Komori et al., 1995).

Some essential oils suppress the release of histamine or leukotriene from mast cells or the production of cytokines. Lavender oil inhibited concentration-dependently cutaneous anaphylaxis induced by anti-dinitrophenyl (DNP) IgE in rats following both topical and intradermal applications, the release of histamine from peritoneal mast cells by anti-DNP IgE, and the anti-DNP IgE-induced release of tumor necrosis factor-alpha (TMFα) from peritoneal mast cells, suggesting the suppression of immediate-type allergic reactions via the inhibition of mast cell degranulation in-vivo and in-vitro. The recruitment of leukocytes into the peritoneal cavity on the intraperitoneal injection of casein into mice was suppressed by intraperitoneal injections of geranium, lemongrass and spearmint oils at a dose of 5 μL/mouse possibly because they suppress neutrophil recruitment (Aoshima et al., 2009; Wada & Yamazaki, 2004).

7.3 Action on TRP receptors

Menthol has specific targets, *i. e.* transient receptor potential vanilloid 3 (TRPV3) and transient receptor potential melastatin subfamily channel 8 (TRPM8) receptors. These receptors are non-selective cation channels with six trans-membrane domains and temperature-sensitive, that is, TRPM8 receptors open at below ca. 25 °C and TRPV3 receptors, at below ca. 32 °C. Thus, menthol acts as a cooling agent. Eucalyptol also opens the channels of TRPM8 receptors, and camphor and thymol open the channels of TRPV3 receptors (Palapoutian et al., 2003).

7.4 Inhibition of acetylcholineesterase

Acetylcholineesterase (AChE) inhibitors have been used in the treatment of Alzheimer's disease. Miyazawa et. al. (1997, 2005) found that monoterpenoids with a *p*-menthane skeleton inhibited AChE. Essential oils of *Mentha* species such as *M. aquatica, M. gentiles*, and *M. avensis* also inhibited AChE (*Ki* = about 50 μg/mL). The treatment of Alzheimer's disease is based on inhibition of the AChE, which hydrolyses acetylcholine, inccreasing acetylcholine available for transmission at the cholinergistic synapse. Therefore, these essential oils and their components may be used for the treatment of Alzheimer's disease.

7.5 Effect on transcription

Hariya (2003) investigated the effect of inhaling odorants (estragon, grapefruit, fennel and pepper oil) on humans and found an increase in relative sympathetic activity (Shen et al., 2005a), which induced noradrenaline release. He found that noradrenaline (0.5 μg/mL) acted on adipose tissue synergistically with percutaneously absorbed caffeine(1 mM) to promote uncoupling protein-3 gene expression of adipose tissue culture. He proposed that inhaling odorants together with caffeine increase noradrenaline release, expression of uncoupling protein-3, which promots thermogenesis from free fatty acids produced by the decomposition of neutral fat, inducing a slimming effect.

Liang et al. (2007) studied the time-dependent effects of ethanol intoxication on $GABA_A$ receptor composition and function in rats, and found decreases in the cell-surface fraction of α4 and δ, but not α1, α5, or γ2, without changes in their total content. Chronic administration of terpinen-4-ol, α-terpineol or linalool to mice also increased the RNA expression of the α4 subunit, while decreasing that of the α1 subunit as did ethanol, when the expression of mRNAs of $GABA_A$ receptor subunits was examined by the real-time RT-

PCR method (Tasaka & Aoshima, 2010). Change of the composition of $GABA_A$ receptor subunits will influence human mind or consciousness, since $GABA_A$ receptors play an important role in neural transmissions as main inhibitory neurotransmittere receptors and the receptors with different composition of subunits have different pharmacological characteristics (Martin & Olsen, 2000).

8. Fragrances and health

8.1 Homeostasis and the hypothalamus

Basic drives such as stress, sleep, temperature regulation, thirst and hunger are essential for survival (Thompson, 2000). They are aroused when internal conditions are less than optimal. For example, when metabolic energy is insufficient, the organism takes foods and optimizes its conditions. This is the basic notion of homeostasis. These basic motives are controlled by the hypothalamus, which makes up only a small proportion of the limbic system of the brain, but exerts profound effects on behavior and experience. Near the hypothalamus, other neural circuits including the medial forebrain bundle and the nucleus accumbens form a reward system. This common pleasure system in the mammalian brain functions for all basic drives, even that which results from addiction.

The hypothalamus controls the endocrine system, the autonomic nervous system and the immune system (Fig. 17) (Thompson, 2000). These systems interact with each other. Hans Selye developed the notion of a general adaptation syndrome in the 1930s. The basic idea he proposed is that the body shows a common, integrated set of responses in an attempt to adapt too many different kinds of stressors. The stressors are not only physical such as temperature change or hard exercise, but also psychological such as divorce or death in the immediate family. He showed that different severe stressors cause similar bodily harm, that is, enlargement of the adrenal cortex, shriveling of the thymus and lymphatic gland, and hemorrhaging from the stomach's inner wall. He proposed three stages: a shock phase involving decreased blood pressure, body temperature and muscle tone, then an adaptation response (a stage of resistance) when the body fights back, and finally a stage of exhaustion where the body's defenses break down because the stress is severe and continues for a long period. The third stage causes the bodily harm mentioned above, and a marked impairment of the immune system, which possibly causes severe illnesses such as cancer.

The adrenal gland is the major gland for coping with stress in humans. Stressors act on the hypothalamus, which releases corticotrophin-releasing hormone (CRH) and also enhances sympathetic nervous pathways. Then CRH reaches the pituitary, from which adrenocorticotrophic hormone (ACTH) is released into the general circulation. ACTH reaches the adrenal cortex and causes endocrine gland cells there to release cortisol (corticosterone, which is the rat analogue of human cortisol), a stress hormone, and a small amount of aldosterone. Cortisol enhances the sympathetic nervous activity of the autonomic nervous system, which releases noradrenalin as a neurotransmitter at the neuromuscular junctions between the postganglionic sympathetic neurons and smooth or cardiac muscle fibers. The sympathetic nervous pathway also releases adrenalin from adrenal medulla. Thus stressors trigger an arousal mechanism for the entire body, which prepares the animal to "fight or flight" in the face of perceived danger.

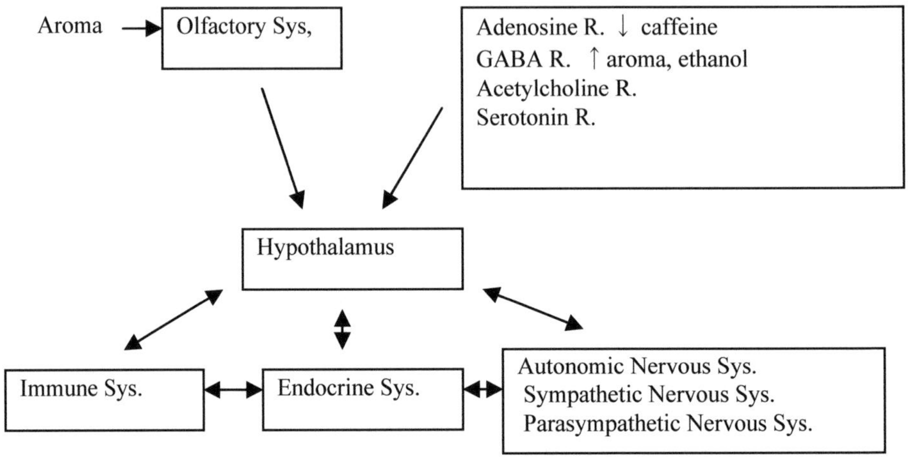

Sys.: system, SNSD: sympathetic nervous system dominant, PNSD: parasympathetic nervous system dominant., R: receptor. ↓: inhibition, ↑: potentiation.

Fig. 17. Schematic representation of interrelations of the olfactory system, the central nervous system, the hypothalamus, and the autonomic nervous, endocrine and immune system and alterations by several food derived components.

8.2 Relationship between stress and health

It has been known for many years that performances in various learning and skill tasks have an inverted-U relation to degree of arousal in both humans and other animals. If you are exhausted or very sleepy, your performances will be poor. If you are aroused properly at some intermediate level of stress, your performances will be optimum. However, if you are extremely aroused under a great deal of stress, your performances will deteriorate. In general, performance is impaired in proportion to the severity of stress *i.e.* the right side of the inverted U (Fig. **18**) (Thompson, 2000).

Health, physicality and mentality also have an inverted-U relation to degree of stress or arousal (Fig. **18**). It is important to maintain a balance between the sympathetic and parasympathetic nervous systems for optimum health. It is also important to keep a balance between granulocyte and lymphocyte numbers in the immune system. Immune function is affected by both the autonomic nervous system and the endocrine system. If the sympathetic nervous system dominates over the parasympathetic system for long periods, granulocyte numbers increase, while lymphocyte numbers decrease, which is likely to induce inflammation or even cancer (Ben-Elivahu. et al., 2007). Granulocytes destroy bacteria through phagocytosis, but produce reactive oxygen species, which may injure near-by cells and could cause inflammation. On the other hand, if the sympathetic nervous system is less dominant than the parasympathetic system for long periods, granulocyte numbers decrease, while lymphocyte numbers increase. Lymphocytes such as natural killer cells are very important to remove cancer cells. However, they can cause allergies when produced in excess. It is important to maintain homeostasis for our physical and mental health.

8.3 Fragrance and health

Mental and physical relaxation is particularly necessary on busy, stressful days. Reportedly, fragrances such as essential oil of lavender, linalool and whiskey enhanced the parasympathetic nervous system in rats by stimulating the olfactory system (Niijima et al., 2009; Shen et al., 2005b). Many fragrant compounds such as terpinen-4-ol and ethanol potentiate the response of $GABA_A$ receptors when they are incorporated into the brain, having tranquilizing, anti-anxiety and anti-stress effects (Aoshima et al., 2001; Wallner et al., 2003). So drinking whiskey or beer enhances the parasympathetic nervous system and causes relaxation even sedation. Herbal teas rich in fragrant compounds may induce similar effects. These beneficial activities will be enhanced even more when combined with other activities such as walking in a forest, massage with essential oils, light sports, bathing at mild temperature, listening to healing music or the intake of a balanced diet.

On the other hand, fragrances such as essential oils of grapefruit and limonene enhance sympathetic nervous activity through the olfactory system (Shen et al., 2005a). Active compounds such as caffeine in tea and coffee, and capsaicin in red pepper also enhance the sympathetic nervous system via the autonomic nervous system. Caffeine and polyphenols such as catechins in tea or chlorogenic acid in coffee inhibited the response of $GABA_A$ receptors (Hossain et al., 2007). On the other hand, Bouayed et al. (2007) have found that intraperitoneally injection of mice by chlorogenic acid resulted in anxiolytic effects. They have also found that chlorogenic acid protected granulocytes from oxidative stress. Fruits and coffees may provide health-promoting advantages to consumers, since chlorogenic acid is one of the most abundant polyphenols in coffees and fruits such as plums, apples and cherries.

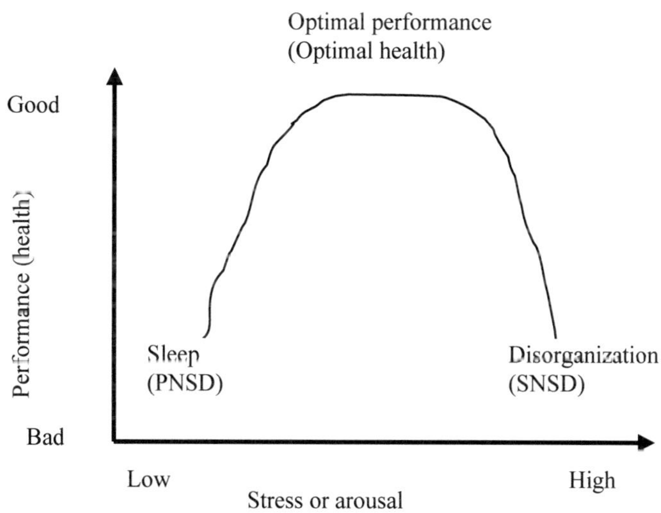

Fig. 18. The inverted-U function relating stress or arousal to performance and health.

Thus, these beverages are useful when we are very sleepy. The relationships among stressors, the autonomic nervous system, the immune system and $GABA_A$ receptors are summarized in Table **5**.

Increase	Stressor	Decrease
Synpathetic N. S. dominant	Autonomic N.S.	Parasympathetic N.S. dominant
Fright. anxiety	Situation	Laugh, peace
Granulocyte increase	Leukocyte	Lymphocyte increase
Increase	Reactive oxygen species	Decrease
Inflammation, cancer	Ills caused by excess	Allergy
Muscle	Increase of blood flow	Digestive organs
Increase	Energy consumption	Decrease
Decrease	Body weight	Increase
Limonene	Fragrance	Linalool
Suppression	$GABA_A$ receptor response	Potentiation

Abbreviation used; N.S.: nervous system

Table 5. Stress, autonomic nervous system, immune system and $GABA_A$ receptor.

9. Conclusion

Fragrant compounds and their physiological activities have been made use of since ancient times. However, relevant scientific studies have been slow to emerge, since these activities are very weak. In the last twenty years, studies on the olfactory system have developed extensively since olfactory receptors were clarified at a molecular level. Fragrant compounds bind to olfactory receptors in olfactory sensory neurons and produce smell images in the olfactory bulb, which are analyzed by the brain. These smell images influence the amygdale and hippocampus in the limbic system, which play an important role in emotion and memory, respectively. The signals through the olfactory system are also transferred to the hypothalamus, which controls the autonomic nervous, endocrine and immune systems. Terpens and higher alcohols such as leaf alcohols produced by trees and grasses or flavors from foods possibly induce attractive feelings for humans, while bad odors such as ammonia and hydrogen sulfide cause evasive behaviors.Thus odors affect mental state and emotion as well as peripheral systems. On the other hand, fragrant compounds are incorporated into the blood stream through the skin, lung, stomach and intestines. Since most are hydrophobic, they cross the blood-brain barrier, enter the brain and are possibly accumulated there. Many of them potentiate the response of $GABA_A$ receptors and induce mental relaxation. Some of them inhibit acetylcholine esterase or affect the immune system directly.

Beverages such as tea, coffee and liquor contain physiologically active compounds such as caffeine and ethanol, which can cause addiction. Moreover, they contain flavors, which stimulate the olfactory system. Many of these flavors come from the beverage's raw materials, such as tea leaves, coffee beans, fruits or grain, others come from additives such as flowers or hops, or storage in oak barrels. Some flavors are produced during processing, such as crumpling and drying for teas, roasting for coffee, and fermentation for liquors. These flavors affect emotion and consciousness together with other active compounds and play an important role in a beverage's qualities. Fragrant compounds in beverages may enhance the effects of active components such as caffeine. Fragrant compounds in liquors increase the potentiation of $GABA_A$ receptors together with ethanol, and induce mental relaxation. On the other hand, flavors in tea and coffee may increase the arousal activity of caffeine, whose mechanism has not been clarified yet. Thus, beverages are useful for balancing the autonomic nervous system in terms of sympathetic and parasympathetic activity, which will help to maintain homeostasis and health. It is necessary to clarify the physiological activities of fragrant compounds at concentrations used under physiological conditions. Essential oils such as lavender, tea tree or eucalyptus have already been used for medical aromatherpy.

10. Acknowledgment

I thank my students and co-researchers, especially Prof. S. J. Hossain in Khulna University, and Dr. H. Koda and Dr. Y. Kiso in Suntory Ltd., for co-operation of these experiments.

11. References

Aoshima, H. & Hamamoto, K. (1999). Potentiation of GABA receptors expressed in *Xenopus* oocytes by perfume and phytoncid. *Biosci Biotchnol Biochem* 63: 743-8

Aoshima, H., Hossain, S.J., Hamamoto, K. *et al.* (2001). Kinetic analyses of alcohol-induced potentiation of the response $GABA_A$ receptors composed of α1 and β1 subunits. *J Biochem*, 130: 703-9

Aoshima, H., Hossain, S.J., Imamura, H. & Shingai, R. (2001). Effects of bisphenol A and its derivatives on the response of $GABA_A$ receptors expressed in *Xenopus* oocytes. *Biosci. Biotechnol Biochem*, 65, 2070-7

Aoshima, H., Hossain, S.J., Koda, H. & Kiso, Y. (2008). Beer and GABA receptors, In *Beer in Health and Disease Prevention*, Preedy, V.R. & Watson, R.R. (Eds.). pp. 193-201. Academic Press (Elsevier), ISBN 978-0123738912, London, UK

Aoshima, H., Oda, K., Orihara, Y. et al. (2009). Effects of essential oils on the response of $GABA_A$ receptors, sleeping time in mice induced by a sleeping drug and plasma adrenocorticotropic levels of rats. *Aroma Res*, 10, 58-64

Aoshima, H., Takeda, K., Okita, Y. et al. (2006) Effects of beer and hop on ionotropic γ-aminobutyric acid receptors. *J Agric Food Chem*, 54, 2514-19

Aoshima, H. & Tenpaku, Y. (1997). Modulation of GABA receptors expressed in Xenopus oocytes by 13-L-hydroperoxylinoleic acid and food additives. *Biosci Biotechnol Biochem*, 61, 2051-7

Ben-Elivahu, S., Page, G.G. & Schleifer, S.J. (2007). Stress, NK cells, and cancer: Still a promissory note. *Brain Behav Immun*, 21, 881-7

Bouayed, J., Rammal, H., Dicko, A., Younos, C. & Soulimani, R. (2007) Chlorogenic acid, a polyphenol from *Prunus domestica* (Mirabelle), with coupled anxiolytic and antioxidant effects. *J Neurol Sci*, 262, 77–84.

Buck, L.B. (2000). The molecular architecture of odor and pheromone sensing in mammals. *Cell*, 100, 611-8

Buck, L. & Axel, R.A. (1991). A novel multigene family may encode odorant receptors: a molecular basis for odor recognition. *Cell*, 65, 175-87

Cal, K. (2006). Skin penetration of terpenes from essential oils and topical vehicles. *Planta Med*, 72, 311-6

Carey, A.F., Wang, G., Su, C.Y., Zwiebel, L.J. & Carlson, J.R. (2010). Odorant reception in the malaria mosquito *Anopheles gambiae*. *Nature*, 464, 66-72

Chebib, M. & Johnston, G.A.R. (2000). GABA-activated ligand gated ion channels: medicinal chemistry and molecular biology. *J Med Chem*, 43, 1427-47

Chu, S. & Dawnes, J.J. (2000) Review; Odour-evoked Autobiographical memories: Psychological investigations of Proustian Phenomena. *Chem Senses*, 25, 111-6

Chuyen, N.V. & Ishikawa, T. (Eds.). (2008). *Science and function of coffee*. IK Co, Tokyo, Japan ISBN 978-4874922538

Dayawansa, S., Umeno, K., Takakura, H. et al. (2003). Autonomic responses during inhalation of natural fragrance of Cedrol in humans. *Auton Neurosci*, 108, 79-86

Dewick. P.M. (2002). *Medicinal Natural Products*. 2ed ed., John Wiley & Sons, ISBN 978-0470741672, Oxford, UK

Field, T., Diego, M., Hernandez-Reif, M. et al. (2005). Lavender fragrance cleansing gel effects on relaxation. *Int J Neurosci*, 115, 207-22

Franks, N.P. & Lieb, W.R. (1994). Molecular and cellular mechanisms of general anaesthesia. *Nature*, 367, 507-14

Fujiwara, R., Komori, T., Noda, Y. et al. (1998). Effect of a long-term inhalation of fragrances on the stress-indiced immunosuppression in mice. *Neuroimmunomodulation*, 5, 318-22

Hariya, T. (2003). The possibility regulating the function of adipose cells by odorants. *Aroma Res*, 6, 72-8

Hatanaka, A. (1993). The biogeneration of green odour by green leaves. *Phytochemistry*, 35, 1201-18

Hess, G.P., Cash, D.J. & Aoshima, H. (1983). Acetylcholine receptor-controlled ion translocation: chemical kinetic investigations of the mechanism. *Ann Rev of Biophy Bioeng*, 12, 443-73

Hongratanaworakit, T. & Buchbauer, G. (2006). Relaxing effect of ylang ylang oil on humans after transdermal absorption. *Phytother Res*, 20, 758-63

Hossain, S.J. & Aoshima, H. (2008). Fragrant compounds in foods and beverages enhance the $GABA_A$ receptor response, In: *Amino Acid Receptor Research*, Paley, B.F. & Warfield, T.E. (Eds.). pp. 269-291, Nova Science Pub Inc, New York, USA ISBN 978-1604562835

Hossain, S.J., Aoshima, H., Koda, H. & Kiso, Y. (2002a). Potentiation of the inototropic GABA receptor response by whiskey fragrance. *J Agric Food Chem*, 50, 6828-34

Hossain, S.J., Aoshima, H., Koda, H. & Kiso, Y. (2003). Effects of coffee components on the response of $GABA_A$ receptor expressed in *Xenopus* oocytes. *J Agric Food Chem*, 51, 7568-75

Hossain, S.J., Aoshima, H., Koda, H. &, Kiso, Y. (2004). Fragrances in oolong tea that enhance the response of $GABA_A$ receptor. *Biosci Biotechnol Biochem*, 68, 1842-8

Hossain, S.J., Aoshima, H., Koda, H. & Kiso, Y. (2007). Review of functional studies of beverage components acting on the recombinant $GABA_A$ neuroreceptor, and Na^+/glucose cotransporter-response using the *Xenopus* oocyte expression system and electrophysiological measurements. *Food Biotechnol*, 21, 237-70

Hossain, S.J., Hamamoto, K., Aoshima, H. & Hara, Y. (2002b). Effects of tea components on the response of $GABA_A$ receptor expressed in *Xenopus* oocytes. *J Agric Food Chem*, 50, 3954-60

Inoue, S. (2002). *Microorganisms and Aroma*, Flagrance Journal Ltd., ISBN 4-89479-057-2, Tokyo, Japan

Inoue, N., Kuroda, K., Sugimoto, A., Kakuda, T. & Fushiki, T. (2003). Autonomic nervous responses according to preference for the odor of jasmine tea. *Biosci Biotechnol Biochem*, 67, 1206-14

Johnson, B.A., Farahbod, H. & Leon, M. (2005). Interactions between odorant functional group and hydrocarbon structure influence activity in glomerular response modules in the rat olfactory bulb. *J Comp Neurol*, 483, 205-16

Johnston, G.A.R., Hanrahan, J.R., Chebib, M. et al. (2006). Moduration of ionotropic GABA receptors by natural products of plant origin. *Advances in Pharmacol*, 54, 285-316

Julius, D. & Katz, L.C. (2004). A Nobel for smell. *Cell*, 119, 747-52

Kobayakawa, K., Kobayakawa, R., Matsumoto, H. et al. (2007). Innate versus learned odor processing in the mouse olfactory bulb. *Nature*, 450, 503-50

Koda, H., Hossain, S.J., Kiso, Y. & Aoshima, H. (2003). Aging of whiskey increases the potentiation of $GABA_A$ receptor response. *J Agric Food Chem*, 51, 5238-44

Komori, T., Fujiwara, R., Tanida, M., Nomura, J. & Yokoyama, M.M. (1995). Effect of citrus fragrance on immune function and depressive state. *Neuroimmunomodulation*, 2, 174-80

Kurahashi, T. & Menini, A. (1997). Mechanism of odorant adaptation in the olfactory receptor cell. *Nature*, 385, 725-9

Kurihara, H. (1993). Enhancement of the behavioral toxicity induced by combined administration of ethanol with methylxanthines: evaluation by discrete avoidance in mice. *J Toxicol Sci*, 18, 95-101

Kurihara, H., Asami, T. Ida, I. & Tadokoro S. (1992). Characteristics of the ambulation-increasing effect of the noncompetitive NMDA antagonist MK-801 in mice: assessment by the coadministration with central-acting drugs. *Jpn J Pharmacol*, 58, 11-8

Kuroda, K., Inoue, N., Ito, Y., et al. (2005). Sedative effects of the jasmine tea odor and (R)-(-) linalool, one of its major odor components, on autonomic nerve activity and mood states. *Eur J Appl Physiol*, 95, 107-14

Li, Q., Kobayashi, M., Wakayama, Y. et al. (2009). Effect of phytoncide from trees on human natural killer cell function. *Int J Immunopathol Pharmacol*, 22, 951-9

Li, Q., Morimoto, K., Kobayashi, M. et al. (2008) A forest bathing trip increases human natural killer activity and expression of anti-cancer proteins in female subjects. *J Biol Regul Homeost Agents*, 22, 45-55

Li, Q., Nakadai, A., Matsushima, H. et al. (2006) Phytoncides (wood essential oils) induce human natural killer cell activity. *Immunopharmacol Immunotoxicol*, 28, 319- 33

Liang, J., Suryanarayanan, A., Abriam, A. et al. (2007). Mechanisms of reversible $GABA_A$ receptor plasticity after ethanol intoxication. *J Neurosci*, 27, 12367-77

Linster, C., Menon, A.V., Singh, C.Y. & Wilson, D.A. (2009). Odor-specific habituation arises from interaction of afferent synaptic adaptation and intrinsic synaptic potentiation in olfactory cortex. *Learn Mem*, 16, 452-9

Maarse, H. (Ed.). (1991). *Volatile compounds in foods and beverages*. Marcel Dekker Inc, ISBN 0-8247-8390-5, New York, USA

Martin, D.L. & Olsen, R.W. (Eds.). (2000). *GABA the Nervous System:The View at Fifty Years*. Lippincott Williams & Wolkins, ISBN 0-7817-2267-5, Philadelphia, USA

Matsui, K. (2006). Green leaf volatiles: hydroperoxide lyase pathway of oxylipin metabolism. *Curr Opinion in Plant Biol* 9, 274-80

Mihic, S.J., Ye, Q., Wick, M.J. et al. (1997). Sites of alcohol and volatile anesthetic action on $GABA_A$ and glycine receptors. *Nature*, 389, 385-9

Miyazawa, M., Watanabe, H. & Kameoka, H. (1997). Inhibition of acetylcholinesterase activity by monoterpenoids with a *p*-menthene skeleton. *J Agric Food Chem*, 45, 677-9

Miyazawa, M. & Yamafuji, C. (2005). Inhibition of acetylcholinesterase activity by tea tree oil and constituent terpenoids. *Flavour Fragr J*, 20, 617-20

Mubassara, S., Ushijima, A., Tan, N. & Aoshima, H. (2008). Effects of essential oils on the response of $GABA_A$ receptors and sleeping time in mice induced by a sleeping drug. *Aroma Res*, 9, 257-63

Nagashima, T. (2010). *Herbal tea*. Fraegurance Journal Ltd., ISBN 978-4-89479-189-3, Tokyo, Japan

Nakashima, T., Akamatsu, M., Hatanaka, A. & Kiyohara, T. (2004). Attenuation of stress-induced elevations in plasma ACTH level and body temperature in rats by green odor. *Physiol Behav*, 80, 481-8

Nicholls, D. G. (1994). *Proteins, Transmitters and Synapses*,. Blackwell Scientific Publications, ISBN 0-632-03661-3, Oxford, UK

Niijima, A. (2008). Effect of scent stimulation on the autonomic nervous system and metabolic regulation of lipids: physiological base of aromatherapy. *Jiritushinnkei (Autonomic nervous system)*, 45, 178-86

Niijima, A., Koda, H., Kiso, Y. & Nagai, K. (2009). Moduration of autonomic nerve activity by whiskey aroma. *Aroma Res*, 10, 256-9

Patapoutian, A., Peier, A.M., Story, G.M. & Viswanath, V. (2003). ThermoTRP channels and beyond: mechanisms of temperature sensation. *Nat Rev Neurosci*, 4, 529-39

Sasabe, T., Kobayashi, M., Kondo, Y. et al. (2003). Activation of the anterior cingulate gyrus by 'Green Odor': a positron emission tomography study in the monkey. *Chem Senses*, 28, 565-72

Schellenberg, R., Sauer, S., Abourashed, E.A., Koetter, U. & Brattström, A. (2004). The fixed combination of valerian and hops (Ze91019) acts via a central adenosine mechanism. *Planta Med*, 70, 594-7

Semmelhack, J.L. & Wang, J.W. (2009). Select *Drosophila* glomeruli mediate innate olfactory attraction and aversion. *Nature*, 459, 218-23

Shen, J., Niijima, A., Tanida, M. et al. (2005a). Olfactory stimulation with scent of grapefruit oil affects autonomic nerves, lipolysis and appetite in rats. *Neurosci lett*, 380, 289-94

Shen, J., Niijima, A., Tanida, M. et al. (2005b). Olfactory stimulation with scent of lavender oil affects autonomic nerves, lipolysis and appetite in rats. *Neurosci lett*, 383, 188-93

Shepherd, G.M. (2000). Smell images and the flavour system in the human brain. *Nature*; 444, 316-21

Stewart, W.B., Kauer, J.S. & Shepherd, G.M. (1979). Functional organization of rat olfactory bulb anaoyzed by the 2-deoxyglucose method. *J Comp Neurol*,185, 715-34

Takahashi, Y.K., Nagayama, S. & Mori, K. (2004). Detection and masking of spoiled food smells by odor maps in the olfactory bulb. *J Neurosci*, 24, 8690-4

Takeuchi, H., Ishida, H., Hikichi, S. & Kurahashi, T. (2009). Mechanism of olfactory masking in the sensory cilia. *J Gen Physiol*, 133, 583-600

Tanida, M., Niijima, A., Shen, J. et al. (2005). Olfactory stimulation with scent of essential oil of grapefruit affects autonomic neurotransmission and blood pressure. *Brain Res*, 1058, 44-55

Tasaka, T. & Aoshima, H. (2010). Effect of monoterpene alcohols on the expression of subunits in the brain of mice. *Aroma Res*, 11, 110-6

Thompson, R.F. (2000). *The brain* (third edition), Worth Pub., ISBN 0-7167-3226-2, New York, USA

Tsang, H.W. & Ho, T.Y. (2010). A systematic review on the anxiolytic effects of aromatherapy on rodents under experimentally induced anxiety models. *Rev Neurosci*, 21, 141-52

Tsuchiya, T., Tanida, M., Uenoyama, S. & Nakayama, Y. (1992). Effects of olfactory stimulation with jasmine and its component chemicals on the duration of pentobarbital-induced sleep in mice. *Life Sci*, 50, 1097-102

Turner, S.L. & Ray, A. (2009). Modification of CO_2 avoidance behaviour in *Drosophila* by inhibitory odorants. *Nature*, 461, 277-81

Uesato, S., Taniguchi, K., Kumagai, A. et al. (2003). Inhibitory effects of 3-*O*-acyl-(+)-catechins on Epstein-Barr virus activation. *Chem Pharm Bull*, 51, 1448-50

Umezu, T. (1999). Anti-conflict effects of plant-derived essential oils. *Pharmacol Biochem Behav*, 64, 35-40

Umezu, T. (2000). Behavioral effects of plant-derived essential oils in the Geller type conflict test in mice. *Jpn J Pharmacol*, 83, 150-3

Umezu, T. (2009). Evidence for dopamine involvement in ambulation promoted by menthone in mice. *Pharmacol Biochem Behav*, 91, 315-20

Umezu, T., Sakata, A. & Ito, H. (2001). Ambulation-promoting effect of peppermint oil and identification of its active constituents. *Pharmacol Biochem Behav*, 69, 383-90

Vignes, M., Maurice, T., Lante, F. et al. (2006) Anxiolytic properties of green tea polyphenol (-)-epigallocatechin gallate (EGCG). *Brai Res, 1110, 102-15*

Wada, M. & Yamazaki, K. (Eds.). (2004). *Aroma and Medicine*, Fragrance Journal Ltd., ISBN 978-4894790766, Tokyo, Japan

Wallner, M., Hanchar, H.J. & Olsen, R.W. (2003). Ethanol enhances $\alpha_4\beta_3\delta$ and $\alpha_6\beta_3\delta$ γ-aminobutyric acid type A receptors at low concentrations known to affect humans. *Proc Natl Acad Sci USA*, 100, 15218-23

Xu, F., Liu, N., Kida, I. et al. (2003). Odor maps of aldehydes and esters reveled by functional MRI in the glomerular layer of the mouse olfactory bulb. *Proc Natl Acad Sci USA*, 100, 11029-34

Yamada, K., Miura, T., Mimaki, Y. & Sashida, Y. (1996). Effect of inhalation of chamomile oil on plasma ACTH level in ovari-ectomized-rat under restriction stress. *Biol Pharm Bull*, 19, 1244-6

Yamada, K., Mimaki, Y. & Sashida, Y. (1994). Anticonvulsive effects of inhaling lavender oil vapour. *Biol Pharm Bull*, 17: 359-60

Yamanishi, S. (1992) *Science of tea*, Shokabo, ISBN 978-4- 7853-8567-5, Tokyo, Japan

Yatagai, M. (2010). *Physiological activities of plant aroma*, Fragrance Journal Ltd., ISBN 978-4-89479-184-8, Tokyo, Japan

Yokogoshi, H. (Ed.). (2006). *Development and perspectives of anti-stress food*. CMC Pub. Co., ISBN 4-88231-589-0, Tokyo, Japan

Yoshimura, H. (2006). The potential of caffeine for functional modification from cortical synapses to neuron networks in the brain. *Curr Neuropharmacol*, 3, 309-16

Zanoli, P., Rivasi, M., Zavatti, M., Brusiani, F. & Baraldi, M. (2005). New insight in the neuropharmacological activity of Humulus lupulus L. *J Ethnopharmacol*, 102, 102-6

7

The Therapeutic Benefits of Essential Oils

Abdelouaheb Djilani[1] and Amadou Dicko[2]
[1]*LSBO, BADJI MOKHTAR-Annaba University,*
[2]*LCME, Metz University,*
[1]*Algeria*
[2]*France*

1. Introduction

Since ancient times, essential oils are recognized for their medicinal value and they are very interesting and powerful natural plant products. They continue to be of paramount importance until the present day. Essential oils have been used as perfumes, flavors for foods and beverages, or to heal both body and mind for thousands of years (Baris et al., 2006; Margaris et al., 1982; Tisserand, 1997; Wei & Shibamoto 2010). Record findings in Mesopotamia, China, India, Persia and ancient Egypt show their uses for many treatments in various forms. For example, in the ancient Egypt, the population extracted oils by infusion. Later; Greeks and Romans used distillation and thus gave aromatic plants an additional value. With the advent of Islamic civilization, extraction techniques have been further refined. In the era of the Renaissance, Europeans have taken over the task and with the development of science the composition and the nature of essential oils have been well established and studied (Burt, 2004; Peeyush et al., 2011; Steven, 2010; Suaib et al., 2007). Nowadays, peppermint, lavender, geranium, eucalyptus, rose, bergamot, sandalwood and chamomile essential oils are the most frequently traded ones.

2. Definition and localization of essential oils

Essential oils (also called volatile or ethereal oils, because they evaporate when exposed to heat in contrast to fixed oils) are odorous and volatile compounds found only in 10% of the plant kingdom and are stored in plants in special brittle secretory structures, such as glands, secretory hairs, secretory ducts, secretory cavities or resin ducts (Ahmadi et al., 2002; Bezić et al., 2009; Ciccarelli et al., 2008; Gershenzon et al., 1994; Liolios et al., 2010, Morone-Fortunato et al., 2010; Sangwan et al., 2001; Wagner et al., 1996). The total essential oil content of plants is generally very low and rarely exceeds 1% (Bowles, 2003), but in some cases, for example clove (*Syzygium aromaticum*) and nutmeg (*Myristica fragrans*), it reaches more than 10%. Essential oils are hydrophobic, are soluble in alcohol, non polar or weakly polar solvents, waxes and oils, but only slightly soluble in water and most are colourless or pale yellow, with exception of the blue essential oil of chamomile (*Matricaria chamomilla*) and most are liquid and of lower density than water (sassafras, vetiver, cinnamon and clove essential oils being exceptions) (Gupta et al., 2010; Martín et al., 2010). Due to their molecular structures (presence of olefenic double bonds and functional groups such as hydroxyl, aldehyde, ester); essential oils are readily oxidizable by light, heat and air (Skold et al., 2006; Skold et al., 2008). Some examples of oxidations are illustrated in figure 1.

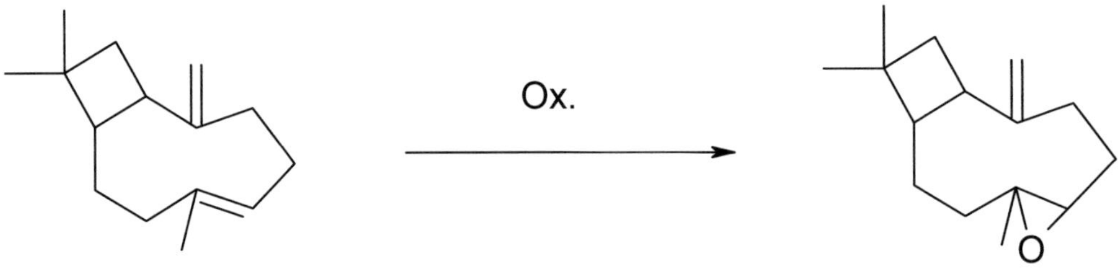

β-caryophyllene caryophyllene oxide

Fig. 1. a. Oxidation (ox.) of β-caryophyllene by air at room temperature.

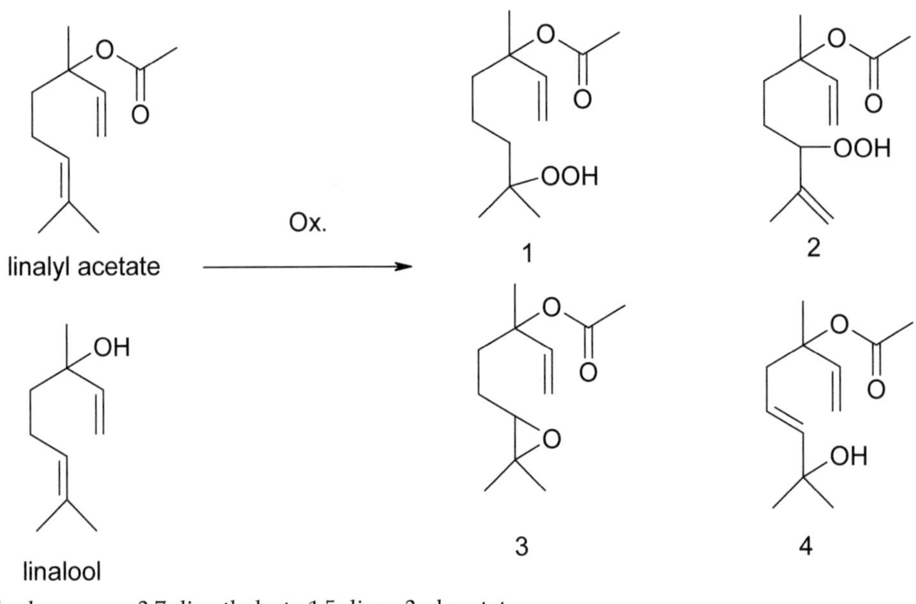

1: 7-hydroper-oxy-3,7-dimethylocta-1,5-diene-3-yl acetate
2: 3,6-hydroperoxy-3,7-dimethylocta-1,7-diene-3-yl acetate
3: 6,7-epoxy-3,7-dimethyl-1-octene-3-yl acetate
4:7-hydroxy-3,7-dimethylocta-1,5-diene-3-yl acetate

Fig. 1.b. Oxidation (ox.) of linalyl acetate and linalool by air at room temperature.

3. Extraction of essential oils

Oils contained within plant cells are liberated through heat and pressure from various parts of the plant matter; for example, the leaves, flowers, fruit, grass, roots, wood, bark, gums and blossom. The extraction of essential oils from plant material can be achieved by various methods, of which hydro-distillation, steam and steam/water distillation are the most common method of extraction (Bowles, 2003; Margaris et al., 1982; Surburg & Panten, 2006). Other methods include solvent extraction, aqueous infusion, cold or hot pressing, effleurage, supercritical fluid extraction and phytonic process (Da Porto et al., 2009; Hunter, 2009; Lahlou, 2004; Martínez, 2008; Pourmortazavi & Hajimirsadeghi, 2007; Surburg & Panten, 2006). This later process has been newly developed; it uses refrigerant hydrofluorocarbons solvents at low temperatures (below room temperature), resulting in good quality of the extracted oils. Thus, the chemical composition of the oil, both quantitative and qualitative, differs according to the extraction technique. For example, hydro-distillation and steam-distillation methods yield oils rich in terpene hydrocarbons. In contrast, the super-critical extracted oils contained a higher percentage of oxygenated compounds (Donelian et al., 2009; Eikani et al., 2007; Reverchon, 1997; Wenqiang et al., 2007).

Essential oils are highly complex mixtures of volatile compounds, and many contain about 20 to 60 individual compounds, albeit some may contain more than 100 different components (Miguel, 2010; Sell, 2006; Skaltsa et al., 2003; Thormar, 2011), such as jasmine, lemon and cinnamon essential oils.

The major volatile constituents are hydrocarbons (e.g. pinene, limonene, bisabolene), alcohols (e.g. linalol, santalol), acids (e.g. benzoic acid, geranic acid), aldehydes (e.g. citral), cyclic aldehydes (e.g. cuminal), ketones (e.g. camphor), lactones (e.g. bergaptene), phenols (e.g. eugenol), phenolic ethers (e.g. anethole), oxides (e.g. 1,8 cineole) and esters (e.g. geranyl acetate) (Deans, 1992). All these compounds may be classified into two main categories: terpenoids and phenylpropanoids (Andrade et al., 2011; De Sousa, 2011; Griffin et al., 1999; Lis-Balchin, 1997; Sangwan et al., 2001) or also into hydrocarbons and oxygenated compounds (Akhila, 2006; Halm, 2008; Hunter, 2009; Margaris et al. 1982; Pourmortazavi and Hajimirsadeghi, 2007; Shibamoto, 2010). This latter classification seems less complex, and for the current book chapter, we have adopted it. The fragrance and chemical composition of essential oils can vary according to the geo-climatic location and growing conditions (soil type, climate, altitude and amount of water available), season (for example before or after flowering), and time of day when harvesting is achieved, etc (Andrade et al., 2011; Deans et al., 1992; Margaris et al., 1982; Pengelly, 2004; Sangwan et al., 2001). In addition, there is another important factor that influences the chemical composition of essential oils, namely the genetic composition of the plant. Therefore, all these biotope factors (genetic and epigenetic) influence the biochemical synthesis of essential oils in a given plant. Thus, the same species of plant can produce a similar essential oil, however with different chemical composition, resulting in different therapeutic activities. These variations in chemical composition led to the notion of chemotypes. The chemotype is generally defined as a distinct population within the same species (plant or microorganism) that produces different chemical profiles for a particular class of secondary metabolites. Some examples of various chemotypes are given in Table 1:

Plant	Chemotype 1	Chemotype 2	Chemotype 3
Thyme (*Thymus vulgaris L.*)	Thymol	Thujanol	Linalool
Peppermint (*Mentha piperita L.*)	Menthol	Carvone	Limonene.
Rosemary (*Rosmarinus officinalis L.*)	Camphor	1,8 cineole	Verbenone
Dill (*Anethum graveolens L.*)	Carvone	Limonene	Phellandrene
Lavender (*Lavandula angustifolia Mill.*)	Linalool	Linalyl acetate	β-Caryophyllene

Table 1. Main chemotypes of some aromatic plants

4. Trade of essential oils

The knowledge of composition of essential oils and their therapeutic properties have contributed to the development of their cultivation and markets. Although only 100 species are well known for their essential oils, there are over 2000 plant species distributed over 60 families such as *Lamiaceae, Umbelliferae* and *Compositae* which can biosynthesize essential oils. They are about 3,000 essential oils, out of which approximately 300 are commercially important and are traded in the world market (Baylac and Racine, 2003; Burt, 2004; Delamare et al., 2007; Sivropoulou et al., 1995; 1996; 1997).

Essential oils constitute a major group of agro-based industrial products and they find applications in various types of industries, such as food products, drinks, perfumes, pharmaceuticals and cosmetics (Anwar et al., 2009a; 2009b; Burt, 2004; Celiktas et al., 2007; Hammer et al., 2008; Hay & Svoboda, 1993; Hussain et al., 2008; Teixeira da Silva, 2004).

The world production and consumption of essential oils is increasing very fast (Lawless, 1995). Despite their high costs (due to the large quantity of plant material required), essential oil production has been increasing. The estimates of world production of essential oils vary from 40,000 to 60,000 tonnes per annum and represent a market of approximately 700 million US $ (Verlet, 1994).

The predominately produced essential oils for industry purposes are from orange, cornmint, eucalyptus, citronella, peppermint, and lemon (Hunter, 2009) but the more commonly domestically used ones include lavender, chamomile, peppermint, tea tree oil, eucalyptus, geranium, jasmine, rose, lemon, orange, rosemary, frankincense, and sandalwood. The countries that dominate the essential oils market worldwide are Brazil, China, USA, Indonesia, India and Mexico. The major consumers are the USA, EU (especially Germany, United Kingdom and France) and Japan.

5. Bioavailability of essential oils

The term bioavailability, one of the principal pharmacokinetic properties of drugs, is used to describe the fraction of an administered dose of unchanged drug that reaches the systemic circulation and can be used for a specific function and/or stored. By definition, when a drug is administered intravenously, its bioavailability is 100%. However, when a drug is administered via other routes (such as oral), it has to pass absorption and metabolic barriers, before it reaches the general circulation system, and its bioavailability is prone to decrease (due to gastro-intestinal metabolism, incomplete absorption or first-pass metabolism). Bioavailability is measured by pharmacokinetic analysis of blood samples taken from the systemic circulation and reflects the fraction of the drug reaching the systemic circulation. If a compound is poorly absorbed or extensively metabolised beforehand, only a limited fraction of the dose administered will reach the systemic circulation. Thus, in order to achieve a high bioavailability, the compound must be of sufficiently high absorption and of low renal clearance (measurement of the renal or other organ excretion ability).

Various factors can affect bioavailability such as biochemical, physiological, physicochemical interactions; habitual mix of the diet; individual characteristics (life-stage and life-style) as well as the genotype. In the case of essential oils, the comprehension of their bioavailability by studying their absorption, distribution, metabolism and excretion in the human body is necessary. Unfortunately, there exists only limited data on the bioavailability of essential oils, and most studies are based on animal models.

All findings confirm that most essential oils are rapidly absorbed after dermal, oral, or pulmonary administration and cross the blood-brain barrier and interact with receptors in the central nervous system, and then affect relevant biological functions such as relaxation, sleep, digestion etc. .....

Most essential oil components are metabolized and either eliminated by the kidneys in the form of polar compounds following limited phase I enzyme metabolism by conjugation with glucuronate or sulfate, or exhaled via the lungs as CO_2. For example, after oral administration of (-)-menthol, 35% of the original menthol content was excreted renally as menthol glucuronide (Bronaugh et al., 1990; Buchbauer, 1993; Hotchkiss et al., 1990; Jirovetz et al., 1992; Kohlert et al., 2000).

The same happens with thymol, carvacrol, limonene and eugenol. After their oral administration, sulphate and glucuronide forms have been detected in urine and in plasma, respectively (Buchbauer et al., 1993; Guénette et al., 2007; Michiels et al., 2008). The fast metabolism and short half-life of active compounds has led to the belief that there is a minimum risk of accumulation in body tissues (Kohlert et al., 2002).

6. Therapeutic benefits of essential oils

The feeding with aromatic herbs, spices and some dietary supplements can supply the body with essential oils. There are a lot of specific dietary sources of essential oils, such as example orange and citrus peel, caraway, dill; cherry, spearmint, caraway, spearmint, black pepper and lemongrass. Thus, human exposure to essential oils through the diet or environment is widespread. However, only little information is available on the estimation of essential oil intake. In most cases, essential oils can be absorbed from the food matrix or as pure products and cross the blood brain barrier easily. This later property is due to the lipophilic character of volatile compounds and their small size.

The action of essential oils begins by entering the human body via three possible different ways including direct absorption through inhalation, ingestion or diffusion through the skin tissue.

6.1 Absorption through the skin

Essential oil compounds are fat soluble, and thus they have the ability to permeate the membranes of the skin before being captured by the micro-circulation and drained into the systemic circulation, which reaches all targets organs (Adorjan & Buchbauer, 2010; Baser & Buchbauer, 2010).

6.2 Inhalation

Another way by which essential oils enter the body is inhalation. Due to their volatility, they can be inhaled easily through the respiratory tract and lungs, which can distribute them into the bloodstream (Margaris et al., 1982; Moss et al, 2003). In general, the respiratory tract offers the most rapid way of entry followed by the dermal pathway.

6.3 Ingestion

Oral ingestion of essential oils needs attention due to the potential toxicity of some oils. Ingested essential oil compounds and/or their metabolites may then be absorbed and delivered to the rest of the body by the bloodstream and then distributed to parts of the body. Once essential oil molecules are in body, they interrelate with physiological functions by three distinct modes of action:

- Biochemical (pharmacological): Interacting in the bloodstream and interacting chemically with hormones and enzymes such as farnesene.
- Physiological: By acting (for example phytohormones) on specific physiological function. For example, the essential oil of fennel contains a form of estrogen-like compounds that may be effective for female problems such as lactation and menstruation.
- Psychological: by inhalation, the olfactory area of the brain (limbic system) undergoes an action triggered by the essential oil molecules and then, chemical and neurotransmitter messengers provide changes in the mental and emotional behavior of the person (Buchbauer, 1993; Johnson, 2011; Shibamoto et al, 2010). Lavender and lemon essential oils are examples for their sedative and relaxant properties.

Biological activity of essential oils may be due to one of the compounds or due to the entire mixture. In the following, we present effects of different classes of compounds present in essential oils together with their major properties and we give some examples of essential oils and their potential therapeutic activities.

7. Classes of essential oil compounds and their biological activities

7.1 Hydrocarbons

The majority of essential oils fall into this category; these contain molecules of hydrogen and carbon only and are classified into terpenes (monoterpenes: C10, sesquiterpenes: C15, and diterpenes: C20). These hydrocarbons may be acyclic, alicyclic (monocyclic, bicyclic or tricyclic) or aromatic. Limonene, myrcene, p-menthane, α-pinene, β-pinene, α-sabinene, p-cymene, myrcene, α–phellandrene, thujane, fenchane, farnesene, azulene, cadinene and sabinene are some examples of this family of products. These compounds have been associated with various therapeutic activities (Table 2). Some structures of these compounds are given in figure 2.

7.2 Esters

Esters are sweet smelling and give a pleasant smell to the oils and are very commonly found in a large number of essential oils. They include for example, linalyl acetate, geraniol acetate, eugenol acetate and bornyl acetate (Figure 3). Esters are anti-inflammatory, spasmolytic, sedative, and antifungal (Table 2).

7.3 Oxides

Oxides or cyclic ethers are the strongest odorants, and by far the most known oxide is 1,8-cineole, as it is the most omnipresent one in essential oils. Other examples of oxides are bisabolone oxide, linalool oxide, sclareol oxide and ascaridole (Figure 4). Their therapeutic benefits are expectorant and stimulant of nervous system (Table 2).

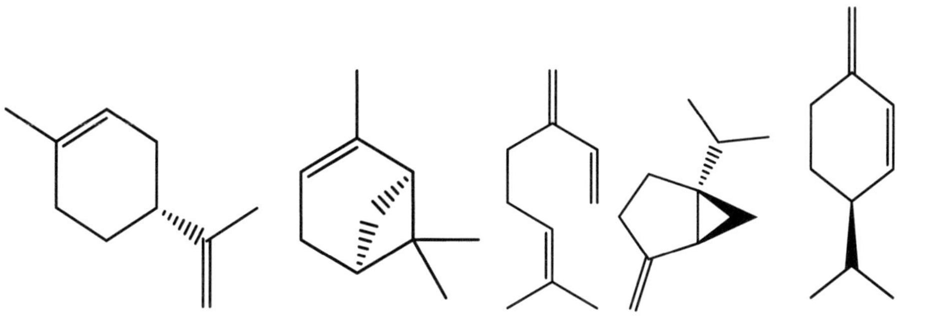

(+)-limonene (-)-α-pinene myrcene sabinene β–phellandrene

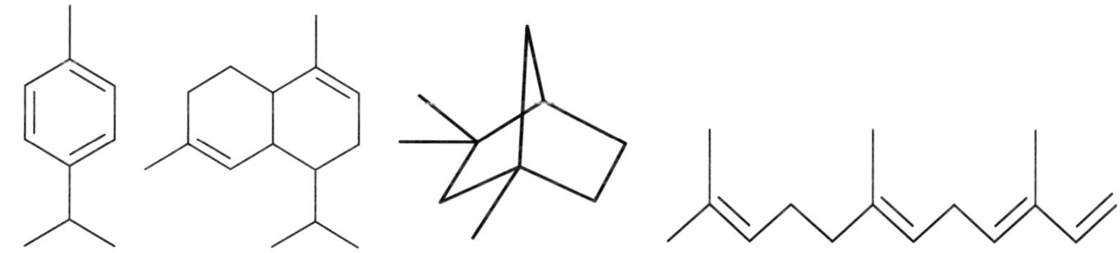

p-Cymene Cadinene Fenchane Farnesene

Fig. 2. Structures of some hydrocarbons commonly found in essential oils.

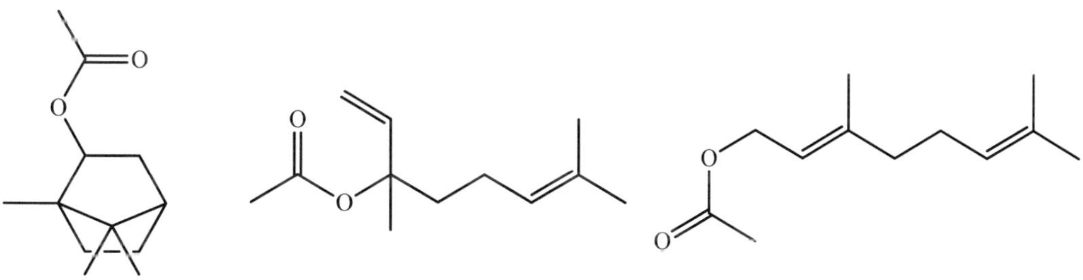

bornyl acetate linalyl acetate geranyl acetate

Fig. 3. Structures of some esters commonly found in essential oils.

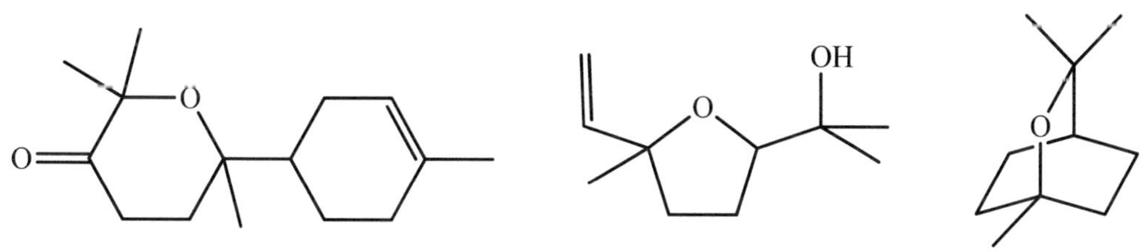

bisabolone oxide linalool oxide 1,8-cineole

Fig. 4. Structures of some oxides commonly found in essential oils.

Class of compounds	Example	Bioactivities	Literature
Hydrocarbons	Limonene, myrcene, pinene, pinene, sabinene, cymene, myrcene, phellandrene.	Stimulant, antiviral, antitumour, decongestant, antibacterial, hepatoprotective	Ozbek, 2003; Pengelly, 2004; Bowles, 2003; Svoboda & Hampson, 1999; Deans et al., 1992; Griffin et al., 1999; Edris, 2007; Baser & Buchbauer, 2010
Esters	linalyl acetate, geraniol acetate, eugenol acetate, bornyl acetate	spasmolytic, sedative, antifungal, anaesthetic, anti-inflammatory.	Pengelly, 2004; De Sousa et al., 2011; Sugawara et al., 1998; Peana et al., 2002 ; Ghelardini et al., 1999; De Sousa, 2011.
Oxides	bisabolone oxide, linalool oxide, sclareol oxide, ascaridole	anti-inflammatory, Expectorant, stimulant	Pengelly, 2004; Ghelardini et al., 2001; De Sousa, 2011.
Lactones	nepetalactone, bergaptene, costuslactone, dihydronepetalactone, alantrolactone.	Antimicrobial; antiviral; Antipyretic, sedative, hypotensive; analgesic	Pengelly, 2004; De Sousa, 2011; Miceli et al., 2005 ; Gomes et al., 2009.
Alcohols	linalol, menthol, borneol, santalol, nerol, citronellol, geraniol	Antimicrobial, antiseptic, tonifying, balancing, spasmolytic, anaesthetic; anti-inflammatory.	Pengelly, 2004; Sugawara et al., 1998; De Sousa, 2011; Ghelardini et al., 1999; Peana et al., 2002.
Phenols	thymol, eugenol, carvacrol, chavicol	antimicrobial,spasmolytic, anaesthetic, irritant, immune stimulating	Pengelly, 2004; Ghelardini et al., 1999; De Sousa, 2011.
Aldehydes	citral, myrtenal, cuminaldehyde, citronellal, cinnamaldehyde, benzaldehyde	Antiviral, antimicrobial, tonic, vasodilators, hypotensive, calming, antipyretic, sedative, spasmolytic	Dorman & Deans, 2000; Pengelly, 2004;
Ketones	carvone, menthone, pulegone, fenchone, camphor, thujone, verbenone	mucolytic, cell regenerating, sedative, antiviral, neurotoxic, analgesic, digestive, spasmolytic	Pengelly, 2004; De Sousa et al. 2008; De Sousa, 2011; Gali-Muhtassib et al., 2000

Table 2. Different classes of essential oils compounds and their bioactivities.

7.4 Lactones

Lactones are of relatively high molecular weight and are usually found in pressed oils. Some examples of lactones are nepetalactone, bergaptene, costuslactone, dihydronepetalactone, alantrolactone, epinepetalactone, aesculatine, citroptene, and psoralen (Figure 5). They may be used for antipyretic, sedative and hypotensive purposes, but their contraindication is allergy, especially such involving the skin (Table 2).

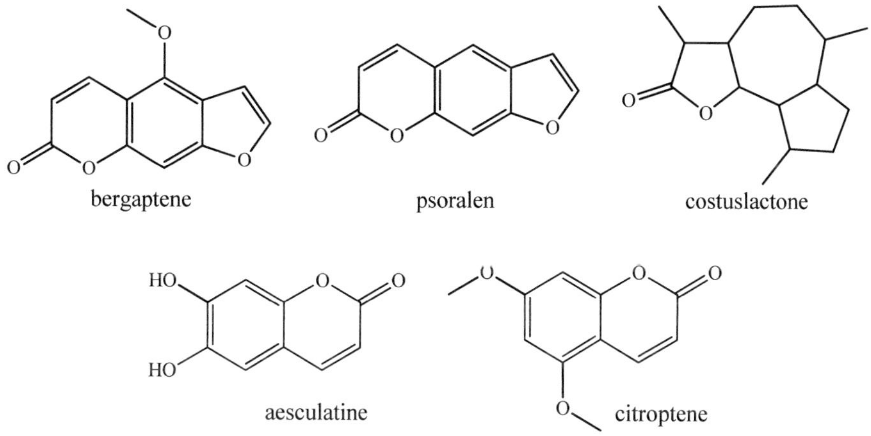

Fig. 5. Structures of some lactones commonly found in essential oils.

7.5 Alcohols

In addition to their pleasant fragrance, alcohols are the most therapeutically beneficial of essential oil components with no reported contraindications. They are antimicrobial, antiseptic, tonifying, balancing and spasmolytic (Table 2). Examples of essential oil alcohols are linalol, menthol, borneol, santalol, nerol, citronellol and geraniol (Figure 6).

7.6 Phenols

These aromatic components are among the most reactive, potentially toxic and irritant, especially for the skin and the mucous membranes. Their properties are similar to alcohols but more pronounced. They possess antimicrobial, rubefacient properties, stimulate the immune and nervous systems and may reduce cholesterol (Table 2). Phenols are often found as crystals at room temperature, and the most common ones are thymol, eugenol, carvacrol and chavicol (Figure 7).

7.7 Aldehydes

Aldehydes are common essential oil components that are unstable and oxidize easily. Many aldehydes are mucous membrane irritants and are skin sensitizers. They have characteristically sweet, pleasant fruity odors and are found in some of our most well known culinary herbs such as cumin and cinnamon. Therapeutically, certain aldehydes have been described as: antiviral, antimicrobial, tonic, vasodilators, hypotensive, calming, antipyretic and spasmolytic (Table 2). Common examples of aldehydes in essential oils include citral (geranial and neral), myrtenal, cuminaldehyde, citronellal, cinnamaldehyde and benzaldehyde (Figure 8).

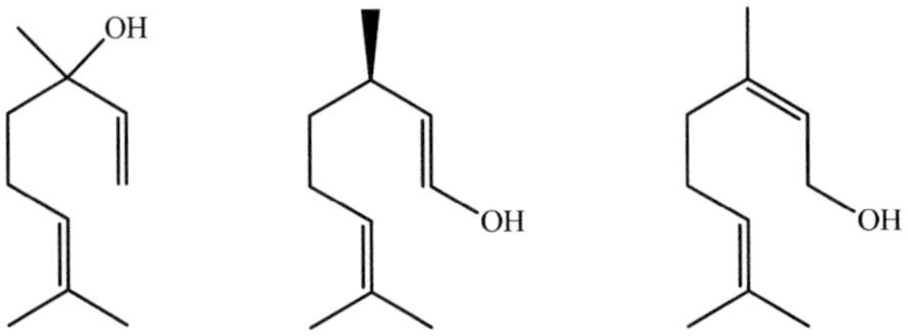

linalool (+)-citronellol nerol

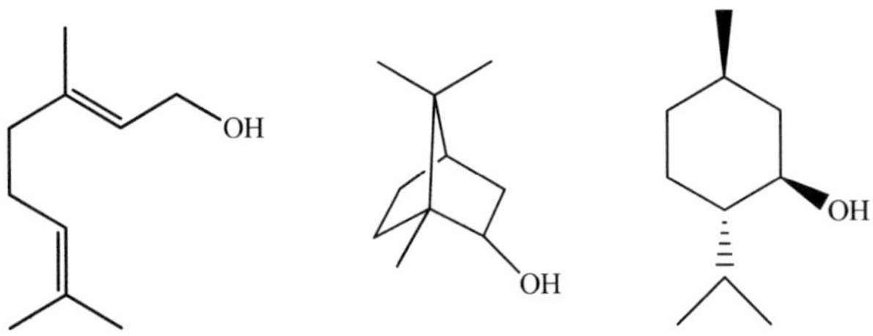

geraniol borneol menthol

Fig. 6. Structures of some alcohols commonly found in essential oils.

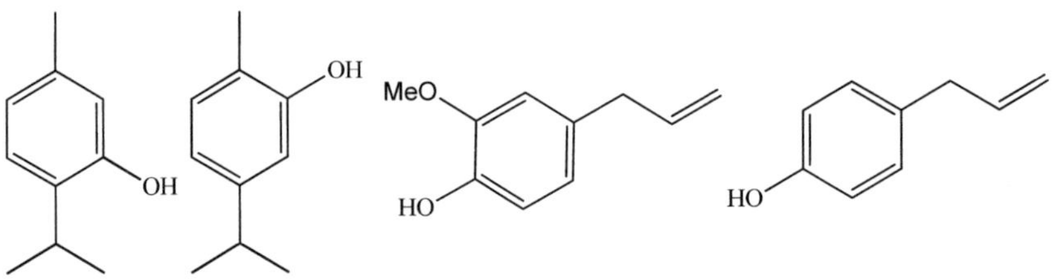

thymol carvacrol eugenol chavicol

Fig. 7. Structures of some phenols commonly found in essential oils.

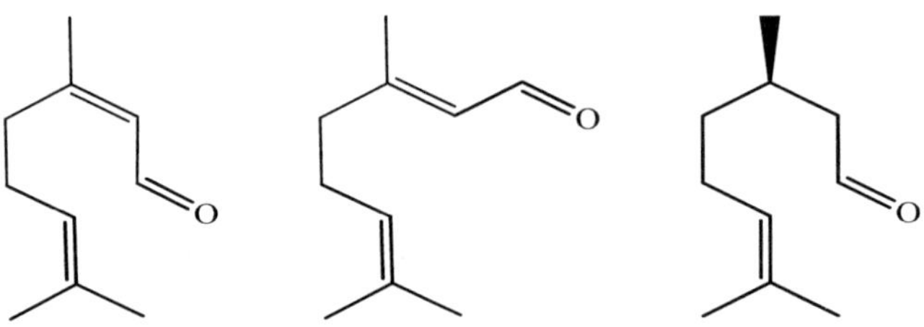

neral geranial citronellal

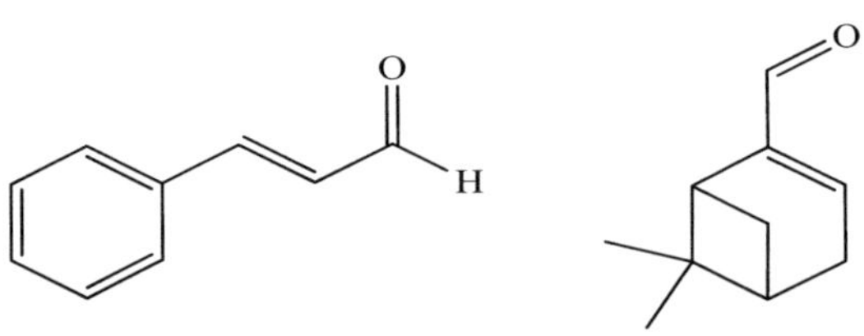

cinnamaldehyde myrtenal

Fig. 8. Structures of some aldehydes commonly found in essential oils.

7.8 Ketones

Ketones are not very common in the majority of essential oils; they are relatively stable molecules and are not particularly important as fragrances or flavor substances. In some cases, ketones are neurotoxic and abortifacients such as camphor and thujone (Gali-Muhtassib et al., 2000) but have some therapeutic effects. They may be mucolytic, cell regenerating; sedative, antiviral, analgesic and digestive (Table 2). Due to their stability, ketones are not easily metabolized by the liver. Common examples of ketones found in essential oils include carvone, menthone, pulegone, fenchone, camphor, thujone and verbenone (Figure 9).

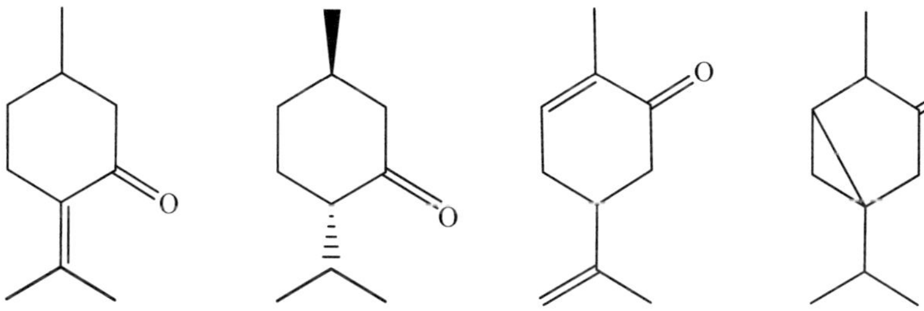

pulegone menthone carvone thujone

camphor verbenone fenchone

Fig. 9. Structures of some ketones commonly encountered in essential oils.

In Table 2; the different classes of these compounds are summarized with their bioactivities based on various biological studies cited in literature.

8. Mechanism of the biological activities of essential oils

So far, there is no study that can give us a clear idea and be accurate on the mode of action of the essential oils. Given the complexity of their chemical composition, everything suggests that this mode of action is complex, and it is difficult to identify the molecular pathway of action. It is very likely that each of the constituents of the essential oils has its own mechanism of action.

8.1 Antibacterial and antifungal action

Because of the variability of amounts and profiles of the components of essential oils, it is likely that their antimicrobial activity is not due to a single mechanism, but to several sites of action at the cellular level. Then, different modes of action are involved in the antimicrobial activity of essential oils.

One of the possibilities for action is the generation of irreversible damage to the membrane of bacterial cells, that induce material losses (cytoplasmic), leakage of ions, loss of energy substrate (glucose, ATP), leading directly to the lysis of bacteria (cytolysis) and therefore to its death. Another possibility of action is inhibition of production of amylase and protease which stop the toxin production, electron flow and result in coagulation of the cell content (Bakkali et al., 2008; Burt 2004; Di Pasqua et al., 2007; Hammer et al., 2008).

Antifungal actions are quite similar to those described for bacteria. However, two additional phenomena inhibiting the action of yeast are worth mentioning: the establishment of a pH gradient across the cytoplasmic membrane and the blocking of energy production of yeasts which involve the disruption of the bacterial membrane.

8.2 Antiviral activity

The complex mixture of essential oils usually shows a higher antiviral activity than individual compounds (due probably to synergism phenomena); with exception of β-caryophyllene which is the most famous antiviral compounds found in many different essential oils from different plant families. Different mechanisms of antiviral activity of different essential oils and their constituents seem to be present. The antiviral activity of the essential oil is principally due to direct virucidal effects (by denaturing viral structural proteins or glycoproteins). Proposed mechanisms suggest that essential oils interfere with the virus envelope by inhibiting specific processes in the viral replication cycle or by masking viral components, which are necessary for adsorption or entry into host cells, thus, they prevent the cell-to-cell virus diffusion (Saddi et al., 2007).

9. Therapeutic properties of some essential Oils

9.1 Chamomille essential oil (*Matricaria chamomilla*):

9.1.1 Main active compounds: Bisabolol and chamazulene (Cemek et al.; 2008; Kamatou & Viljoen, 2010).

9.1.2 Properties: anti-inflammatory, anti-allergic, anti-pruritic, healing, decongestive (decongest the skin) and antispasmodic (Bnouham, 2010; Tolouee et al., 2010, Alves et al., 2010; Mckay & Blumberg, 2006).

9.2 Anise essential oil (*Pimpinella anisum*):

9.2.1 Main active compound: Anethole (Andrade et al., 2011; Mata et al., 2007;)

9.2.2 Proprieties: antispasmodic, emmenagogue, stomachic, carminative, diuretic, general cardiac stimulant. (Jaiswal et al., 2009; Muchtaridi et al., 2010; Nerio et al., 2010; Tabanca et al., 2006).

9.3 Nutmeg essential oil (*Myristica fragrans*):

9.3.1 Main active compounds: Sabinene, 4-terpineol and myristicin (Muchtaridi et al., 2010).

9.3.2 Properties: Antimicrobial, pesticidal activity, general tonic, brain and circulatory, hepatoprotective, aphrodisiac, Stimulating the digestive, carminative and digestive systems Analgesic, Emmenagogue, Antiseptic, anti-parasitic (Sankarikutty & Narayanan, 1993; Spricigo et al., 1999; Tomaino et al., 2005).

9.4 Cedar essential oil (*Cedrus libani*):

9.4.1 Main active compound: Limonene (Cetin et al., 2009).

9.4.2 Properties: Larvicidal, Lymphotonic, draining powerful diuretic, Regenerative blood, Healing, astringent, Scalp Tonic, Antifungal, Anti-mosquito and anti-moth Decongestant and antiseptic respiratory Relaxing and comforting (Dharmagadda et al., 2005; Kizil et al., 2002; Loizzo et al., 2008; Svoboda et al., 1999)

9.5 Dill essential oil (*Anethum graveolens*):

9.5.1 Main active compound: Carvone (Lazutka et al., 2001; Kishore et al.,1993)

9.5.2 Properties: Antispasmodic in gastrointestinal disorders, fluidity of bronchial secretions. (Bakkali et al., 2008; Edris, 2007; Jirovetz et al., 2003; Sridhar et al., 2003.)

9.6 Garlic essential oil (*Allium sativum*):

9.6.1 Main active compound: Diallylle disulfide (Kendler, 1987; Thomson & Ali, 2003)

9.6.2 Properties: Protects and maintains the cardiovascular system, hypoglycemic, Regulates blood pressure vermifuge, antimicrobial, antiviral, anti-fungal and anti-parasitic, insecticidal and larvicidal, antioxidant (Klevenhusen et al., 2011; Lazarević et al., 2011; Lau et al., 1983; Park & Shin, 2005)

9.7 Clove essential oil (*Syzygium aromaticus*):

9.7.1 Main active compound: Eugenol and eugenyle acetate (Silva & Fernandes, 2010; Fichi et al., 2007)

9.7.2 Properties: Antiviral, antimicrobial, antifungal, general stimulating, hypertensive aphrodisiac, light stomachic, carminative, anesthetic. (de Paoli et al., 2007; Koba et al., 2011; Machado et al., 2011; Politeo et al., 2010).

9.8 Cinnamon essential oil (*Cinnamomum cassia*):

9.8.1 Main active compound: Cinnamaldehyde (Hseini & Kahouadji, 2007; Vyawahare et al., 2009).

9.8.2 Properties: Powerful, antibacterial, antiviral, antifungal and parasiticide, uterine tonic, anticoagulant, insecticide. (Cheng et al., 2004; Geng et al., 2011; Unlu *et al.*, 2010).

9.9 Sweet orange essential oil (*Citrus sinensis*):

9.9.1. Main active compound: Limonene (Hosni et al., 2010; Viudamartos et al., 2008)

9.9.2. Properties: Antiseptic, sedative, stomachic, carminative, tonic, excellent food flavoring (Anagnostopoulou et al., 2006; Ezeonu et al., 2001; Singh et al., 2010).

9.10. Eucalyptus essential oil (*Eucalyptus globulus*):

9.10.1. Main active compound: 1,8-cineole (Nerio et al., 2009; Vilela et al., 2009)

9.10.2. Properties: Anticatarrhale, expectorant and mucolytic, antimicrobial, Antiviral (Ben-Arye et al., 2011; Ben Hadj et al., 2011; Caballero-Gallardo et al., 2011; Gende et al., 2010).

9.11. Peppermint essential oil (*Mentha piperita*):

9.11.1. Main active compound: menthol and menthone (Sala, 2011; Alexopoulos et al., 2011).

9.11.2. Properties: Tonic and stimulant, decongestant, anesthetic and analgesic antipruritic, refreshing, antimicrobial, anti-inflammatory, expectorant, mucolytic, emmenagogue (De Sousa, 2011; Kumar et al., 2011; Sabzghabaee et al., 2011; Singh et al., 2011).

9.12. Lavender essential oil (*Lavandula officinalis*):

9.12.1. Main active compound: Linalol and linalyle acétate (Hajhashemi et al., 2003; Lee et al., 2011).

9.12.2. Properties: antispasmodic, sedative, relaxing, analgesic, anti-inflammatory, antimicrobial (Kloucek et al., 2011; Pohlit et al., 2011; Woronuk et al., 2011; Zuzarte et al., 2011).

9.13. Tea tree essential oil (*Melaleuca alternifolia*):

9.13.1. Main active compound: Terpinène-1-ol-4. (Van Vuuren et al., 2009 ; Hammer et al., 2008)

9.13.2. Properties: Antimicrobial, antiviral, antiasthenic, neurotonic, lymphatic, decongestant, radioprotective, antispasmodic (Garozzo et al., 2009; Lobo et al., 2011; Mickienė et al., 2011).

9.14. Lemon essential oil (*Citrus limonum*):

9.14.1. Main active compound: limonene (Fisher & Phillips, 2008; Kim et al., 2003)

9.14.2. Properties: Strengthen natural immunity, metabolism regulator, tonic nervous system, antimicrobial, antiviral, digestive tonic carminative and purgative (Koul et al., 2008; Pavela *et al.*, 2005; Pavela et al., 2008; Ponce et al., 2004).

10. Conclusion

According to literature, we can say that the essential oils and their components have many uses, both in pharmacology and in food. In addition, they are endowed with interesting biological activities and have a therapeutic potential. For example, essential oils exhibit antimicrobial activities, antiviral activities with broad spectrum, and may be useful as natural remedies and it seems that essential oils can be used as a suitable therapy for many pathologies. In the cosmetic and in the food industry, essential oils uses are an integral part, as they may play different roles. Therefore, economic importance of essential oils is indisputable. It appears therefore imperative to preserve our natural, diverse flora and support its protection in order to keep this inexhaustible source of molecules destined for multiple targets.

11. References

Adorjan, B. & Buchbauer, G. (2010). Biological properties of essential oils: an updated review. *Flavour Fragr. J.*, Vol.25, pp. 407-426.

Ahmadi, L.; Mirza, M. & Shahmir, F. (2002). The volatile constituents of Artemisia marschaliana Sprengel and its secretory elements. *Flavour Fragr. J.*, Vol.17, pp. 141-143.

Akhila, A. (2006). *Essential Oil-Bearing Grasses, The genus Cymbopogon; Medicinal and Aromatic Plants-Industrial Profiles*, CRC Press, New York.

Alexopoulos, A.; Kimbaris, A.C.; Plessas, S.; Mantzourani, I.; Theodoridou, I.; Stavropoulou, E.; Polissiou, M.G. & Bezirtzoglou, E. (2011). Antibacterial activities of essential oils from eight Greek aromatic plants against clinical isolates of *Staphylococcus aureus. Anaerobe in press.*

Alves, A.M.H.; Gonçalves, J.C.R.; Cruz, J.S. & Araújo, D.A.M. (2010). Evaluation of the sesquiterpene (−)-α-bisabolol as a novel peripheral nervous blocker. *Neuroscience Letters*, Vol.472, pp.11-15.

Anagnostopoulou, M.A.; Kefalas, P.; Papageorgiou, V.P.; Assimopoulou, A.N. & Boskou, B. (2006). Radical scavenging activity of various extracts and fractions of sweet orange peel (*Citrus sinensis*). *Food Chemistry*, Vol.94, pp.19-25.

Gali-Muhtasib, H. ; Hilan, C. & Khater, C. (2000). Traditional uses of *Salvia libanotica* (East Mediterranean sage) and the effects of its essential oils. *Journal of Ethnopharmacology*, Vol. 71 (2000) pp. 513–520.

Andrade, E.H.A.; Alves, C.N.; Guimarães, E.F.; Carreira, L.M.M. & Maia, J.G.S. (2011). Variability in essential oil composition of *Piper dilatatum* L.C. Rich. *Biochem. Syst. Ecol.*, Vol.39, pp.669-675.

Anwar, F.; Ali, M.; Hussain, A.I.; & Shahid, M. (2009). Antioxidant and antimicrobial activities of essential oils and extracts of fennel (*Foeniculum vulgare* Mill.) seeds from Pakistan. *Flavour Fragr. J.*, Vol.24, pp.170-176.

Anwar, F.; Hussain, A.I.; Sherazi, S.T.H. & Bhanger, M.I. (2009). Changes in composition and antioxidant and antimicrobial activities of essential oil of fennel (*Foeniculum vulgare* Mill.) fruit at different stages of maturity. *Journal of Herbs, Spices and Medicinals Plants*, Vol.15, pp.1-16.

Bakkali, F.; Averbeck, S.; Averbeck, D.; & Idaomar, M. (2008). Biological effects of essential oils. *Food and Chemical Toxicology*, Vol.46, pp.446-475.

Baris, O.; Güllüce, M.; Sahin, F.; Ozer, H.; Kılıc, H.; Ozkan, H.; Sökmen, M. and Ozbek, T. (2006). Biological activities of the essential oil and methanol extract of *Achillea biebersteini* Afan Afan (Asteraceae). *Turkish J. Biol.*, Vol.30, pp.65-73.

Baser, K.H.C. & Buchbauer, G. (2010). *Handbook of essential oils: Science, Technology, and Applications*, CRC Press NW.

Baylac, S. & Racine, P. (2003). Inhibition of 5-lipoxygenase by essential oils and other natural fragrant extracts. *International Journal of Aromatherapy*, Vol.13, pp.138-142.

Ben Hadj Ahmed, S.; Sghaier, R.M.; Guesmi, F.; Kaabi, B.; Mejri, M.; Attia, H.; Laouini, D.; Smaalı, I. (2011). Evaluation of antileishmanial, cytotoxic and antioxidant activities of essential oils extracted from plants issued from the leishmaniasis-endemic region of Sned (Tunisia). *Natural Product Research*, Vol.25, pp. 1195-1201.

Ben-Arye, E.; Dudai, N.; Eini, A.; Torem, M.; Schiff, E. & Rakover, Y. (2011). Treatment of Upper Respiratory Tract Infections in Primary Care: A Randomized Study Using Aromatic Herbs. *Evidence-Based Complementary and Alternative Medicine*, Vol.2011, Article ID 690346, 7 pages.

Bezić, N.; Šamanić, I.; Dunkić, V.; Besendorfer, V. & Puizina, J. (2009). Essential Oil Composition and Internal Transcribed Spacer (ITS) Sequence Variability of Four South-Croatian *Satureja* Species (Lamiaceae). *Molecules*, Vol.14, pp. 925-938.

Bnouham, M. (2010). Medicinal Plants with Potential Galactagogue Activity Used in the Moroccan Pharmacopoeia. *Journal of Complementary and Integrative Medicine*. Vol.7, Iss. 1, Article 52.

Bowles, E.J. (2003). *The Chemistry of Aromatherapeutic Oils*; 3rd Edition Griffin Press.

Bronaugh, R.L.; Wester, R.C.; Bucks, D.; Maibach, H.I.; & Sarason, R. (1990). *In vivo* percutaneous absorption of fragrance ingredients in rhesus monkeys and humans. *Food and Chemical Toxicology*, Vol.28, pp. 369-374.

Buchbauer, G. (1993). Molecular interaction: biological effects and modes of action of essential oils. *International Journal of Aromatherapy*, Vol.5, pp. 11-14.

Buchbauer, G.; Jirovetz, L.; Jager, W.; Plank, C. & Dietrich, H. (1993). Fragrance compounds and essential oils with sedative effects upon inhalation. *J. Pharm. Sci.*, Vol.82, pp. 660-664.

Burt, S. (2004). Essential oils: their antibacterial properties and potential applications in foods. *International Journal of Food Microbiology*, Vol.94, pp. 223-253.

Caballero-Gallardo, K.; Olivero-Verbel, J. & Stashenko, E.E. (2011). Repellent Activity of Essential Oils and Some of Their Individual Constituents against *Tribolium castaneum* Herbst. *J. Agric. Food Chem.*, Vol.59, pp. 1690-1696.

Celiktas, O. Y.; Kocabas, E. E. H.; Bedir, E.; Sukan, F. V.; Ozek, T. & Baser, K.H.C. (2007). Antimicrobial activities of methanol extracts and essential oils of *Rosmarinus officinalis*, depending on location and seasonal variations. *Food Chemistry*, Vol.100, pp. 553-559.

Cemek, M.; Kaga, S.; Simsek, N.; Buyukokuroglu, M.E. & Konuk, M. (2008). Antihyperglycemic and antioxidative potential of *Matricaria chamomilla* L. in streptozotocin-induced diabetic rats. *J. Nat. Med.* Vol.62, pp. 284-293.

Cetin, H.; Kurt, Y.; Isik, K. & Yanikoglu, A. (2009). Larvicidal effect of *Cedrus libani* seed oils on mosquito *Culex pipiens*. *Pharmaceutical Biology*, Vol.47, pp. 665-668.

Cheng, S.S.; Liu, J.Y.; Tsai, K.H.; Chen, W.J. & Chang, S.T. (2004). Chemical composition and mosquito larvicidal activity of essential oils from leaves of different *Cinnamomum osmophloeum* provenances. *J. Agric. Food Chem.*, Vol.52, pp. 4395-4400.

Ciccarelli, D.; Garbari, F. & Pagni, A.M. (2008). The flower of *Myrtus communis* (Myrtaceae): Secretory structures, unicellular papillae, and their ecological role. *Flora*, Vol.203, pp. 85-93.

Da Porto, C.; Decorti, D. & Kikic, I. (2009). Flavour compounds of *Lavandula angustifolia* L. to use in food manufacturing: Comparison of three different extraction methods. *Food Chemistry*, Vol.112, pp. 1072-1078.

de Paoli, S.; Giani, T.S.; Presta, G.A.; Pereira, M.O.; da Fonseca, A.D.; Brandão-Neto, J.; Medeiros, A.D.; Santos-Filho, S.D. & Bernardo-Filho, M. (2007). Effects of Clove (*Caryophyllus aromaticus* L.) on the Labeling of Blood Constituents with Technetium-99m and on the Morphology of Red Blood Cells. *Brazilian archives of biology and technology*, Vol.50, pp. 175-182.

De Sousa, D.P. (2011). Analgesic-like Activity of Essential Oils Constituents. *Molecules*, Vol.16, pp. 2233-2252.

De Sousa, D.P.; Júnior, G.A.S.; Andrade, L.N. & Batista, J.S. (2011). Spasmolytic Activity of Chiral Monoterpene Esters. *Rec. Nat. Prod.*, Vol.5, pp. 117-122.

De Sousa, D.P.; Júnior, G.A.; Andrade, L.N.; Calasans, F.R.; Nunes, X.P.; Barbosa-Filho, J.M. & Batista, J.S. (2008). Structure and spasmolytic activity relationships of analogues found in many aromatic plants. *Z. Naturforsch. C.* Vol.63, pp. 808-812.

Deans, S. G.; Svoboda, K.P.; Gundidza, M. & Brechany, E.Y. (1992). Essential oil profiles of several temperate and tropical aromatic plants: their antimicrobial and antioxidant activities. *Acta Hortic.*, Vol.306, pp. 229-232.

Delamare, A.P.L.; Moschen-Pistorello, I.T.; Artico, L.; Atti-Serafini, L.; Echeverrigaray, S. (2007). Antibacterial activity of the essential oils of *Salvia officinalis* L. and *Salvia triloba* L. cultivated in South Brazil. *Food chemistry,* Vol.100, pp. 603-608.

Dharmagadda, V.S.S.; Naik, S.N.; Mittal, P.K.; Vasudevan, P. (2005). Larvicidal activity of *Tagetes patula* essential oil against three mosquito species. *Bioresource Technology,* Vol.96, pp. 1235-1240.

Di Pasqua, R.; Betts, G.; Hoskins, N.; Edwards, M.; Ercolini, D.; Mauriello, G. (2007). Membrane toxicity of antimicrobial compounds from essential oils. *J. Agric. Food Chem.,* Vol.55, pp. 4863-4870.

Donelian, A.; Carlson, L.H.C.; Lopes, T.J. & Machado, R.A.F. (2009). Comparison of extraction of patchouli (*Pogostemon cablin*) essential oil with supercritical CO2 and by steam distillation. *The Journal of Supercritical Fluids,* Vol.48, pp. 15-20.

Dorman, H.J.D. & Deans, S.G. (2000). Antimicrobial agents from plants: antibacterial activity of plant volatile oils. *Journal of Applied Microbiology,* Vol.88, pp. 308-316.

Edris, A.E. (2007). Pharmaceutical and Therapeutic Potentials of Essential Oils and Their Individual Volatile Constituents. *Phytotheray Research,* Vol.21, pp.308-323.

Eikani, M. H.; Golmohammad, F. & Rowshanzamir, S. (2007). Subcritical water extraction of essential oils from coriander seeds (*Coriandrum sativum* L.). *Journal of Food Engineering,* Vol.80, pp. 735-740.

Ezeonu, F.C.; Chidume, G.I. & Udedi, S.C. (2001). Insecticidal properties of volatile extracts of orange peels. *Bioresource Technology,* Vol.76, pp. 273-274.

Fichi, G.; Flamini G.; Giovanelli, F.; Otranto, D. & Perrucci, S. (2007). Efficacy of an essential oil of *Eugenia caryophyllata* against *Psoroptes cuniculi. Experimental Parasitology,* Vol.115, pp. 168-172.

Fisher, K. & Phillips, C. (2008). Potential antimicrobial uses of essential oils in food: is citrus the answer. *Trends in Food Science & Technology,* Vol.19, pp. 156-164.

Gali-Muhtasib, H.; Hilan, C. & Khater, C. (2000). Traditional uses of *Salvia libanotica* (East Mediterranean sage) and the effects of its essential oils. *Journal of Ethnopharmacology,* Vol.71, pp. 513-520.

Garozzo, A.; Timpanaro, R.; Bisignano, B.; Furneri, P.M.; Bisignano, G. & Castro, A. (2009). *In vitro* antiviral activity of *Melaleuca alternifolia. Letters in Applied Microbiology,* Vol.49, pp. 806-808.

Gende, L.; Maggi, M.; Van Baren, C.; Di Leo Lira, A.; Bandoni, A.; Fritz, R. & Eguaras, M. (2010). Antimicrobial and miticide activities of *Eucalyptus globulus* essential oils obtained from different Argentine regions. *Spanish Journal of Agricultural Research,* Vol.8, pp. 642-650.

Geng, S.; Cui, Z.; Huang, X.; Chen, Y.; Xu, D. & Xiong, P. (2011). Variations in essential oil yield and composition during *Cinnamomum cassia* bark growth. *Industrial Crops and Products*, Vol.33, pp. 248-252.

Gershenzon, J. (1994). Metabolic costs of terpenoid accumulation in higher plants. *Journal of Chemical Ecology*, Vol.20, pp. 1281-1328.

Ghelardini, C.; Galeotti, N. & Mazzanti, G. (1999). Local anaesthetic activity of the essential oil of *Lavandula angustifolia*. *Planta Medica*, Vol.65, pp. 700–703.

Ghelardini, C.; Galeotti, N. & Mazzanti, G. (2001). Local anaesthetic activity of monoterpenes and phenylpropanes of essential oils. *Planta Medica*, Vol.67, pp. 564–566.

Gomes, N.G.M; Campos, M.G.; Órfão, J.M.C. & Ribeiro, C.A.F. (2009). Plants with neurobiological activity as potential targets for drug discovery. *Progress in Neuro-Psychopharmacology & Biological Psychiatry*, Vol.33, pp. 1372-1389.

Griffin, S.G.; Wyllie, S.G.; Markham, J.L. & Leach, D.N. (1999). The role of structure and molecular properties of terpenoids in determining their antimicrobial activity. *Flavour and Fragrance Journal*, Vol.14, pp. 322-332.

Guénette, S.A.; Ross, A.; Marier, J.F. ; Beaudry, F. & Vacho, P. (2007). Pharmacokinetics of eugenol and its effects on thermal hypersensitivity in rats. *Eur. J. Pharmacol.*, Vol.562, pp. 60-67.

Gupta, V.; Mittal, P.; Bansal, P.; Khokra, S.L. & Kaushik, D. (2010). Pharmacological Potential of *Matricaria recutita*. *International Journal of Pharmaceutical Sciences and Drug Research*, Vol.2, pp. 12-16.

Hajhashemi, V.; Ghannadi, A. & Sharif, B. (2003). Anti-inflammatory and analgesic properties of the leaf extracts and essential oil of *Lavandula angustifolia* Mill. *Journal of Ethnopharmacology*, Vol.89, pp. 67-71.

Halm, M.A. (2008). Essential Oils for Management of Symptoms in Critically Ill Patients. *Am. J. Crit. Care*, Vol.17, pp. 160-163.

Hammer, K.A.; Carson, C.F.; Dunstan, J.A.; Hale, J.; Lehmann, H.; Robinson, C.J.; Prescott, S.L. & Riley, T.V. (2008). Antimicrobial and anti-inflammatory activity of five *Taxandria fragrans* oils *in vitro*. *Microbiology and immunology*, Vol.52, pp. 522-530.

Hay, R.K.M. & Svoboda, K.P. (1993). Botany. In Volatile oil crops: their biology, biochemistry and production, Hay, R.K.M. and Waterman, P.G. (eds.), Longman Scientific & Technical, Harlow, pp. 5-22.

Hosni, K.; Zahed, N.; Chrif, R.; Abid, I.; Medfei, W.; Kallel, M.; Ben Brahim, N. & Sebei, H. (2010). Composition of peel essential oils from four selected *Tunisian Citrus* species: Evidence for the genotypic influence. *Food Chemistry*, Vol.123, pp. 1098-1104.

Hotchkiss, S.A.; Chidgey, M.; Rose, S. & Caldwell, J. (1990). Percutaneous absorption of benzyl acetate through rat skin in vitro. Validation of an in vitro model against in vivo data. *Food and Chemical Toxicology*, Vol.28, pp. 443-447.

Hseini, S. & Kahouadji, A. (2007). Étude ethnobotanique de la flore médicinale dans la région de Rabat (Maroc occidental); LAZAROA Vol.28, pp.79-93.

Hunter, M. (2009). Essential oils: Art, Agriculture, Science, Industry and Entrepreneurship. Nova Science Publishers, Inc., New York.

Hussain, A.I.; Anwar, F.; Sherazi, S.T.H.; Przybylski, R. (2008). Chemical composition, antioxidant and antimicrobial activities of basil (*Ocimum basilicum*) essential oils depends on seasonal variations. *Food Chem.*, Vol.108, pp. 986-995.

Jaiswal, P.; Kumar, P.; Singh, V.K. & Singh, D.K. (2009). Biological Effects of *Myristica fragrans*. *Annual Review of Biomedical Sciences*, Vol.11, pp. 21-29.

Jirovetz, L.; Buchbauer, G.; Jager, W.; Woidich, A. & Nikiforov, A. (1992) Analysis of fragrance compounds in blood samples of mice by gas chromatography, mass spectrometry, GCFTIR and GCAES after inhalation of sandalwood oil. *Biomedical Chromatography*, Vol.6, pp. 133-134.

Jirovetz, L.; Buchbauer, G.; Stoyanova, A.S.; Georgiev, E.V. & Damianova, S.T. (2003). Composition, quality control, and antimicrobial activity of the essential oil of long-time stored dill (*Anethum graveolens* L.) seeds from Bulgaria. *J. Agric. Food Chem.*, Vol.51, pp. 3854-3857.

Johnson, A.J. (2011). Cognitive Facilitation Following Intentional Odor Exposure. *Sensors*, Vol.11, pp. 5469-5488.

Kamatou, G.P.P. & Viljoen A.M. (2010). A Review of the Application and Pharmacological Properties of α-Bisabolol and α-Bisabolol-Rich Oils. *J. Am. Oil Chem. Soc.*, Vol. 87, pp. 1–7.

Kendler, B.S. (1987). Garlic (*Allium sativum*) and onion (*Allium cepa*): a review of their relationship to cardiovascular disease. *Preventive Medicine*, Vol.16, pp. 670-685.

Kim, E.H.; Kim, H.K. & Ahn, Y.J. (1993). Acaricidal Activity of Plant Essential Oils against *Tyrophagus putrescentiae* (Acari: Acaridae). *Journal of Asia-Pacific Entomology*, Vol.6, pp.77-82.

Kishore, N.; Mishra, A.K. & Chansouria, J.P.N. (1993). Fungitoxicity of essential oils against dermatophytes. *Mycoses*, Vol.36, pp. 211-215.

Kizil, M.; Kizil, G.; Yavuz, M. & Aytekin, C. (2002). Antimicrobial activity of resins obtained from the roots and stems of *Cedrus libani* and *Abies Cilicia*. *Applied Biochemistry and Microbiology*, Vol.38, pp. 144-146.

Klevenhusen, F.; Zeitz, J.O.; Duval, S.; Kreuzer, M. & Soliva, C.R. (2011). Garlic oil and its principal component diallyl disulfide fail to mitigate methane, but improve digestibility in sheep. *Animal Feed Science and Technology*, Vol.166-167, pp. 356-363.

Kloucek, P.; Smid, J.; Frankova, A.; Kokoska, L.; Valterova, I. & Pavela, R. (2011). Fast screening method for assessment of antimicrobial activity of essential oils in vapor phase. *Food Research International*, Article in Press.

Koba, K.; Nenonene, A.Y.; Raynaud, C.; Chaumont, J.P. & Sanda, K. (2011). Antibacterial Activities of the Buds Essential Oil of *Syzygium aromaticum* (L.) 42 Merr. & Perry from Togo. *Journal of Biologically Active Products from Nature*, Vol.1, pp. 42-51.

Kohlert, C.; Schindler, G.; Marz, R.W.; Abel, G.; Brinkhaus, B.; Derendorf, H.; Grafe, E.U. & Veit, M. (2002). Systemic availability and pharmacokinetics of thymol in humans. *J. Clin. Pharmacol.* Vol.42, pp. 731-737.

Kohlert, C.; van Rensen I.; Marz, R.; Schindler, G.; Graefe, E.U. & Veit, M. (2000). Bioavailability and pharmacokinetics of natural volatile terpenes in animals and humans. *Planta Medica*, Vol. 66, pp. 495-505.

Koul O., Walia S., Dhaliwal G.S. (2008). Essential Oils as Green Pesticides: Potential and Constraints. *Biopestic. Int.*, 2008, Vol.4, pp. 63-84.

Kumar, P.; Mishra, S.; Malik, A.; Satya, S. (2011). Insecticidal properties of *Mentha* species. *Industrial Crops and Products*, Vol.34, pp. 802-817.

Lahlou, M. (2004). Methods to Study the Phytochemistry and Bioactivity of Essential Oils. *Phytother. Res.*, Vol.18, pp. 435-448.

Lau, B.H.S.; Adetumbi, M.A. & Sanchez, A. (1983). *Allium sativum* (garlic) and atherosclerosis. *Nutrition Research* Vol.3, pp. 119-128.

Lawless, J. (1995). The illustrated encyclopedia of essential oils: the complete illustrated guide to the use of oils in aromatherapy and herbalism. Shaftesbury, Dorset, UK: Element.

Lazarević, J.S.; Đorđević, A.S.; Zlatković, B.K.; Radulović, N.S. & Palić, R.M. (2011). Chemical composition and antioxidant and antimicrobial activities of essential oil of *Allium sphaerocephalon* L. subsp. *sphaerocephalon* (Liliaceae) inflorescences. *Journal of the Science of Food and Agriculture* Vol.91, pp. 322-329.

Lazutka, J.R; Mierauskiene J.; Slapsyte, G. & Dedonyte, V. (2001). Genotoxicity of dill (*Anethum graveolens* L.), peppermint (*Mentha piperita* L.) and pine (*Pinus sylvestris* L.) essential oils in human lymphocytes and *Drosophila melanogaster*. *Food Chem. Toxicol.* Vol.39, pp.485-492.

Lee, Y.L.; Wu, Y.; Tsang, W.H.; Leung, A.Y. & Cheung, W.M. (2011). A Systematic Review on the Anxiolytic Effects of Aromatherapy in People with Anxiety Symptoms. *The Journal of Alternative and Complementary Medicine*. Vol.17, pp. 101-108.

Liolios, C.C.; Graikou, K.; Skaltsa, E. & Chinou, I. (2010). Dittany of Crete: A botanical and ethnopharmacological. *Journal of Ethnopharmacology*, Vol.131, pp. 229-241.

Lis-Balchin, M. (1997). Essential oils and 'aromatherapy': their modern role in healing. *The Journal of the Royal Society for the Promotion of Health*, Vol.117, pp. 324-329.

Lobo, R.; Prabhu, K.; Shirwaikar, A.; Shirwaikar, A. & Ballal, M. (2011). Formulation and Evaluation of Antiseptic Activity of the Herbal Cream Containing *Curcuma longa* and Tea Tree Oil. *Journal of Biologically Active Products from Nature*, Vol.1, pp. 27-32.

Loizzo, M.R.; Saab, A.; Tundis, R.; Statti, G.A.; Lampronti, I.H.; Menichini, F.; Gambari, R.; Cinatl, J. & Doerr, H.W. (2008). Phytochemical analysis and in vitro evaluation of the biological activity against herpes simplex virus type 1 (HSV-1) of *Cedrus libani* A. Rich. *Phytomedicine*, Vol.15, pp. 79-83.

Machado, M.; Dinis, A.M.; Salgueiro, L.; Custódio; J.B.A.; Cavaleiro, C. & Sousa, M.C. (2011). Anti-Giardia activity of *Syzygium aromaticum* essential oil and eugenol: Effects on growth, viability, adherence and ultrastructure. *Experimental Parasitology*, Vol.127, pp. 732-739.

Margaris, N.; Koedam A. & Vokou, D. (1982). Aromatic Plants: basic and applied aspects. The Hague, London, Boston, Martinus Nijhoff Publishers.

Martín, A.; Varona, S.; Navarrete, A. & Cocero, M.J. (2010). Encapsulation and Co-Precipitation Processes with Supercritical Fluids: Applications with Essential Oils. *The Open Chemical Engineering Journal*, Vol. 4, pp. 31-41.

Martínez, J.L. (2008). Supercritical Fluid Extraction of Nutraceuticals and Bioactive Compounds, CRC Press, NW.

Mata, A.T.; Proenc, C.; Ferreira, A.R.; Serralheiro, M.L.M., Nogueira, J.M.F. & Araujo M.E.M. (2007). Antioxidant and antiacetylcholinesterase activities of five plants used as Portuguese food spices. *Food Chemistry*, Vol.103, pp. 778-786.

McKay, D.L. & Blumberg, J.B. (2006). A Review of the bioactivity and potential health benefits of chamomile tea (*Matricaria recutita* L.). *Phytotherapy Research*, Vol.20, pp. 519–530.

Miceli, N.; Taviano, M.F.; Giuffrida D.; Trovato, A.; Tzakou, O. & Galati E.M. (2005). Anti-inflammatory activity of extract and fractions from *Nepeta sibthorpii* Bentham. *Journal of Ethnopharmacology*, Vol.97, pp. 261-266.

Michiels, J.; Missotten, J.; Dierick, N.; Fremaut, D.; Maene, P. & De Smet, S. (2008). In vitro degradation and in vivo passage kinetics of carvacrol, thymol, eugenol and transcinnnamaldehyde along the gastrointestinal tract of piglets. *J Sci Food Agric*. Vol.88, pp. 2371-2381.

Mickienė, R.; Bakutis, B. & Baliukonienė, V. (2011). Antimicrobial activity of two essential oils. *Ann. Agric. Environ. Med.*, Vol.18, pp. 139-144.

Miguel, M.G. (2010). Antioxidant and Anti-Inflammatory Activities of Essential Oils. *Molecules*, Vol.15, pp. 9252-9287.

Morone-Fortunato, I.; Montemurro, C.; Ruta, C.; Perrini, R.; Sabetta, W.; Blanco, A.; Lorusso, E. & Avato, P. (2010). Essential oils, genetic relationships and in vitro establishment of *Helichrysum italicum* (Roth) G. Don ssp. *italicum* from wild Mediterranean germplasm. *Industrial Crops and Products*, Vol.32, pp. 639-649.

Moss, M.; Cook, J.; Wesnes, K. & Duckett, P. (2003). Aromas of rosemary and lavender essential oils differentially affect cognition and mood in healthy adults. *International Journal of Neuroscience*, Vol.113, pp. 15-38.

Muchtaridi; Subarnas, A.; Apriyantono, A. & Mustarichien R. (2010). Identification of Compounds in the Essential Oil of Nutmeg Seeds (*Myristica fragrans* Houtt.) That Inhibit Locomotor Activity in Mice. *Int. J. Mol. Sci.*, Vol.11, pp. 4771-4781.

Nerio, L.S.; Olivero-Verbel, J. & Stashenko, E.E. (2009). Repellent activity of essential oils from seven aromatic plants grown in Colombia against *Sitophilus zeamais* Motschulsky (*Coleoptera*). *Journal of Stored Products Research*, Vol.45, pp. 212-214.

Nerio, L.S.; Olivero-Verbel, J. & Stashenko, E. (2010). Repellent activity of essential oils. *Bioresource Technology*, Vol.101, pp. 372-378.

Ozbek, H.; Ugras, S.; Dulger, H.; Bayram, I.; Tuncer, I.; Ozturk, G. & Öztürk, A. (2003). Hepatoprotective effect of *F. vulgare* essential oil. *Fitoterapia*, Vol.74, pp. 317-319.

Park, I.K. & Shin, S.C. (2005). Fumigant activity of plant essential oils and components from garlic (*Allium sativum*) and clove bud (*Eugenia caryophyllata*) oils against the Japanese termite (*Reticulitermes speratus* Kolbe). *J. Agric. Food Chem.*, Vol.53, pp. 4388-4392.

Pavela, R. (2008). Insecticidal properties of several essential oils on the house fly (*Musca domestica* L.). *Phytotherapy Research*, Vol.22, pp. 274-278.

Pavela, R.; (2005). Insecticidal activity of some essential oils against larvae of *Spodoptera littoralis*. *Fitoterapia*, Vol.76, pp. 691-696.

Peana, A.T.; D'Aquila, P.S.; Panin, F.; Serra; G.; Pippia, P. & Moretti, M.D.L. (2002). Anti-inflammatory activity of linalool and linalyl acetate constituents of essential oils *Phytomedicine*, Vol.9, pp. 721-726.

Peeyush, K.; Sapna, M.; Anushree, M. & Santosh, S. (2011). Insecticidal properties of *Mentha* species. *Industrial Crops and Products*, Vol.34, pp. 802-817.

Pengelly, A. (2004). The Constituents of Medicinal Plants; Allen & Unwin.

Pohlit, A.M.; Lopes, N.P.; Gama, R.A.; Tadei, W.P. & Neto, V.F.D. (2011). Patent Literature on Mosquito Repellent Inventions which Contain Plant Essential Oils. *Planta Med.* Vol.77, pp. 598-617.

Politeo, O.; Jukic, M. & Milos, M. (2010). Comparison of chemical composition and antioxidant activity of glycosidically bound and free volatiles from clove (*Eugenia caryophyllata* Thunb.). *Journal of Food Biochemistry*, Vol.34, pp. 129-141.

Ponce, A.G.; del Valle, C.E. & Roura, S.I. (2004). Natural essential oils as reducing agents of peroxidase activity in leafy vegetables. *Lebensmittel-Wissenschaft und-Technologie*, Vol.37, pp. 199-204.

Pourmortazavi, S.M. & Hajimirsadeghi, S.S. (2007). Supercritical fluid extraction in plant essential and volatile oil analysis. *Journal of Chromatography A*, Vol.1163, pp. 2-24.

Reverchon, E. (1997). Supercritical fluid extraction and fractionation of essential oils and related products. *J. Supercrit. Fluids*, Vol.10, pp. 1-37.

Sabzghabaee, A.M.; Nili, F.; Ghannadi, A.; Eizadi-Mood, N.; Maryam Anvari M. (2011). Role of menthol in treatment of candidial napkin dermatitis. *World Journal of Pediatrics*, Vol.7, pp. 167-170.

Saddi, M.; A. Sanna, F.; Cottiglia, L.; Chisu, L.; Casu, L.; Bonsignore, A.; De Logu, A. (2007). *Ann. Clin. Microbiol. Antimicrob.* Vol.6, pp. 1-10.

Sala, H.; (2011). Aromatherapy: Current and Emerging Applications. *Alternative and Complementary Therapies*, vol.17, pp. 26-31.

Sangwan, N.S.; Farooqi, A.H.A.; Shabih, F.; Sangwan, R.S. (2001). Regulation of essential oil production in plants. *Plant Growth Regulation*, Vol.34, pp.3-21.

Sankarikutty, B. & Narayanan, C.S. (1993), Essential Oils Isolation and Production; Encyclopedia of Food Sciences and Nutrition Academic Press. (Second Edition). pp. 2185-2189.

Sell, C.S. (2006). The Chemistry of Fragrance. From Perfumer to Consumer. 2nd edition. *The Royal Society of Chemistry*. Cambridge. 329 p.

Shibamoto, K.; Mochizuki, M. & Kusuhara, M. (2010). Aroma Therapy in Anti-Aging Medicine. *Anti-Aging Medicine*, Vol.7, pp. 55-59.

Silva, N.C.C. & Fernandes, J.A. (2010). Biological properties of medicinal plants: a review of their antimicrobial activity. *The Journal of Venomous Animals and Toxins including Tropical Diseases*, Vol.16, pp. 402-413.

Singh, P.; Shukla, R.; Prakash, B.; Kumar, A.; Singh, S.; Kumar, P.; Mishra, P.K. & Dubey, N.K. (2010). Chemical profile, antifungal, antiaflatoxigenic and antioxidant activity of *Citrus maxima* Burm. and *Citrus sinensis* (L.) Osbeck essential oils and their cyclic monoterpene, dl-limonene; *Food and Chemical Toxicology*, Vol.48, pp. 1734-1740.

Singh, R.; Shushni, M.A.M. & Belkheir, A. (2011). Antibacterial and antioxidant activities of *Mentha piperita* L; *Arabian Journal of Chemistry*, in press.

Sivropoulou, A.; Kokkini, S.; Lanaras, T. & Arsenakis, M. (1995). Antimicrobial activity of mint essential oils. *J. Agric. Food Chem.* Vol.43, pp. 2384-2388.

Sivropoulou, A.; Nikolau, C.; Papanikolau, E.; Kokkini, S.; Lanaras, T. & Arsenakis, M. (1997). Antimicrobial, cytotoxic, and antiviral activities of *Salvia fruticosa* essential oil. *Journal of Agricultural & Food Chemistry* Vol.45, pp.3197-3201.

Sivropoulou, A.; Papanikolaou, E.; Nikolau, C.; Kokkini, S.; Lanaras, T. & Arsenakis, M. (1996). Antimicrobial and cytotoxic activities of *origanum* essential oils. *J. Agric. Food Chem.* Vol.44, pp. 1202-1205.

Skaltsa, H.D.; Demetzos, C.; Lazari, D. & Sokovic, M. (2003). Essential oil analysis and antimicrobial activity of eight *Stachys* species from Greece. *Phytochemistry*, Vol.64, pp.743-752.

Skold, M.; Karlberg, A.T.; Matura, M. & Borje, A. (2006). The fragrance chemical β-caryophyllene—air oxidation and skin sensitization. *Food and Chemical Toxicology*, Vol.44, pp. 538–545.

Skold, M.; Hagvall, L. & Karlberg, A.T. (2008). Autoxidation of linalyl acetate, the main compound of lavender oil, creates potent contact allergens. *Contact Dermatitis*, Vol.58, pp. 9-14.

Spricigo, C.B.; Pinto, L.T.; Bolzan, A. & Novais, A.F. (1999). Extraction of essential oil and lipids from nutmeg by liquid carbon dioxide; *The Journal of Supercritical Fluids*, Vol.15, pp. 253-259.

Sridhar, S.R.; Rajagopal, R.V.; Rajavel, R.; Masilamani, S. & Narasimhan, S. (2003). Antifungal activity of some essential oils. *J. Agric. Food Chem.*, Vol. 51, pp. 7596-7599.

Steven, B.K. (2010). Traditional Medicine: A global perspective. Edited by Steven B. K.; Pharmaceutical Press.

Suaib, L.; Dwivedi, G.R.; Darokar, M.P.; Kaira, A. & Khanuja, S.P.S. (2007). Potential of rosemary oil to be used in drug-resistant infection. *Alternative therapies*, Vol.13, pp. 54-59.

Sugawara, Y.; Hara, C.; Tamura, K., Fujii, T.; Nakamura, K.; Masujima, T. & Aoki, T. (1998). Sedative effect on humans of inhalation of essential oil of linalool: Sensory evaluation and physiological measurements using optically active linalools. *Analytica Chimica Acta* Vol.365, pp. 293-299.

Surburg, H. & Panten J. (2006). Common Fragrance and Flavor Materials. Preparation, Properties and Uses. 5th Ed. WILEY-VCH, Weinheim.

Svoboda, K.; Hampson, J. & Hunter, T. (1999). Secretory tissues: Storage and chemical variation of essential oils in secretory tissues of higher plants and their bioactivity. *International Journal of Aromatherapy*, Vol.9, pp. 124-131.

Tabanca, N.; Demirci, B.; Ozek, T.; Kirimer, N.; Baser, K.H.C.; Bedir, E.; Khan, I.A. & Wedge, D.E. (2006). Gas chromatographic–mass spectrometric analysis of essential oils from *Pimpinella* species gathered from Central and Northern Turkey. *Journal of Chromatography A*, Vol.1117, pp. 194-205.

Teixeira da Silva, J.A. (2004). Mining the essential oils of the Anthemideae. *African Journal of Biotechnology*, Vol.3, pp. 706-720.

Thomson, M. & Ali, M. (2003). Garlic [*Allium sativum*]: a review of its potential use as an anti-cancer agent. *Current Cancer Drug Targets*, Vol.3, pp. 67-81.

Thormar, H. (2011). Lipids and Essential Oils as Antimicrobial Agents; John Wiley & Sons; Chichester.

Tisserand, R. B. (1997). In The Art of Aromatherapy; Healing Arts Press: Rochester, VT.

Tolouee, M.; Alinezhad, S.; Saberi, R.; Eslamifar, A.; Zad, S.J.; Jaimand, K.; Taeb, J.; Rezaee, M.B.; Kawachi, M.; Ghahfarokhi, M.S. & Abyaneh, M.R. (2010). Effect of *Matricaria chamomilla* L. flower essential oil on the growth and ultrastructure of *Aspergillus nigervan* Tieghem. *International Journal of Food Microbiology*, Vol.139, pp. 127-133.

Tomaino, A.; Cimino, F.; Zimbalatti, V.; Venuti, V.; Sulfaro, V.; De Pasquale, A. & Saija, A. (2005). Influence of heating on antioxidant activity and the chemical composition of some spice essential oils. *Food Chemistry*, Vol.89, pp. 549-554.

Unlu, M.; Ergene, E.; Unlu, G.V.; Zeytinoglu, H.S. & Vural, N. (2010). Composition, antimicrobial activity and in vitro cytotoxicity of essential oil from *Cinnamomum zeylanicum* Blume (Lauraceae). *Food and Chemical Toxicology,* Vol.48, pp. 3274-3280.

Van Vuuren, S.F.; Suliman, S. & Viljoen, A.M. (2009). The antimicrobial activity of four commercial essential oils in combination with conventional antimicrobials. *Letters in Applied Microbiology,* Vol.48, pp. 440-446.

Verlet, N. (1994). Huiles essentielles: production mondiale, échanges internationaux et évolution des prix. *Res. Mediterranea Magazine,* Vol.1, pp. 4-9.

Vilela, G.R.; de Almeida, G.S.; Moraes, M.H.D.; Brito, J.D.; Da Silva, M.F.; Silva, S.C.; De Stefano Piedade, S.M.; Calori-Domingues, M.A. & Da Gloria, E.M. (2009). Activity of essential oil and its major compound, 1,8-cineole, from *Eucalyptus globulus Labill.*, against the storage fungi *Aspergillus flavus* Link and *Aspergillus parasiticus* Speare. *Journal of Stored Products Research,* Vol.45, pp. 108-111.

Viudamartos, M.; RuizNavajas, Y.; Fernándezlópez, J. & Pérezalvarez, J. (2008). Antifungal activity of lemon (*Citrus lemon* L.), mandarin (*Citrus reticulata* L.), grapefruit (*Citrus paradisi* L.) and orange (*Citrus sinensis* L.) essential oils. *Food Control,* Vol.19, pp. 1130-1138.

Vyawahare, N.S.; Deshmukh, V.V.; Gadkari, M.R. & Kagathara, V.G. (2009). Plants with Antiulcer Activity. *Phcog Rev.*, Vol.3, pp. 118-125.

Wagner, G.J. (1996). Secreting glandular trichomes: More than just hairs. *Plant Physiol.* Vol.96, pp. 675-679.

Wei, A. & Shibamoto, T. (2010). Antioxidant/Lipoxygenase Inhibitory Activities and Chemical Compositions of Selected Essential Oil. *J. Agric. Food Chem.* Vol.58, pp. 7218-7225.

Wenqiang, G.; Shufen, L.; Ruixiang, Y.; Shaokun, T. & Can, Q. (2007). Comparison of essential oils of clove buds extracted with supercritical carbon dioxide and other three traditional extraction methods Chemistry of essential oils. *Food Chemistry,* Vol.101, pp.1558-1564.

Woronuk, G.; Demissie, Z.; Rheault, M. & Mahmoud, S. (2011). Biosynthesis and Therapeutic Properties of *Lavandula* Essential Oil Constituents. *Planta Med.,* Vol.77, pp. 7-15.

Zuzarte, M.; Gonçalves, M.J.; Cavaleiro, C.; Canhoto, J.; Vale-Silva, L.; Silva, M.J.; Pinto, E. & Salgueir, L. (2011). Chemical composition and antifungal activity of the essential oils of *Lavandula viridis* L'Hér. *Journal of Medical Microbiology,* Vol.60, pp. 612-618.

8

Functional Foods Based on Traditional Chinese Medicine

Chunyan Yao[1], Ruiwen Hao[2],
Shengli Pan[3] and Yin Wang[1]
[1]*Zhejiang Academy of Medical Sciences, Hangzhou, Zhejiang Province,*
[2]*Shanghai Aipu Food Industry Co. Ltd., Baoshan District, Shanghai,*
[3]*Fudan University, Zhangjiang, Shanghai,*
China

1. Introduction

The primary role of foods is to satisfy hunger and provide people with necessary nutrients, while today more and more foods or phytochemical compounds from edible plants are used to prevent nutrition-related diseases and to increase physical and mental well-being of people. These products are called functional foods, which were introduced first in Japan in the early 1980s. In the last decades, functional food industries have been developed rapidly, and functional foods have been accepted in many countries by the public due to the demand for healthier foods and products. However, there is no clear definition regarding functional foods so far at global level (Menrad, 2003; Kaushik and Kaushik, 2010). Different countries have a different regulatory framework for functional foods. The modern concept of functional food was proposed first by the Japanese academic society in the early 1980s. The legislation for functional foods was first implemented as "Foods for Special Health Use (FOSHU)" in 1991 (Arai et al., 2001). According to FOSHU, functional foods refer to natural or formulated foods containing ingredients that aid specific body functions in addition to being nutritious. Functional foods consist of food- and drink-based formulations. This first legislation highlighted the functional food as a food which should be consumed as part of the daily diet (not a capsule, tablet, or powder). A working definition of functional foods has been developed by Functional Food Science in Europe (FUFOSE) in the 1990s, stating that this is a food item that beneficially affects one or more target functions in the body beyond adequate nutritional effects in a way that is relevant to either an improved state of health and well being and /or reduction of risk of disease. The American Dietetic Association (ADA, 1999) described functional foods as any potentially healthful food or food ingredient that may provide a health benefit beyond the traditional nutrients that it contains. This is a simple definition for spanning the vast topic of food and health. The International Food Information Council (IFIC, 1998) defines functional foods as foods that provide health benefits beyond basic nutrition. Generally, functional foods may be classified as conventional food, probiotics and nutraceuticals. Conventional food includes whole foods with functional effects at some physiological level, for example, nuts and tomatoes (Kaushik and Kaushik, 2010). Probiotics are live microorganisms (in most cases, bacteria) that are similar to beneficial microorganisms found in the human gut. They are also called "friendly bacteria" or "good bacteria." Probiotics are available to consumers mainly in the form of dietary supplements and foods. They can also be used as complementary and alternative medicine. Most often, the bacteria come from two groups, *Lactobacillus* or *Bifidobacterium*. The probiotics can alleviate lactose intolerance, treat diarrhoea, enhance immune functions, etc. De Vrese et al (2001) discussed the influence of colonic flora, the colonic milieu (e.g. pH), and gas production (hydrogen) on symptoms of lactose intolerance. With reference to diarrhoea treating, although the evidence supporting the prevention of travellers' diarrhoea by probiotics is weak, there exists an overall protective effect on the prevention and treatment of diarrhoea from antibiotic-associated diarrhoea, with especially strong evidence on the efficacy of *Lactobacillus* in treating diarrhoea from rotavirus infection (Santosa et al., 2006; Guarino et al., 2009). As for probiotics and their role in immunomodulation, although several in vitro and in vivo studies on probiotics effects on immunity have been reported, the specific mechanisms of the observed changes remain unclear. Generally, an enhanced sIgA production and splenocyte proliferation were observed during the probiotics treatment. Moreover, regarding the cytokine production, several studies have shown that cytokine production by cells of the immune system can be altered by probiotic use (Erickson & Hubbard, 2000).

Nutraceuticals are foods or food ingredients that have defined physiological effects. They are derived from plants or foods, and are usually packaged and labelled similar to drugs. Functional foods and nutraceuticals are synonymous in many media and literature. Dr. Stephen De Felice, M.D., the founder and chairman of the foundation for innovation in Medicine, defined a nutraceutical as a food or part of a food that provide medical or health benefits, including the prevention and treatment of diseases. According to this concept, the nutraceuticals cover everything, including dietary supplements, fortified foods, functional foods and medical foods (Brower, 1998; Hardy, 2000; Kalra, 2003). Thus, the definition of nutraceuticals from the USA is a diet supplement that delivers a concentrated form of a presumed bioactive agent originating from a food, presented in a non-food matrix, and used to enhance health in dosages that typically exceed those that could be obtained from normal foods (Zeisel, 1999). Both in Canada and in Great Britain, functional foods are essentially a food, while a nutraceutical is an isolated or concentrated form. Thus, nutraceuticals are also naturally occurring dietary substances, with partly rather pharmaceutical dosage. Vitamins, minerals and herbal supplements fall into this scope. Currently, dozens of functional food components from plants as well as animals are under investigation for their potential role in disease prevention and health promotion.

With increasing needs and demand for a healthier life, functional foods are becoming more and more popular, while from the market perspective, functional foods are difficult to quantify because different definitions are used in the world. Therefore, more work should be done to establish a proper regulatory and harmonization of regulation and guidelines on functional foods at a global level, to ensure the quality and safety for active utilization of functional foods in different countries.

From the point of view of specific products, the difference between a functional food and a drug is not clearly defined and is not easy to classify (Pletscher, 2004). Generally, foods are usually defined as food and drink, while drugs are usually intended for use in the diagnosis, cure, treatment, or prevention of disease. There has been no common shared ground between foods and drugs in the past, and health-related claims were not permitted on foods. However, regulatory authorities and new legislation in some countries have opened up the possibility of properly substantiated health-related claims for foods and their ingredients in several countries and at the international level (Lupien, 2002). However, some of the functional foods may not be without side effects. For example, Bachmann and Hoffmann (2004) reported the interaction of L-Carnitin (a functional food) with acenocoumarol, an anticoagulant. In the event of such an incident, diagnosis becomes more difficult. Therefore, we have to face the reality that there is often no clear defined border between a functional food and a drug, and much more research work should be done on the safety, pharmacological effects and potential risks of the interaction of functional foods or functional foods and drugs.

The goal of this chapter is to offer a comprehensive review of developments and current status of the functional food area, as well as of medicated diets based on Traditional Chinese Medicines (TCM).

2. Chinese functional foods

2.1 Definition of functional food in China

Modern functional food appeared in the 1980's in China. Presently, the newly emerged functional food industry is under rapid development. In June 1996, the Ministry of Health of the People's Republic of China promulgated "the provision of functional foods administration". In this document, a functional food is defined as a food that has special health functions. It is suitable for consumption by special groups of people and has the function of regulating human body functions, but is not used for therapeutic purposes (Ministry of Heath, 1996).

A similar definition is given in "General Standard for Health (Functional) Foods (GB 16740−1997)", proclaimed by the General Administration of Quality Supervision, Inspection and Quarantine of the People's Republic of China (used to be the State Bureau of Quality and Technical Supervision) in May 1997. Functional foods are foods which help regulating body functions by special groups of chemical compounds, but not for therapeutic purposes (General Administration of Quality Supervision, Inspection and Quarantine of the People's Republic of China, 1997).

According to the guideline of registration for functional foods, promulgated by the State's Food and Drugs Administration in July 2005, health (functional) food means that a food has special health functions or is able to supply vitamins or minerals. It is suitable for consumption by special groups of people and has the function of regulating human body functions, but is not used for therapeutic purposes. It should also not cause any harm, whether acute or sub-acute or chronic (State Food and Drug Administration, 2005). These three regulations contain all the information when a functional food is registered, such as the definition, the materials needed, laboratory testing data, the suitable consumers, the package, labelling and advertisement regulations.

2.2 The history and development of functional foods in China

China is the treasure house of traditional Chinese medicine (TCM). The functional food, which initiated from Japan, is still based on the ancient Chinese saying "Medicine and food are isogenic". The documented use of functional foods in the history of TCM began as early as 1000 B.C. in the West Zhou Dynasty. The basic theories and knowledge of foods and nutrition had been established and developed in the process of looking for foods in ancient times. Hundreds of functional foods and corresponding recipes were recorded in many Chinese classical TCM publications, such as the medical classic of the yellow emperor (Han Dynasty), ShenNongBenCaoJing (Han Dynasty), Compendium of Materia Medica (Ming Dynasty), etc. These basic nutrition theories of TCM gave us some information and kinds of ideas for the application of the TMC/compounds of TCM in functional foods. Many present used products came from these monographs. Also, these medical volumes have accumulated and rendered valuable experiences for later nutrition researchers to develop superior functional foods and medical dietary foods in China. Over the past decades, functional foods have becoming one of the most important industries in China. Since the guideline of registration for functional foods were promulgated in 2005, the new functional foods were encouraged by the Food and Drug Administration (FDA) of the People's Republic of China. In the National Development Plan of Science and Technology for the Twelfth Five-year Program (2011-2015), research institutes and corporations are expected to develop safer functional foods with higher quality and better activities to increase the physiological functions, especially for old people, pregnant women, children and population in special working environment.

The reports from WTO showed that among all the population in the world, about 70% is in sub-optimal health situation, which means there is more than 1 billion of the population in sub-optimal health condition in China. Currently, China is becoming one of the most important and developed markets for functional foods, which are based on traditional dietary cultures and the rapid economic development among individuals and communities (Tee, 2002). With the globalization, a lot of famous international brands of functional foods garrisoned the market of China mainland in quick succession. In 1995, the first foreign company was Amway, since then, the number of abroad companies increased by more than 12% annually. Up to now, there are about 20 foreign food companies in China mainland, such as Amway, P&G, Jamieson, and VIVA. It is predicted by some international authoritative institutions that the sales volume of functional foods in China had reached 50 billion Chinese Yuan in 2005. By the end of July, 2011, 11,456 products were registrated in the Ministry of Health (MOH) and the State' FDA (SFDA) in China, including 10,807 domestic products and 649 foreign/international ones. Of the approved products, only about 30% of them are still currently on the market (Wu, 2004).

There are 27 categories of product specific health claims according to Technical Standards for Testing and Assessment of Health Food in China (2003), which are function related or refer to reduction of disease risk (Table 1) (Ministry of Health, 2003). However, an adjustment scheme of the health claims of functional food (a draft for discussing) has been published on the website of the SFDA on August 1st, 2011. According to the adjustment scheme, some health claims will likely be cancelled, e.g. the claim of assisting in blood pressure reduction, or combined, e.g. improving skin water content and improving skin oil content (SFDA, 2011). Most functional foods are found in 11 different formats, including capsule, tablet, powder, granule, liquid and food forms to fit people's busy lifestyle.

Allowed Claim FOR FUNCTIONAL FOODS IN CHINA	
1) Enhanced immunity	14) Improves nutritional anemia
	15) Assists in protecting against chemical injury to the liver
2) Antioxidative	16) Eliminates acne
3) Assists in memory improvement	17) Eliminates skin chloasma
4) Alleviates eye fatigue	18) Improves skin water content
5) Facilitates lead excretion	19) Improves skin oil content
6) Moisten and clean throat	20) Regulates gastrointestinal tract flora
7) Improves sleep	21) Facilitates digestion
8) Facilitates milk secretion	22) Facilitates faecal excretion
9) Alleviates physical fatigue	23) Assists in protecting against gastric mucosa damage
10) Enhances anoxia endurance	24) Weight loss
11) Assists in irradiation hazard protection	25) Assists in blood lipid reduction
12) Improves child growth and development	26) Assists in blood sugar reduction
13) Increases bone density	27) Assists in blood pressure reduction

Table 1. Permitted health claims of functional foods in China

2.3 Examples of functional foods based on the TCM

Nowadays, with the rapid development of science and technology, the functional food industries have become a rising industry in the 21st century. People also have refocused their attention from disease treatment towards disease prevention and health improvement, from drugs towards functional foods, for achieving a high quality of life. Below are a few selected examples of functional foods based on TCM. The examples of 4 kinds of functional foods with good activities and long history in TCM culture will be introduced. This is not only because raw materials of 4 functional foods are mainly produced in China, but also these functional foods are still precious and expensive till now.

2.3.1 Oviductus Ranae

The Chinese forest frog was first documented in the Compendium of Materia Medica and was called "Shange". The original Shange recorded in the works of Chinese materia medica is *Rana temporaria chensinensis* David. Oviductus Ranae is made from the oviduct of the female forest frog and was recorded by Ben-Cao-Tu-Jing (Susong of Song Dynasty), Compendium of Materia Medica and Chinese Pharmacopoeia for its effective health tonic. In ancient China, Oviductus Ranae was only used by high officials as a precious invigorant. The market price of Oviductus Ranae is about 3,000 $/kg at present and it is also known as "soft gold". The main nutritious elements of Oviductus Ranae are proteins (54.93%), amino acids (tryptophan 15.69 mg/100g, lysine 7.20 mg/100g), human chorionic gonadotrophin (HCG) (107.50 μg/g), estradiol (52.3±5.89 pg/100 mg), progesterone (187.9±19.4 pg/100 mg), testosterone (15.3±1.4 pg/100 mg), and diverse trace elements (K 1.65%, Ca 5.71%, Na 3.56%, Mg 0.53%, Cu 0.01%, Zn 0.02%, Fe 0.47%, Mn 0.06%, Se 0.001%, Cr 0.01%, Mo 0.002%, Sr 0.004%, Co 0.63%, Cd 1.09%, La 1.53%, Ba 3.73%, B 0.37%, Ni 0.65% and V 1.56%) and vitamins. It was reported that Oviductus Ranae can enhance immunity of mice (Gao et al., 1996), improve sleep (Hua et al., 2009) and relieve physical fatigue (Zhang et al., 2011). Oviductus Ranae at the dose of 68.34 mg/100 g culture medium markedly prolonged the mean life span of drosphila metanogaster (Liu et al., 1998). And Oviductus Ranae capsules have the potential protective effect on the reproductive organs of aged mice (Liang et al., 2008). Though Oviductus Ranae has been used for hundreds of years, there exist still no controlled human studies. More detailed researches about the activities and safety about Oviductus Ranae are needed.

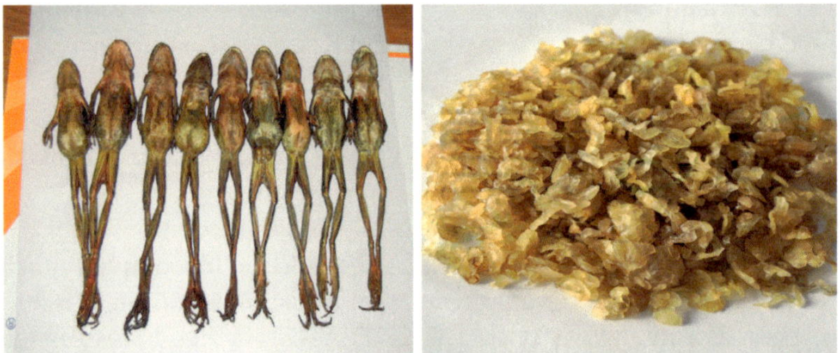

Fig. 1. *Rana temporaria chensinensis* and Oviductus Ranae

2.3.2 TiepiFengdou

TiepiFengdou is processed by the stems of *Dendrobium officinale* Kimura et Migo (Orchidaceae). In many papers, the synonym *Dendrobium candidum* was used. *D. officinale* Kimura et Migo is an endangered species in China and ranked "the first of the nine Chinese fairy herbs" (Ji, 1999). It is mainly distributed in southern China, e.g. Zhejiang, Jiangxi, Guangxi, Guizhou, Yunnan provinces and Tibet. The stems of *D. officinale* have been used as a traditional Chinese tonic medicine for hundreds of years, and have been recorded in many medical monographs such as ShenNongBenCaoJing, Compendium of Materia Medica, Supplement to the Compendium of Materia Medica, etc. According to the ancient literature and modern pharmacological research results, it can benefit human health in many aspects, such as nourishing yin and clearing away unhealthy heat, benefiting the stomach, promoting the production of body fluids, resisting cancer and prolonging life (Ji, 1999; The Pharmacopoeia Commission of PRC, 2010). It contains dendrobium polysaccharides (23%) (Huang et al., 1994), alkaloids (0.02%-0.04%) (Zhu et al., 2010), amino acids (133 mg/g dried materials) (Huang et al., 1994), as well as several trace mineral elements, including Fe 292 μg/g, Zn 12 μg/g, Mn 53μg/g, Cu 3.6μg/g (Weng, 2003). Modern pharmacological studies (Liu et al., 2011) showed that *D. officinale* and its polysaccharides can significantly enhance cellular immunity and nonspecific immunity in mice. Humoral immunity was also enhanced after oral administration of *D. officinale*, but the polysaccharides had no influence. The antioxidant activity of total polysaccharide and purified polysaccharide DCPP3c-1 from suspension-cultured protocorms of *D. candidum* had been studied, which showed the good antioxidant activity in vitro (He et al., 2007). Wu et al. (2004) indicated that the extract of *D. candidum* has obvious anti-hyperglycemic effects in adrenaline-induced hyperglycosemia mice and streptozotocin-diabetic rats, and the mechanisms are stimulating the secretion of insulin from βcells and inhibiting the secretion of glucagons from αcells, and it can probably decrease the decomposition of live glucogen and increase the synthesis of the liver glycogen. Moreover, *D. officinale* Kimura et Migo is one of the five species which contain chrysotoxene, erianin and confusarin. These compounds present good antitumour and anti-oxidation activities in vitro and in vivo (Chen et al., 2006; Gong et al., 2004).

Fig. 2. *Dendrobium officinale* and TiepiFengdou

2.3.3 Poria

Poria cocos (Polyporaceae), the Chinese name being Fu-Ling, is an oriental rotten pine-tree fungus. It has been widely used as a Chinese traditional herbal medicine for its diuretic, sedative, and tonic effects for centuries and was first documented by ShenNongBenCaoJing, the first Chinese medical pharmacopoeia written in the Han Dynasty. Various studies of this fungus have demonstrated its marked multiple activities in different experimental models, such as the antioxidant activity in mice (Wu et al., 2004; Lin et al., 2011), anti-hyperglycemic activity in streptozocin- treated mice (Li et al., 2010), anti-aging activity in mice (Hou et al., 2004), anti-inflammatory activities in human leukemic U937 and HL-60 cells (Chen & Chang, 2004), and in the models of xylene-induced ear swelling in mice and the chronic inflammatory reaction of rat granuloma induced by cotton (Hou et al., 2003), and at last the antitumor activities in S180 and K525 cells (Wu et al., 1994). *P. cocos* is widely distributed in Hubei, Henan, Anhui and Zhejiang provinces. Several triterpenes, pachyman, and pachymaran have been identified from *P. cocos* (Tai et al., 1992; Cheung, 1997; Mizushina et al., 2004; Zjawiony, 2004). Dehydrotrametenolic acid, one triterpene constituent of *P. cocos*,

Fig. 3. *Poria cocos* and Poria

was shown to reduce hyperglycemia in db/db mice, which can activate peroxide proliferator-activated receptor-γ (PPAR-γ) (Sato et al., 2002) and enhance the insulin sensitivity (Li et al., 2010). Yu et al. (2004) suggested that pachyman could improve cell-medicated immunity and could have anti-tumor function in the chickens infected with marek's disease virus (MDV).

2.3.4 Bitter gourd buccal tablet

Momordica charantia L., commonly known as bitter gourd, bitter melon, bitter lemon, balsam pear, or karela, is an economically important medicinal plant belonging to the family cucurbitaceae. It is widely cultivated in Asia, Africa, and South America, and has been used in various parts of the world to treat diabetes. The immature fruits are eaten as vegetables and are a good source of vitamin C, vitamin A, phosphorus and iron (Sultana & Bari Miah, 2003; Paul et al., 2009). The vitamin C content of Chinese bitter gourd varies from 440 mg/kg to 780 mg/kg per edible portions (Behera et al., 2010). Among the secondary bioactive metabolites of *M. charantia*, are cucurbitane-type triterpenoids. These compounds and their aglycones showed some biological effects beneficial to diabetes and obesity (Lee et al., 2009). Leung et al. (2009) reviewed the anti-diabetic and hypoglycaemic effects of *M. charantia* in animal studies and clinical studies. Fruits and seeds of bitter gourd possess medicinal properties such as anti-HIV, anti-ulcer, anti-inflammatory, anti-leukemic, antimicrobial, antitumor and antidiabetic properties (Taylor, 2002). The Compendium of Materia Medica records that *M. charantia* has the function of eliminating heat, making people vigorous, clearing people's mind and eyes, tonifying the kidney and so on. Freeze-dried bitter melon capsules are widely available and marketed in health food stores across North America and Western European countries. And in China, the bitter gourd buccal tablet was made from the concentrated solution of bitter gourd, maltitol, sorbitol and isomaltooligosaccharide (Zhang et al., 2004). After the processes of mixing, prilling, drying and tabletting, a kind of healthy bitter gourd buccal tablet of low calories was prepared, which is a functional food and usually used as a auxiliary medical service for treating diabetic (Sun et al., 2000; Liu et al., 2002). This production can be found in many supermarkets and functional food stores.

Fig. 4. *Momordica charantia* L. and Bitter gourd buccal tablet

3. Chinese medicated diets

In the theories of TCM, food and medicine are of equal importance in preventing and treating diseases. Influenced by these theories, Chinese people prefer to foods made of food materials and edible herbal medicines offered by restaurants or home made by themselves to regulate the body physiology, prevent diseases or promote recovery. This kind of food is called medicated diet/food in China. In the third part of the chapter, the definition, the history and the current status of medicated diets are discussed.

3.1 Definition of Chinese medicated diets

Medicated food/diet is delicious foods made of food materials and herbal medicines under TCM theories, nutrition and Chinese cooking technologies guidance, with a good colour, aroma and taste. It has not only the efficiency of medicine but also the delicacy of food, and has been used to prevent and cure diseases, build up one's health and prolong one's life.

3.2 The history of Chinese medicated diets

Chinese medicated diet has a long history in Chinese culture. Dietotherapy is the way to cure and prevent disease by eating food with curative effects, which is one of the parts of science of health preserving of TCM. In China, its history can be dated back to the Shang Dynasty 3,500 B. C. In the Yellow Emperor's Internal Classic, a medical classic in TCM, which appeared approximately in the Warring States Period, several medicated diet prescriptions were recorded. Sun Simiao, a well-known doctor in the Tang Dynasty, listed and discussed dietetic treatment, dietetic treatment for senile health care and health preservation, etc, in his books of BeiJiQianJinYaoFang and QianJinYiFang. It is emphasized in these books that "dietetic therapy should come first for any senile diseases, and then followed by medicine if they are not cured." These two books were substantial in medicated diet prescriptions, which were important marks of the establishment of the dietotherapy in China. The book ShiLiaoBenCao (Dietotherapy of Medical Material) by Meng Xian, a student of Sun Simiao and a famous doctor of the Tang Dynasty 618~907 A. C., has a great influence on later generations. It is the earliest extant monograph of dietetic treatment. Later, similar kinds of books on dietetic treatment have been published, including Peaceful Holy Benevolent Prescriptions (Wang Huaiyin and some others, Song Dynasty), Principles of Correct Diet (Husihui, Yuan Dynasty), the Compendium of Materia Medica (Li Shizhen, Ming Dynasty), among others. In Recipe of Suixiju by Wang Shixiong (Qing Dynasty), 331 species belonging to 7 phyla of medicated food and drink were introduced. With the development of economy and the continuous increase of the people's living standard, medicated diet is more and more valued by the people, and a number of scientific works on medicated diet have been published recently.

Medicated diet has some therapy effect. It can be prepared either from Chinese herbs alone, or from Chinese herbs and food according to certain prescriptions, by processing and cooking. In light of its form and process, the medicated diet can be divided into 3 types.

1. **Liquid diet:**
a. Fresh juice

This is the juice extracted from edible Chinese fruits alone, such as fresh fruits, or together with some fresh, clean-washed Chinese herbs. Fresh lotus leaf juice from lotus leaves (leaves of *Nelumbo nucifera* Gaertn) is a good example, documented in Compendium of Materia Medica and Pharmacopoeia of the People's Republic of China (2005). Studies on the main chemical constituents found that the content of flavonoids from lotus leaf reached 36mg/g (Zhang et al., 2005) and thus constitutes a good source of polyphenols. Research results of Lin et al. (2009) indicated that the antioxidant capacity of lotus leaves is partially relevant to its flavonoids. Huang et al. (2011) suggest that the lotus leaf methanolic extract and its active constituent catechin are useful in the control of hyperglycemia in non-insulin-dependent diabetes mellitus through their action as insulin secretagogues in vitro and in vivo. The total alkaloids extracts of lotus leaf have the function of regulating the lipids of the hyperlipidemia rats (Zhu and Li, 2010). The clinical observations showed that the lotus leaves have the significant effect on reducing blood lipid after 3 months treatment (Xie, 2010).

b. Medicated tea

This is the mixed powder of drugs with tea or without tea. Chinese herbs such as fruits, flowers and vegetables are often used as ingredients of medicated tea, while some drastic or extremely bitter herbs are not used. Generally, this kind of medicated tea is taken frequently as a common tea. Ginger and Sugar Tea, for instance, which were recorded in General Records of Holy Universal Relief (also called ShengJiZongLu, Zhaoji and others, Song Dynasty) and were used to treat wind-cold type of common cold, is made from fresh ginger (*Zingiber officinale* Roscoe) and brown sugar. Ginger, from the rhizome of *Zingiber officinale* Rosco (Zingiberaceae), is a common condiment for foods and beverages in China. It is a well-known spice and herbal medicine used to treat diseases such as the common cold, cough and gastrointestinal problems. Hiroshi et al. (2010) suggested that the repeated administration of the aqueous constituents of ginger augmented the serum corticosterone level and that this may have gradually induced anti-inflammatory activity. Gao and Zhang (2010) revealed that crude polysaccharides and favonoids from *Z. offcinale* have antibacterial activities. The volatile oil of ginger has protective effects on antioxidation activity of carbon tetracholoride damaged mice (Sun, 2010). Gingerols are the major pungent constituents of ginger, which has the significant antioxidant activity (Masuda et al., 2004) and have antiplatelet activity in rats (Jiang et al., 2010). Results from Liu et al. (2009) suggest that hypolipidemic effects of mulberry extract are via an enhancement of Low-density lipoprotein receptor gene expression and the clearance ability of low-density lipoprotein and a decrease in the lipid biosynthesis.

c. Medicated soup

Medicated soup is a kind of thick soup, which is usually stewed or braised in water with edible Chinese herbs alone, or along with meat, chicken, spareribs and so on. It is to be taken as a common soup to prevent diseases and strengthen health. Four Ingredients Decoction (SIWU Decoction), for example, is such a soup which contains four individual herbs, Radix Rehmanniae (*Rehmannia glutinosa* Libosch.), Radix Paeoniae Alba (*Paeonia lacti flora* Pall.), Rhizoma Chuanxiong (*Ligusticum chuanxiong* Hort.) and Radix Angelicae Sinensis (*Angelica Sinensis* (*Oliv.*) Diels.). This decoction is used to invigorate qi and promote blood circulation (Yu et al., 2003). Zhang et al. (2000) confirmed the effect of Siwu Decoction on red blood cell immunoadhesive function and stem cell multiplication in mice. Moreover, Lu et al. (2001) reviewed the pharmacological effects of Siwu Decocotion in immune, blood, and cardiovascular systems and Siwu Decocotion have significant radioprotective effect (Lu et al., 2001; Guo et al., 2004).

d. Medicated wine

This is a liquid made by combining wine with Chinese herbs. It can be made by either infusing or brewing, for example, Chinese wolfberry wine. The major active components from wolfberry (*Lycium chinense* Mill.) are polysaccharides. Ho et al (2009) demonstrated that a fraction of polysaccharides from wolfberry can antagonize glutamate excitotoxicity in rat cortical neurons, which provided remarkable neuroprotective effects of wolfberry. And the extract of Chinese wolfberry had a significant anti-fatigue effect in mice (Qin et al., 2009; Yin & Wang, 2010). Some research results showed that this anti-fatigue activity of wolfberry might relate to polysaccharide extract (Sheng & Fan, 2011).

2. **Semi-liquid diet:**
a. Medicated gruel

This is a kind of gruel prepared by cooking rice, together with herbs, or the decoctions, which have the function for health care. For example, gruel of Pipayc (Folium Eriobotryae) was cooked by rice and Folium Eriobotryae. Folium Eriobotryae is dry leaves of *Eriobotrya japonica* (Thunb) Lindl (Rosaceae), which was first documented in MingYiBieLu by Tao Hongjing. It has a long history of application for relieving cough and reducing sputum. Modern pharmacological research proved that Folium Eriobotryae has the anti-inflammatory and antitussive effects (Wang et al., 2004), and it was used clinically to treat acute and chronic respiratory diseases. Moreover, recent results showed that flavonoid fraction from Folium Eriobotryae has the hypoglycemic effect (Lv et al., 2009)

b. Medicated paste

This is a kind of paste, made by the powders of herbs and round-grained rice flour or wheat flour. Mulberry Gao is a tonic for enriching blood and nourishing yin. Mulberry is a deciduouds tree native to China and Korea, belonging to the Moraceae family. The fruit of mulberry has a tonic effect on kidney energy, and thus, it is used as an antiphlogistic, a diuretic and an expectorant (Koyuncu, 2004). Jeong et al. (2010) indicated that mulberry fruit extract induced human glioma cell death in vitro through ROS-dependent mitochondrial pathway and inhibits glioma growth in vivo via reduction of tumor cell proliferation and induction of apoptosis.

3. **Solid diet:**
a. Medicated cake

This is a cooked cake or noodle prepared by the mixture of Chinese herbs and rice flour, or wheat flour, or bean flour. It is processed by either steaming or baking. Take Eight-Ingredient Cake for example, may be used to strengthen the spleen (Liu et al., 2009). We will have a detailed knowledge to Eight- Ingredient Cake in the following context.

b. Cuisines

This is a large group of medicated diet, including varieties of meat and vegetable dishes. It is prepared by cooking chicken, or duck, or fish, or vegetable, etc. along with vegetables and TCM. It can be cooked in many ways: stewed, braised, steamed, boiled, stir-fried, roasted, and fried and so on. For example, crucian carp (*Carassius auratus* (L.)) steamed with tea, stewed duck with aweto (*Cordyceps sinensis*).

Most medicated diets are in conventional food form. Generally, people would like to cook it at home or buy it from some local restaurants. The medicated diets are especially popular in Southern China, e.g. in Zhejiang, Guangdong and Guangxi provinces. People take different medicated diets according to the point of view of "the unification of humanity and nature" in Chinese traditional medicine theories. The health claims of medicated diets are listed in Table 2 and Table 3. The health claims of medicated diets contain 2 categories, medicated diets for health preservation and common diseases coordination (Peng, 2010).

Allowed Claims for Medicated Diets in China	
Health preservation 1) Preserving five organs	Preserving heart, lung, liver, kidney, conditioning spleen and stomach
2) Facial beautification and figure shaping	Blacking hair, improving eyesight, skin moistening, skin lightener, spots removing, anti-acne, removing wrinkle and emaciating bodies
3) Nourishing	Nourishing brain, improving memory, tranquilizing the mind, enriching blood, regulating qi*, nourishing yin* and strengthening yang*

* The theory of Yin and Yang is one part of the basic theories in TCM, which is directly originated from the traditional Chinese philosophical thought of "the way of Heaven at unity with the course of Man's affairs". The relationship between Yin and Yang include opposition, interdependent, ebb and flow, and transformation. Yang means the outside, the upper, and the back, whereas Yin means the inside, the lower and the front of the human body structure. For example, the hair and skin are Yang, while the organs, heart, spleen, kidney, etc, are Yin. The imbalance of Yin Yang will lead to the occurrence of diseases. Thousands of TCM are classified into Yin and Yang according to the herbal nature. TCMs with Yang nature are prescribed to the patients of Yin deficiency. On the contrary, the patients of Yang deficiency get the TCMs with Yin nature. Of course, the explanation above is only the simple example to let the Yin Yang theory understood easily.

According to the TCM theory, Qi is the basic element of the human body. Qi is divided into 4 types: the most important is Yuan Qi from the kidney, then Zong Qi comes from the chest, Ying Qi generated from the spleen and stomach, flowing into vessels, and the last one is Wei Qi from the spleen and stomach, staying between vessels.

Table 2. Health claims of medicated diets in China for health preservation.

3.3 Some examples of medicated diets

Medicated diets from TCM have a long history in China. In ancient time, medicated diets, especial medicated cake and medicated soup were usually offered only to the noblemen in the Palace. Eight Ingredients cake, for example, was ever the favourite snack of the famous Qianlong emperor and Empress Dowager Cixi. Up till now, Chinese people still like these diets for preserving the health. Four selected medicated diets are discussed here.

3.3.1 Five-Juice Drink

Five-Juice Drink was first documented in Detailed Analysis of Epidemic Warm Diseases by Wu Jutong (Qing Dynasty). It is made from the juices of the water chestnut (Bulbus

Claims	
Common diseases coordination	
1) Pediatric diseases	Salivation, anorexia, infantile malnutrition, intestines parasite infection, diarrhoea, enuresis, obesity
2) Nervous system and mental disorders	Depression, insomnia, schizophrenia, neurasthenia, senile dementia, neuralgia, neuritis and palsy
3) Sense organ diseases	Conjunctivitis, glaucoma, cataract, night blindness, osteopathy, dermatosis
4) Digestive system diseases	Esophagus cancer, gastritis, gastric ulcer, duodenal ulcer, gastric cancer, gastroptosis, hyperchlorhydria, enteritis, rectal cancer, dysentery, constipation, hepatitis, fatty liver, liver cirrhosis, liver cancer
5) Respiratory system diseases	Common cold, pharyngitis, parotitis, relieving cough and eliminating phlegm, bronchitis, pneumonia, lung cancer, pulmonary tuberculosis
6) Circulatory system diseases	Coronary heart disease, arrhythmia, hypertension, hyperlipidemia, anemia virus, angiosclerosis, hemorrhage disease, leukemia
7) Endocrine system disease	Diabetes mellitus, thyroid enlargement, thyroid carcinoma, gout disease
8) Urinary diseases	Diuresis, nephritis, kidney calculi, urinary tract infection, bladder cancer
9) Reproductive system diseases	Dysmenorrhea, menoxenia, mastitis, infertility, pregnant reaction, menopause syndrome, impotence, emission, premature ejaculation, ejaculatory incompetence, prostatitis
10) Motor related system disease	Hyperosteogeny, osteoporosis, rheumatism, Eliminating wind and dampness, rheumatoid

Table 3. Health claims of medicated diets in China for common disease coordination

Heleocharis Tuberosae), fresh reed rhizome (Rhizoma Phragmitis), fresh lotus root (Rhizoma Nelumbinis), pear (Malum Piri) and fresh lilyturf root (Radix Ophiopogonis). Five-Juice Drink is used for satisfying one's thirst and relieving heat in rabbits (Jiang et al., 2007). Wu and Qiu (2009) revealed that water chestnut peel extract has strong inhibitory effect on both bacteria and fungi, especially on bacteria. The effects of liver protection of Rhizoma Phragmitis have been reported in mice and rats (Zhang et al., 2002; Zhang et al., 2002; Li et al., 2005).

3.3.2 Eight ingredients cake

Eight Ingredients cake was invested first by the famous archiater (private doctor of emperor) Chen Shigong (Ming Dynasty), and became more and more popular since the Qing Dynasty. It is said that Eight Ingredients cake was the favourite refection of Empress Dowager Cixi. This cake is made of 8 medicinal herbs, including Ginseng (*Panax ginseng*), Chinese yam (*Dioscorea oppositae* thunb.), Tuckahoe (Poria cocos, Fu-ling), Gordon Euryale Seed (Semen Euryales), lotus seed (Semen Nelumbinis), Rhizoma Atractylodis Macrocephalae, hyacinth bean (Semen Dolichoris Album), Semen Coicis, and high-quality glutinous rice flour. There exist multiple recipes for eight ingredients cake with fine distinctions and it has gradually become a famous snack in China with a sustained reputation of over 100 years. Though this snack has been in popular for many year, these is no related modern scientific literatures about the pharmacological effects recorded in ancient monographs, except the research papers on the individual herbs. More studies on the safety and efficacy are needed.

3.3.3 Herbal teas

Herbal teas, which are also known as tisanes or herbal infusions, are herbal or plant infusions. Unlike actual teas, herbal teas do not come from *Camellia sinensis* plant but come from other sources, such as blossoms of linden, leaves of peppermint and fruits of hibiscus (Trevisanato et al., 2009). In China, herbal tea is commonly known as Liangcha. The documents of traditional Chinese herbal tea are dating back to as early as ancient China. With incomes growing steadily, consumers began looking for healthier drinks. Herbal teas, with its medicinal properties, are becoming one of the most popular soft drinks in China. With the sales of Wang Laoji's Liangcha for example, 9 billion Chinese Yuan have been generated in 2007 in China. The formula of Wang Laoji's Liangcha contains *Mesona chinensis, Plumeria rubra* cv. *Acutifolia, Microcos paniculata,* Flos Chrysanthemi (*Dendranthema morifolium*), *Lonicera japonica,* Spica Prunellae (*Prunella vulgaris* L.), Radix Glycyrrhizae (*Glycyrriza Uralensis*), white granulated sugar and water. Other tea-based soft drinks such as bottled ptisan, Ku Ding tea (leaves of *Ilex latifolia* Thunb) and lemon tea also have become popular in China. Generally, herbal teas are used for clearing hectic heat, removing dampness, improving eyesight and detumescence. Li et al. (2010) reported the effects of Wanglaoji Herbal Tea on cytotoxic T lymphocyte activity in spleen of restraint stress mice. A comparative research of antioxidant activities of 20 herbal teas saled in the market were performed (He et al., 2010). The lipid metabolic dysfunction can be improved by Wanglaoji Herbal Tea in restrained mice, and the mechanism might be related to the amelioration of oxidative situation in plasma (He et al., 2008). But these kinds of beverages are not fit for pregnant women and children as the herbals used in the drink are cool in nature (Boullate et al., 2000). Moreover Wang (2008) evaluated the safety activity of Fuxing Tang Guangdong Herbal Tea in the master dissertation, and the results showed that there are no observed indicators after 30 days feeding on rats.

3.3.4 Gruel of Hetaoren (Semen Juglandis)

In fact, medicated gruels are almost the most popular medicated diet in China because gruels are easy and convenient to cook. Different kind of medicated gruels are provided for different people such as old people, children, women, patients.

Gruel of Hetaoren is made from Semen Juglandis (*Juglans regia* L.) and a proper amount of rice. The recipe is the following: Pound Hetaoren into pieces. Rice, together with a proper quantity of water and Hetaoren pieces are simmered for half an hour. Then the food is ready. Gruel of Hetaoren is used for nourishing the heart, tranquilizing the mind and tonifying the brain to benefit intelligence, and the acetone extract of Semen Juglandis could scavenge oxygen free redicals and have antiaging effect in mice (Bi & Yin, 2006). The phenolic compounds from walnut kernels have the significant antioxidant activities (Zhang et al., 2009). Zhao et al. (2004) indicated the semen juglands extract could improve learning and memory of mice.

4. Conclusion

With the remarkable improvement of people's living standard, in addition to the large aging population, more and more parts of the population is becoming interested in achieving and maintaining well-being and a high quality of life, and the demand for health-preserving foods is growing rapidly.

The functional food industry has been developing rapidly for 30 years in China. It is undeniable that in this industry, a great progress has been made, especially in the number, the claims related and the quality of the functional foods. Versatile products can be provided for different customers. The sales volume of functional foods in China in 2009 have reached 13.4 billion dollars, which means that China has become the second biggest functional food market in the world after the USA.

However, there are new opportunities and challenges in this industry up to now. Functional foods that we find on the market today are often based on the general discoveries in nutritional science, and less on a deliberate research strategy to develop functional food. For future research on functional foods, it would be of interest to focus on the mechanisms by which various food components, such as phytochemicals found in fruits, vegetables, whole grains, and herbs, positively affect health and whether these components work independently or synergistically. New functional foods with precise pharmacological mechanisms are needed.

The elderly population aged 60 has reached 144 million, 11% of the total population in China in 2006 and the number of old person will increase continuously. It is vital to improve the health care work of the aged and prevent occurrence of aging-related diseases, such as cardiovascular disease, stroke, cancer, diabetes mellitus, etc. Functional foods and nutraceuticals constitute a great promise to improve health and prevent aging-related chronic diseases (Ferrari, 2004).

With China being a leading producer of agricultural products, there is a great potential for many commodities to be processed into functional foods for domestic and global markets. More attention should be directed to potential nutraceuticals from agricultural products.

At last, the universal definition of functional food in the world should be stipulated. With the continuous development of food science and technology research, some assessment methods need to be re-evaluated and modified. More effective and safer functional foods should enter the markets for the health of people of the world.

5. References

Arai, S.; Osawa, T.; Ohigashi, H.; Yoshikawa, M.; Kaminogawa, S.; Watanabe, M.; Ogawa, T.; Okubo, K.; Watanabe, S.; Nishino, H.; Shinohara, K.; Esashi, T. & Hirahara, T. (2001). A Mainstay of Functional Food Science in Japan - History, Present Status, and Future Outlook. *Bioscience, Biotechnology, and Biochemistry*, Vol.65, No.1, (Janurary 2010), pp.1-13

Bachmann, H. U. & Hoffmann, A. (2004). Interaction of Food Supplement L-Carnitine with Oral Anticoagulant Acenocoumarol. *Swiss Medical Weekly*, Vol.134, No.25-26, (June 2004), pp. 385

Behera, T. K.; Behera, S. & Bharathi, L. K. (2010). Bitter Gourd: Botany, Horticulture, Breeding, In: *Horticultural Reviews*, Volume 37, J. Janick, (ed.), 101-141, John Wiley & Sons, Inc., ISBN 978-047-1333-74-6, Oxford, UK

Bi, M. & Yin, Z. (2006). Study on Antiaging Effect of Brain from the Extracts of Semen Juglandis. *Research and Practice on Chinese Medicines*, Vol.20, No.3, (May-June 2006), pp. 35-37

Boullate, J. I. & Nace, A. M. (2000). Safety Issues with Herbal Medicine. *Pharmacotherapy*, Vol.20, No.3, (March 2000), pp. 257-269

Brower, V. (1998). Nutraceuticals: Poised for Healthy Slice of Healthcare Market? *Nature Biotechnology*, Vol.16, No.8, (August 1998), pp. 728-731

Chen, Y. G.; Liu, S. H. & Jin, Y. (2006). Chemical Constituents and Pharmaceutical Activity of Eria Plants. *Acta Botanica Boreali-Occidentalia Sinica*, Vol.26, No.4, (April 2006), pp. 854-857

Chen, Y. Y. & Chang, H. M. (2004). Antiproliferative and Differentiating Effects of Polysaccharide Fraction from Fu-Ling (*Poria cocos*) on Human Leukemic U937 and HL-60 Cells. *Food and Chemical Toxicology*, Vol.42, No.5, (May 2004), pp. 759–769

Cheung, P. C. K. (1997). Dietary Fibre Content and Composition of Some Edible Fungi Determined by Two Methods of Analysis. *Journal of the Science of Food and Agriculture*, Vol.73, No.2, (February 1997), pp. 255–260

De Vrese, M.; Stegelmann, A.; Richter, B.; Fenselau, S.; Laue, C. & Schrezenmeir, J. (2001). Probiotics—Compensation for Lactase Insufficiency. *American Journal of Clinical Nutrition*, Vol.73, No.2Suppl, (February 2001), pp. 421S-429S

Erickson, K. L. & Hubbard, N. E. (2000). Probiotic Immunomodulation in Health and Disease. *The Journal of Nutrition*, Vol.130, No.2Suppl, (February 2000), pp. 403S-409S

Ferrari, C. K. (2004). Functional Foods, Herbs and Nutraceuticals: Towards Biochemical Mechanisms of Healthy Aging. *Biogerontology*, Vol.5, No.5, (May 2004), pp. 275–289

Gao, D. M. & Zhang, Y. Y. (2010). Comparative Antibacterial Activities of Crude Polysaccharides and Favonoids from *Zingiber offcinale* and Their Extraction. *Asian Journal of Traditional Medicines*, pp. (235-238)

Gao, Y. L. (1996). Effect of Ovidutus Ranae from Gansu Province on the Immutity and Irritability of Organism. *Joruna1 of Chinese Medicinal Materials*, Vol.19, No.2, (February, 1996), pp. 90-92

General Administration of Quality Supervision, Inspection and Quarantine of the People's Republic of China. (May 1997). In: General Standard for Health (Functional) Foods (GB 16740—1997), Available from http://jckspaqj.aqsiq.gov.cn/ybzspjyjy/gnxgbz/200703/P020070313373054134338.pdf

Gong, Y. Q.; Fan, Y.; Wu, D. Z.; Yang, H.; Hu, Z. B. & Wang, Z. B. (2004). In Vivo and in Vitro Evaluation of Erianin, A Novel Anti-angiogenic Agent. *European Journal of Cancer*, Vol. 40, No.10, (July 2004), pp. 1554-1565

Guarino. A.; Vecchio. A. L. & Canani. R. B. (2009). Probiotics as Prevention and Treatment for Diarrhea. *Current Opinion in Gastroenterology*, Vol.25, No.1, (January 2009), pp.18–23

Guo, P.; Ma, Z. C.; Li, Y. F.; Liang, Q. D.; Wang, J. F. & Wang, S. Q. (2004). Effects of Siwu Tang on Protein Expression of Bone Marrow of Blood Deficiency Mice Induced by Irradiation. *China Journal of Chinese Materia Medica*, Vol.29, No.9, (September 2004), pp. (893-896)

Hardy, G. (2000). Nutraceutical and Functional Food: Introduction and Meaning. *Nutrition*, Vol.16, No.7-8, (July-August 2000), pp. 688-689

He, Q. T.; Yu, Y. G.; Li, C. C.; Li, J. T. & Wu, H. (2011). A Comparative Study of Antioxidant Activities of 20 Kinds of Herbal Teas Saled in the Market. *Food Science*, Vol. 32, No.7, (April 2011), pp. (47-51)

He, R. R.; Kurihara, H; Bao, L. & Yao, X. S. (2008). The Effect of Wang Laoji Liangcha on Plasma Lipids Metabolism in Restraint Mice. *China Journal of Experimental Traditional Medical Formulae*, Vol.14, No.10, (October 2008), pp. (31-33)

He, T. G.; Yang, L. T.; Li, Y. R. & Wang, C. Q. (2007). Antioxidant Activity of Crude and Purified Polysaccharide from Suspension-Cultured Protocormns of *Dendrobium candidum* in Vitro. *Chinese Traditional Patent Medicine*, Vol.29, No.9, (September 2007), pp.1265-1269

Hiroshi, U.; Katsunari, I. & Atsuko, T. (2010). Repeated Oral Administration of a Squeezed Ginger (*Zingiber officinale*) Extract Augmented the Serum Corticosterone Level and Had Anti-Inflammatory Properties. *Bioscience Biotechnology Biochemistry*, Vol.74, No.11, (November 2010), pp. 2248 2252

Ho, Y. S.; Yu, M. S.; Yik, S. Y.; So, K. F.; Yuen, W. H. & Chang, R. C. C. (2009). Polysaccharides from Wolfberry Antagonizes Glutamate Excitotoxicity in Rat Cortical Neurons. *Cellular and Molecular Neurobiology*, Vol.29, No.8, (December 2009), pp. 1233-1244

Hou, A. J.; Peng, S. P. & Xiang, R. (2003). Anti-inflammatory Activity of Pachyman Polysaccharides. *Pharmacology and Clinics of Chinese Materia Medica*, Vol.19, No.3, (May-June 2003), pp.15-16

Hou, A. J., Chen, T. Y., Peng, S. P. Xiang, R. & Liu, P. (2004). Anti-aging Activity of Pachyman Polysaccharides in Mice. *Pharmacology and Clinics of Chinese Materia Medica*, Vol.20, No.3, (May-June 2004), pp. 10-11

Hua, Y. L.; Zhao, M. M. & Ning, Z. X. (2009). Experimental Research of Ovidutus Ranae for Improving Sleep. *China and Foreign Medical Journal*, Vol.7, No.6, (June, 2009), pp. 9-10

Huang, C. F.; Chen, Y. W.; Yang, C. Y.; Lin, H. Y.; Way, T. D.; Chiang, W. C. & Liu, S. H. (2011). Extract of Lotus Leaf (*Nelumbo nucifera*) and Its Active Constituent Catechin with Insulin Secretagogue Activity. *Journal of Agricultural and Food Chemistry*, Vol.59, No.4, (January 2011), pp. 1087–1094

Jeong, J. C.; Jang, S. W.; Kim, T. H.; Kwon, C. H. & Kim, Y. K. (2010). Mulberry Fruit Extract Induced Human Glioma Cell Death in Vitro through ROS-dependent

Mitochondrial Pathway and Inhibits Glioma Growth in Vivo. *Nutrition and Cancer*, Vol.62, No.3, (November-December, 2010), pp. 402-412

Ji, Z. H. (1999). Vol. 19, Orchidaceae, In : *Flora of China*, the editors committee of flora of China, (Eds.), 117, Science Press, ISBN 978-703-0073-88-4, Beijing, China

Jiang, L. Z.; Xu, S.; Wang, Y. & Wang, X. P. (2007). Research of Five-juice Drink on Yin Impairment Caused by Hyperfever of Eqidemic Febrile Disease. *Chinese Archives of Traditional Chinese Medicine*, Vol.25, No.3, (March 2007), pp. 531-533

Jiang, S. Z.; Mi, S. Q. & Wang, N. S. (2010). Effective Antiplatelet and Pharmacodynamics of Gingerols in Rats. *Pharmacolpgy and Clinics of Chinese Materia Medica*, Vol.26, No.3, (May-June 2010), pp. (10-13)

Kalra, E. K. (2003). Nutraceutical: Definition & Introduction. *AAPS PharmSci*, Vol.5, No.3, (September 2003), pp. 27-28

Kaushik, N. & Kaushik, D. (2010). Functional Foods: Overview and Global Regulation. *International Journal of Pharm Recent Research*, Vol.2, No.2, (June-September 2010), pp. 47-52

Koyuncu, F. (2004). Organic Acid Composition of Black Mulberry Fruit. *Chemistry of Natural Compounds*, Vol.40, No.4, (July-August 2004), pp. 367-369

Lee, S. Y.; Eom, S. H.; Kim, Y. K.; Park, N. I. & Park, S. U. (2009). Cucurbitane-Type Triterpenoids in *Momordica Charantia* Linn. *Journal of Medicinal Plants Research*, Vol.3, No.13, (December 2009), pp. 1264-1269

Leung, L.; Birtwhistle, R.; Kotecha, J.; Hannah, Susan. & Cuthbertson, S. (2009). Anti-diabetic and Hypoglycaemic Effects of *Momordoca charantia* (bitter melon): a mini review. *British Journal of Nutrition*, Vol.102, No.12, (December 2009), pp. 1703-1708

Li, L. H.; Zhang, G. S.; Dai, M.; Liu, X. Y. & Wang, S. (2005). The Liver Protective Effect of Rhizoma Phragmitis Polysaccharides on Rat Liver Fibrosis Induced by Carbon Tetrachloride. *Journal of Anhui Traditional Chinese Medical College*, Vol.24, No.2, (March -April 2005), pp. 24-26

Li, S. B.; He. R. R.; Wang, M.; Xie, G.; Yao, X. S. & Kurihara, H. (2010). Effects of Wanglaoji Herbal Tea on Cytotoxic T Lymphocyte Activity in Spleen of Restraint Stress Mice. *Traditional Chinese Drugs Research and Clinical Pharmacology*, Vol.21, No.3, (May-June 2010), pp. 223-226

Li, T. H.; Hou, C. C.; Chang, C. L. T. & Yang, W. C. (2010). Anti-Hyperglycemic Properties of Crude Extract and Triterpenes from *Poria Cocos*. *Evidence-Based Complementary and Alternative Medicine*, Vol. 2011, (August 2010)

Liang, J.; Zhang, X. H.; Zhou, Y.; Huang, Y. J. & Deng, H. Z. (2008). Protective Effect of *Oviductss Ranae* Capsules on the Reproductive Organs of Aged Mice. *Journal of Southern Medical University*, Vol.28, No.6, (June 2008), pp. 982-985

Lin, H. Y.; Kuo, Y. H.; Lin, Y. L. & Chiang, W. C. (2009). Antioxidative Effect and Active Components from Leaves of Lotus (*Nelumbo nucifera*). *Journal of Agricultural and Food Chemistry*, Vol.57, No.15, (July 2009), pp. 6623–6629

Lin, Z. H.; Zhu, D. N.; Yan, Y. Q.; Yu, B. Y.; Wang, Q. J.; Shen, P. N. & Ruan, K. F. (2011). An Antioxidant Phytotherapy to Rescue Neuronal Oxidative Stress. *Evidence-Based Complementary and Alternative Medicine*, Vol. 2011, (February 2011)

Liu, B. S.; Pang, Z. R.; Liu, X. G. & Jia, C. H. (2002). Research Progresses on Hypoglycemic Action of Bitter Gourd and the Compound Preparations. *Hebei Medicine*, Vol.8, No.2, (February 2002), pp. 184-187

Liu, C. X. & Xiao, P. G. (1992). Recent Advances on Ginseng Research in China. *Journal of Ethnopharmacology*, Vol. 36, No.1, (February 1992), pp. 27-38

Liu, L. K.; Chou, F. P.; Chen, Y. C.; Chyau, C. C.; Ho, H. H. & Wang, C. J. (2009). Effects of Mulberry (Morus alba L.) Extracts on Lipid Homeostasis in Vitro and in Vivo. *Journal of Agricultural and Food Chemistry*, Vol.57, No.16, (July 2009), pp. 7605-7611

Liu, X. F.; Zhu, J.; Ge, S. Y.; Xia, L. J.; Yang, H. Y.; Qian, Y. T. & Ren, F. Z. (2011). Orally Administered *Dendrobium Officinale* and Its Polysaccharides Enhance Immune Functions in BALB/c Mice. *Natural Product Communications*, Vol.6, No.6, (June 2011), pp. 867-870

Liu, Y. L.; Liu, Y. S.; Mu, X. S.; Cui, S. Y.; Qian, L.; Ren, X.; Hou, Z. R. & Zhang, J. (1998). The Anti-aging Effect of *Oviductus Ranae*. *Journal of Shenyang Pharmaceutical University*, Vol.15, No.1, (January 1998), pp.56-58

Lu, X. Q.; Ma, Z. C. & Gao, Y. (2001). The Progress in Pharmacological Research of Siwu Decoction. *Chinese Journal of Experimental Traditional Medical Formulae*, Vol.7, No.4, (July-August 2001), pp. (56-59)

Lupien, J. R. (2002). Implications for Food Regulations of NovelFood: Safety and Labeling. *Asia Pacific Journal of Clinical Nutrition*, Vol.11, No.S6, (October 2002), pp. S224-S229

Lv, H.; Chen, J.; Li, W. L.; Ren, B. R.; Wu, J. L. & Zhang, H. Q. (2009). Hypoglycemic Effect of the Total Flavonoid Fraction from Folium Eriobotryae. *Phytomedicine*, Vol.16, No.10, (October 2009), pp. 967-971

Masuda, Y.; Kikuzaki, H.; Hisamoto, M. & Nakatani, N. (2004). Antioxidant Properties of Gingerol Related Compounds from Ginger. *BioFactors*, Vol.21, No.1-4, pp. (293-296)

Menrad, K. (2003). Market and Marketing of Functional Food in Europe. *Journal of Food Engineering*, Vol.56, No.2-3, (February 2003), pp. 181–188

Ministry of Health. (May 1996). In: *The regulation of Health Food*, Available from http://www.moh.gov.cn/publicfiles/business/htmlfiles/mohwsjdj/s3592/200804/29402.htm

Ministry of Health. (February 2003). Procedures for Functional Assessment of Health Food, In: *Technical Standards for Testing and Assessment of Health Food*, The Ministry of Health of People's Republic of China, (Eds), 4-18, Tsinghua Tongfang Electronic Press, ISBN 790-063-9772/G, Beijing, China

Mizushina, Y.; Akihisa, T.; Ukiya, M., Murakami, C.; Kuriyama, I.; Xu, X. N.; Yoshida, H. & Sakaguchi. K. (2004). A Novel DNA Topoisomerase Inhibitor: Dehydroebriconic Acid, One of the Lanostane-Type TriterpeneAcids from *Poria cocos*. *Cancer Science*, Vol.95, No.4, (April 2004), pp. 354–360

Paul, A.; Mitter, K. & Raychaudhuri, S. S. (2009). Effect of Polyamines in Vitro Somatic Embryogenesis in *Momordica charantia* L. *Plant Cell, Tissue and Organ Culture*, Vol.97, No.3, (June 2009), pp. 303-311

Peng, M. Q. (2010). Compendium of Chinese Medicated Dites, Huawen Press, ISBN 978-750-7528-35-0, Beijing, China

Pletscher, W. (2004). Functional food: A Growing and Not Clearly Controlled Market with A Risk Potential Ranging from a Food to a Drug. *Swiss Medical Weekly*, Vol.134, No.25-26, (June 2004), pp. 367-368

Qin, X. Y.; Tang, L.; Yang, L.; Dai, J. F.; Huang, J. Y. & Chang, J. (2009). Study on the Anti-Fatigue Effect of Extract of Chinese Wolfberry in Mice. *Acta Chinese Medicine and Pharmacology*, Vol.37, No.2, (March-April 2009), pp. 8-10

Santosa, S.; Farnworth, E. & Jones, J. H. P. (2006). Probiotics and Their Potential Health Claims. *Nutrition Reviews*, Vol.64, No.6, (June 2006), pp. 265-274

Sato, M.; Tai, T.; Nunoura, Y.; Yajima, Y.; Kawashima, S. & Tanaka, K. (2002). Dehydrotrametenolic Acid Induces Preadipocyte Differentiation and Sensitizes Animal Models of Noninsulin-Dependent Diabetes Mellitus to Insulin. *Biological & Pharmaceutical Bulletin*, vol.25, No.1, (January 2002), pp. 81–86

Sheng, W. & Fan, W. Y. (2011). Effects of Polysaccaride from Fructus Lycii on Anti-anoxia and Anti-fatigue in Mice. *Journal of Xinxiang Medical College*, Vol.28, No.3, (May-June 2011), pp. (298-300)

State Food and Drug Administration. (July 2005). Chapter 1, In: *The guideline of registration for functional food*, State Food and Drug Administration, (Eds), 1, Available from http://www.sda.gov.cn/WS01/CL0053/24516.html

State Food and Drug Administration. (August 2011). Adjustment scheme of the health claims of health food (a draft for discussing), Available from http://www.sda.gov.cn/WS01/CL0027/64433.html

Sultana, R. S. & Bari Miah, M. A. (2003). In Vitro Propagation of Karalla (*Momordica charantea* Linn.) from Nodal Segment and Shoot Tip. *Journal of Biological Sciences*, Vol.3, No.12, (December 2003), pp. 1134-1139

Sun, F. Q. & Zhang, G. Z. (2000). Clinical Observation of Bitter Gourd Buccal Tablet in Treating Diebetic. *Liaoning Journal of Practical Diabetology*, Vol.8, No.4, (April 2000), pp. 34-35

Sun, X. S. (2010). Effects of Ginger Extract on Lipid Peroxidation in Liver of Carbon Tetracholoride Damaged Mice. *Guide of China Medicine*, Vol.8, No.23, (August 2010), pp. 26-27

Tai, T. ; Akahori, A. & Shingu, T. (1992). A Lanostane Triterpenoid from *Poria cocos*. *Phytochemistry*, Vol.31, No.7, (July 1992), pp. 2548-2549

Taylor, L. (2002). Technical Data Report for Bitter Melon (*Momordica charantia*), In: Herbal Secrets of the Rainforest 2nd ed, Sage Press Inc., ISBN 0761517340, Austin

Tee, E. S. (2002). Nutrition Labeling and Claims: Concerns and Challenges; Experiences from the Asia Pacific Region. *Asia Pacific Journal of Clinical Nutrition*. Vol.11, No.S6, pp. S215-S223

The Pharmacopoeia Commission of PRC (Eds.). (2010). Pharmacopoeia of the People's Republic of China, Chinese medicine science and Technology Press, ISBN 978-750-6744-37-9, Beijing, China

Trevisanato, S. I.; Kim, Y. I. (2000). Tea and Health. *Nutrition Reviews*, Vol.58, No.1, (January 200), pp. 1-10

Wang, L. (2008). Investigation on the Safety Evaluation of Fuxing Tang Guangdong Herbal Tea (Master dissertation).

Wang, L. W.; Liu, X. M.; Yu, S. C.; Xiao, P. G. & Yang, J. S. (2004). Anti-Inflammatory and Antitussive Effects of *Eriobotrya Japonica*. *Chinese Traditional and Herbal Drugs*, Vol.35, No.2, (February 2004), pp. 174-176

Weng, D. (2003). FAAS Determination of Trace Elements in *Dendrobium candidum* Using Suspension Sampling with Ultrasonic Agitation. *Chinese Pharmaceutical Journal*, Vol.38, No.9, (September 2003), pp. 704-706

Wu, B.; Tong, L.; Huang, T. Y.; Liang, N. C. & Li, J. L. (1994). Experimental Study on the Antitumor Activity and Mechanisms of Pachyman Polysaccharides. *Chinese Pharmacological Bulletin*, Vol. No.4, (April 1994), pp. 300-304

Wu, H. S.; Xu, J. H.; Chen, L. Z. & Sun, J. J. (2004). Studies on Anti-hyperglycemic Effect and Its Mechanism of *Dendrobium candidum*. *China Journal of Chinese Materia Medica*, Vol.29, No.2, (February 2004), pp. 160-163

Wu, S. J.; Ng, L. T. & Lin, C. C. (2004). Antioxidant Activities of Some Common Ingredients of Traditional Chinese Medicine, *Angelica sinensis*, *Lycium barbarum* and *Poria cocos*. *Phytoterapy research*, Vol.18, No.12, (December, 2004), pp. 1008-1012

Wu, S. J. & Qiu, P. Q. (2009). Inhibitory Effects of Bioactive Components in Water Chestnut on Bacteria. *Food Science*, Vol.30, No.19, (October 2009), pp. 143-146, ISSN 1002-6630

Wu, Y. M. (2004). Actuality and Countermeasure on Health Foods Authorized in China. *Chinese Journal of Public Health Management*. Vol.20, No.2, (March-April 2004), pp. (32-36)

Xie, J. X. (2010). Clinical Observations of the Effect of Lotus Leaves on Hyperlipidemia. *Chronic Pathematology Journal*, Vol. 12, No. 7, (July 2010), pp. 634

Yin, X. W. & Wang, L. (2010). Effects of Fructus Lycii on Anti-anoxia and Anti-fatigue for Mice. *Journal of Agricultural Sciences*, Vol.31, No.1, (January-March 2010), pp. (40-43)

Yu, J. G.; Jiang, Z. Q.; Yan, H. G. & Zhu, L. Y. (2004). Effect of Pachyman on Cell-Medicated Immunity Activity and Anti-Tumor Function in Chicken Infected with vMDV. *Chinese Journal of Veterinary Science and Technology*, Vol.34, No.11, (November 2004), pp. 70-73

Yu, Y. L.; Zhang, C. X.; Liang, A. J.; Xie, J. Q. & Yuan, C. (2003). Research Progresses on Pharmacology and Application of SiWu Decoction. *Practical Journal of Medicine & Pharmacy*, Vol.20, No.12, (December 2003), pp. 955-956

Zeisel, S. H. (1999). Regulation of "Nutraceuticals". *Science*, Vol.285, No.5435, (September 2003), pp. (1853-1855)

Zhang, G. S.; Fan, M. Y.; Peng, D. Y. & Fang, F. (2002). The Liver Protective Activity of Polysaccharides from Rhizoma Phragmitis. *China Journal of Traditional Chinese Medicine and Pharmacy*, Vol.17, No.7, (July 2002), pp. 416-417

Zhang, G. S.; Fan, M. Y.; Peng, D. Y. & Fang, F. (2002). Protective Effect of P-Poly on Carbon Tetrachloride Induced Liver Damage in Mice. *Chinese Pharmacological Bulletin*, Vol.18, No.3, (March 2002), pp.354-355

Zhang, P.; Ge, H. J.; Lai, Y. J. & Zhang, L. (2011). Experimental Study of Ovidutus Ranae for Relieving Physical fatigue. *Journal of Hygiene Research*, Vol.40, No.2, (March-April, 2011), pp. 231-232

Zhang, T.; Tan, Y. Y. & Pan, Y. S. (2000). Effects of Siwu Decoction on RBC Immunoadhesive Function and Stem Cell Multiplication. *Journal of Beijing University of Traditional Chinese Medicine*, Vol.23, No.1, (January 2000), pp. (36-38)

Zhang, Y.; Zhang, R. F.; Zhang, M. W. & Qiu, Y. Q. (2004). Study on Processing Technology of Healthy Bitter Gourd Buccal Tablet. *Food Science and Technology*, Vol.30, No.9, (September 2004), pp 31-33

Zhang, Y. B.; Li, C. X.; Li, J. & Wu, Y. Q. (2005). Study on Extraction Flavonoid From Lotus Leaf. *Food Research and Development*, Vol.26, No.5, (September- October, 2005), pp. 94-96

Zhang, Z. J.; Liao, L. P.; Moore, J.; Wu, T. & Wang, Z. T. (2009). Antioxidant Phenolic Compounds from Walnut Kernels (*Juglans regia* L.). *Food Chemistry*, Vol.113, No.1, (March 2009), pp. (160-165)

Zhao, H. F.; LI, X. M. & Xiao, R. (2004). Effects of Semen Juglands Extract on Improving Learning and Memory of Mice. *Journal of Shanxi Medical University*, Vol.35, No.1, (January 2004), pp. (20-22)

Zhu, L. Z. & Li, W. (2010). Study of Effect of Total Alkaloids of Lotus Leaf on the Lipids of Hyperlipemia Rats. *Heilongjiang Medicine Journal*, Vol.23, No.3, (March 2010), pp. (363-364)

Zhu, Y.; Zhang, A.; He, B.; Zhang, X.; Yu, Q. & Si, J. (2010). Quantitative Variation of Total Alkaloids Contents in *Dendrobium officinale*. *China Journal of Chinese Materia Medica*, Vol.35, No.18, (September 2010), pp. 2388-2391

Zjawiony, J. K. (2004). Biologically Active Compounds from Aphyllophorales (polypore) Fungi. *Journal of Natural Products*, Vol.67, No.2, (February 2004), pp. 300–310

9

Health-Promoting Food Ingredients and Functional Food Processing

Lizhe Wang and Torsten Bohn
Centre de Recherche Public- Gabriel Lippmann
Luxembourg

1. Introduction

One of the greatest challenges food research is facing in this century lies in maintaining sustainable food production and at the same time delivering high quality food products with an added functionality to prevent life-style related diseases such as, cancer, obesity, diabetes, heart disease, stroke. Functional foods that contain bioactive components may provide desirable health benefits beyond basic nutrition and play important roles in the prevention of life-style related diseases. Polyphenols and carotenoids are plant secondary metabolites which are well recognized as natural antioxidants linked to the reduction of the development and progression of life-style related diseases. This chapter focuses on health-promoting food ingredients (polyphenols and carotenoids), food structure and functionality, and bioavailability of these bioactive ingredients, with examples on their commercial applications, namely on functional foods. Thereafter, in order to support successful development of health-promoting food ingredients, this chapter contributes to an understanding of the relationship between food structures, ingredient functionality, in relation to the breakdown of food structures in the gastrointestinal tract and its impact on the bioavailability of bioactive ingredients. The overview on food processing techniques and the processing of functional foods given here will elaborate novel delivery systems for functional food ingredients and their applications in food. Finally, this chapter concludes with microencapsulation techniques and examples of encapsulation of polyphenols and carotenoids; the physical structure of microencapsulated food ingredients and their impacts on food sensorial properties; yielding an outline on the controlled release of encapsulated bioactive compounds in food products.

2. Overview on health-promoting food ingredients and food processing technology

2.1 Health-promoting food ingredients from plant sources

A great number of research activities in the field of health related dietary aspects have demonstrated a significant link between the regular intake of phytochemicals (e.g. polyphenols, carotenoids, phytosterols), and the prevention of life-style related diseases, such as cancer, obesity, diabetes and cardiovascular complications (Greselea, 2011). As a natural source of these phytochemicals, especially the consumption of whole foods (e.g.

coloured fruits and vegetables), has been suggested to be of health benefit (Liu, 2003). Research studies have shown that a polyphenol mixture including anthocyanins, proanthocyanidins and flavonols, naturally occurring in certain red and blue berry fruits, had a stronger activity against cancer cells than purified polyphenols (Herring & Albrecht, 2005). Similar to this finding, the consumption of purified lycopene resulted in a minimal protection against prostate cancer in rats, whereas the consumption of tomatoes lowered this risk more successfully by 26% (Boileau et al., 2003). The potent antioxidant and anticancer activities can be assumed to be due to the additive and synergistic effects of phytochemicals and nutrients, when occurring in a complex mixture in fruits and vegetables (Liu, 2004). This may also explain why no single antioxidant can replace natural compounds in fruits and vegetables to achieve similar health benefits. Early epidemiological surveys indicated that regular consumption of fruits and vegetables can reduce cancer risk, which has been demonstrated through comprehensive studies linked to cancer prevention (Block et al., 1992). Furthermore, the clinical studies have also shown that commonly applied dietary supplements do not have the same health benefits as a natural diet with a large percentage of fruits and vegetables, which has demonstrated to inversely affect tumor formation and cardiovascular diseases (Hertog et al., 1993; Verhoeven et al., 1996). Functional foods, which aim to amplify these positive attributes by the addition of certain purified phytochemicals, have to undergo a paradigm change based on research findings. The beneficial effect of purified phytochemicals is commonly reduced, in comparison with a mixture of phytochemicals in their natural food matrix. Hence, the isolated compounds either lose their bioactive function or can not react in the same way as when they are present in their natural matrix, possibly due to the extraction and purification processes applied. On the other hand, the application of bioactive ingredients sourced from plants, e.g. extracts rich in isoflavones, lycopene and ingredients from herbs, nuts and fibres, has been documented in cancer prevention, however the market for these supplements and potential health claims is currently still widely unregulated (Watson & Preedy, 2010).

2.1.1 Polyphenols

Polyphenols are a group of dietary antioxidants found naturally in fruits and vegetables. They primarily consist of flavonoids including flavanols, flavones, isoflavones, flavonols, flavonones and anthocyanins, and non-flavonoid polyphenolics including phenolic acids, lignans and stilbenes. Polyphenol functionality lies in the prevention of e.g. oxidative, inflammatory, microbial and viral assaults, and therefore in their potential to reduce chronic diseases (Leifert & Abeywardena, 2008a,b; Rimando & Suh, 2008). The mechanisms of antioxidant activity of polyphenols can be characterized by direct scavenging or quenching of oxygen free radicals and inhibition of oxidative enzymes that generate reactive oxygen species (Terao, 2009). Potential mechanisms for the anticarcinogenic actions of polyphenols have been reviewed by Lea (2010), in which one of the most common recognized mechanism is that polyphenols cause the excessive production of hydrogen peroxide and the defense of the cancer cell may be overwhelmed, leading to an inhibition of proliferation and death in the cell. However, polyphenols may also function mutagenic and thus increase carcinogenesis, as they enhance the formation of reactive oxygen species under some specific circumstances (Watson & Preedy, 2010), causing further DNA damage.

The challenges for applications of polyphenols in food system are the initial protection of the bioactivity of the polyphenols, as they may lose their antioxidative properties or bioactive functionalities during processing of food, due to their sensitivity to oxygen, temperature, light (Ottaway, 2008), and to the gastrointestinal tract environment (pH, enzymes) (Bell, 2001); furthermore the development of appropriate formulations to increase solubility of polyphenols according to a specific food matrix. Examples for polyphenols from food sources, with some of their properties are presented in Table 1. Plant extracts frequently produced from regular foods which are rich in polyphenols, are for example, grape seed or pine bark, lemon balm, green tea, olive, rooibos, and aloe vera, etc. Among these, functionality in the prevention of lipid peroxidation has been reported for grape seed extract and maritime pine bark extract (Buchwald-Werner et al., 2008). However, not every commercially available plant extract complies with the regulatory requirements governing their application in foods. The approval of food status for any ingredient depends on the traditional use of the plant as a food, the form in which it is presented to the consumer (i.e. a pill versus a candy) and its physiological functionality, as well as on the processes which are involved in the extraction of the ingredient. In addition to these legal aspects, other technological properties such as stability and solubility are also important for a successful application in food. The plant extracts should not interfere with the product characteristics, such as colour or taste, and they have to be stable within in the food matrix to prevent unwanted precipitation. When included in a food product, the polyphenolic compounds may impart an astringent or bitter taste, or introduce a degree of brown colouring. For example, grape seed extracts are difficult to incorporate into functional foods due to their dark brown colour and low water solubility. Therefore, it is a crucial step during the product development process, to investigate the sensorial impacts of bioactive compounds on foods before their application. Certain methods can help to avoid these issues, e.g. a modified food formulation using microencapsulation technique could be a possibility to mask an unpleasant flavour from the bioactive extracts. The most successful applications of plant extracts containing polyphenols are beverages, including water or tea-based functional drinks, as well as dairy products or other novel product groups such as "smoothies" while the most popular plant extracts used in this type of beverages is green tea extract followed by rooibos extract.

2.1.2 Carotenoids

Carotenoids are natural occurring pigments in plants. There are more than 600 known carotenoids, which can be categorized in two groups: provitamin A carotenoids (α-carotene, β-carotene, β-cryptoxanthin), and non-provitamin A carotenoids (lycopene, lutein, zeaxanthin, astaxanthin), where the former can be metabolized by humans into retinol, and the later exhibit no function of vitamin A in humans (Bohn, 2008). Carotenoids can not be synthesized by our bodies and have to be obtained from food sources, such as fruits, vegetables, certain types of fish or shell fish (Table 2). The potent antioxidant activities of carotenoids and their health beneficial functionality, such as maintaining eye health, and prevention of chronic diseases have been widely investigated. Although, a recent study showed that β-carotene increases the risk of lung cancer in smokers when it was administered as a purified food supplement (Goodman et al., 2004), the health benefits and cancer preventing properties of carotenoids, when occurring in a natural diet of fruits and vegetables, cannot be denied. For example, data collected from 1993 case subjects with first primary incident of adenocarcinoma of the colon, strongly suggested that lutein was inversely associated with colon cancer in both men and women, when incorporating

Polyphenols	Main compounds	Food sources	Properties
Anthocyanidins	Malvidin, cyanidin, delphinidin, pelargoindin, peonidin, peltunidin and their glycosides	Red wine, all kinds of berries, red grape, orange, aubergine, avocado	Highly sensitive to temperature, oxidation, pH, and light; water soluble
Flavanones	Hesperetin, hesperidin, naringenin, naringin, homoeriodictyol	Citrus, such as, grapefruit juice, orange juice	Sensitive to oxidation, light and pH; aglycones insoluble but glycosides soluble in water
Flavones	Apigenin, luteolin	Fruit and vegetables, such as, celery, oregano, parsley	Sensitive to oxidation and pH; aglycones slightly soluble but glycosides soluble in water
Flavonols	Kaempferol, myricetin, quercetin, and their glycosides	Fruit and vegetables, such as, apples, onions, tea, red wine	Sensitive to oxidation, lights and pH; aglycones slightly soluble but glycosides soluble in water
Flavan-3-ols	Catechin, epicatechin, gallocatechin, epigallocatechin	Tea (green/black), red wine, dark chocolate	Sensitive to oxidation and pH; astringent and bitter; slightly soluble in water
Isoflavones	Daidzein, genistein, genistein	Soy beans, peas, peanuts	Sensitive to temperature, oxidation, lights and pH; astringent and bitter; soy smell; water soluble
Hydroxybenzoic acids	Gallic acid, *p*-hydroxybenzoic, vanillic acid	Berries, tea, wheat	Sensitive to temperature, oxidation, lights and pH; soluble in water
Hydroxycinnamic Acids	Caffeic acid, ferulic acid, *p*-coumaric acid, sinapic acid	Fruits, oats, rice	Sensitive to oxidation and pH; most slightly soluble in water
Lignans	Pinoresinol, podophyllotoxin, steganacin	Flaxseed, sesame, whole grain wheat bread, rice, cashew nuts, sunflower seeds	Relatively stable under normal conditions; unpleasant flavour; water soluble
Tannins	Castalin, pentagalloyl glucose, procyanidins	Tea, berries, wines, chocolate, black walnuts, black/red beans	Sensitive to high temperature and oxidation; astringent and bitter, water soluble

Table 1. Polyphenols from food sources and their properties (adapted from Ottaway, 2008; Watson & Preedy, 2010)

spinach, broccoli, lettuce, tomatoes, oranges and orange juice, carrots, celery, and greens into the diet (Slattery et al., 2000). In vitro studies, it has been suggested that carotenoids (β-cryptoxanthin, lycopene) stimulated bone formation and mineralization, and may prevent osteoporosis development (Yamaguchi & Uchiyama, 2004). Zeaxanthin and lutein might possess the function of preventing age-related macular degeneration (Sajilata et al., 2008). Epidemiological evidence has indicated that lycopene may protect individuals from colorectal cancer and men from prostate cancer (Schwarz et al., 2008).

In the last few decades efforts have been made to develop genetically modified food sources with an increased carotenoid content. Golden Rice is one of the well-known examples, which is a crop product that has 1.6 mg/g total carotenoids in the rice endosperm (Ye et al., 2000). This already posed a significant increase to the values found in natural rice varieties, e.g. 0.02 μg/g for pale varieties and a highest value of 0.13 μg/g for a dark variety (Frei & Becker, 2004). A later developed variety of golden rice (Paine et al., 2005) contains up to 37 mg/g of total carotenoids and recent clinical trials have shown that this functional product is a good source of vitamin A for humans (Tang et al., 2009).

At present, carotenoids are most often used as food colouring substances in commercial products. β-carotene, lycopene, astaxanthin, lutein, and zeaxanthin are also consumed as dietary supplement in human nutrition. Physically, carotenoids are almost insoluble in the aqueous phase, but slightly soluble in the lipid phase (0.2 g/L). However, the stability and solubility of carotenoids can be considerably improved with encapsulation technology during food processing, which will be discussed later in this chapter.

2.2 Food structure, functionality and bioavailability

The functionality and bioavailability of bioactive compounds are strongly affected and determined by their chemical properties, in terms of solubilisation and depolymerisation (Lesmes & McClements, 2009). Also further processing of the food material may dramatically affect the bioavailability of nutrients and phytochemicals, as do the environmental conditions during its passage through the gastrointestinal tract (GI).

The influence of heat and mass transfer in food processing affects food microstructures. The complexity of food matrix determines both food texture and also the release of functional components. It has been suggested that quantitative structure function relationships (QSFR) can help the rational design and efficient production of such functional food system (Lattanzio et al., 2009). However, currently the knowledge base on bioactive ingredients and food structure is very limited. Future studies should provide further data which can aid in the understanding of bioavailability of specific compounds and hence an improved description of food processes (Armand et al., 1997; Lesmes & McClements, 2009), as not only the bioavailability, of selected micronutrients and phytochemicals, but also the general ingredient stability is affected by the food matrix. For example, it has been found that only a minor part of the carotenoids in raw fruits or vegetables is absorbed in the intestines, probably due to the fact that carotenoids in most plant foods exist as crystals or are bound to proteins. In contrast, carotenoids dissolved in vegetable oils show a higher bioavailability (Parker, 1997). Incorporation of carotenoids into micro and nano structures in food matrices may influence their solubility and crystallinity, and thus their absorption. After formulating carotenoids into particulate systems which allow for sufficient solubility and release during digestion, they may be more easily delivered into cellular compartments, improving their bioavailability.

The current knowledge of the effects of processing methods on the bioavailability of individual food components has been reviewed by Faulks & Southon (2008). The understanding of the following considerations are crucial for the development of novel products with added health beneficial value, and they could aid in predicting the absorption rate, metabolism and bioavailability of bioactive compounds within the human organism: (i) original compounds might be present in a form which is not directly available in the human digestive system; (ii) the food matrix has a significant effect on the release and availability of bioactive ingredients; (iii) compounds might need an additional carrier substance to aid solubility; (iv) released ingredients might not be fully absorbed; (v) functional responses to bioactive compounds may vary throughout the population according to their genetic makeup. Faulks & Southon (2008) concluded that interactions of foods with the human body are extremely complex, due to the large variety of physicochemical processes, their effect on food structure and also the individual's metabolism characteristics. There is also still a lack in the understanding of how single food components are digested. Current developments in the area of functional food have already demonstrated that the bioavailability of bioactive ingredients can be improved by the selection and development of a delivery and protection method for bioactive ingredients. Plant sterol absorption, for example, differs greatly throughout various food matrixes, with milk being a good carrier, being up to three times

Carotenoid species	**Food sources**	**Potential health benefits**
β-carotene	Carrots, sweet potatoes, apricots, cantaloupes, peaches, dark green leafy vegetables	Most potent pro-vitamin A form, maintains healthy eyes; vision cycle, reduces the risk of cancers and heart diseases
Lycopene	Tomatoes, watermelon, red peppers, pink Grapefruit	Against many types of cancers, prevention of cardiovascular disease
β-cryptoxanthin	Peppers, pumpkin, squash, peas, chilli, sweet corns	Prevention of vitamin A deficiency, prevention of colon cancer; improves lung function
Lutein	Spinach, kale, collard greens, turnip greens, lettuce, broccoli	Prevention of macular degeneration
α- carotene	Sweet potatoes, carrots, kales, spinach, turnip greens, winter squash, collard greens, cilantro, fresh thyme, cantaloupe, lettuce, broccoli	Prevention of vitamin A deficiency, heart disease prevention
Zeaxanthin	Sweet corn, persimmon, spinach, egg yolk, green peas, brussel sprouts, citrus fruits, peaches, apricots, papayas	Prevention of macular degeneration
Astaxanthin	Salmon, shrimp, trout, seabream, lobster, fish eggs	Maintains healthy eyes

Table 2. Main dietary carotenoids and their major food sources (adapted from Wildman, 2007)

Functional ingredients	A	B	C	D	E	F	G	H	I	J
Carotenoids	X		X		X					
Plant sterols	X							X		
Saponines	X	X			X			X		
Glucosinolates	X	X						X		
Polyphenols	X	X	X	X	X	X	X		X	
Protease inhibitors	X		X						X	
Terpenes	X									
Phyto-estrogens	X		X							
Sulfides	X	X	X	X	X	X	X	X		X
Phytic acid	X		X		X			X	X	
Dietary fibre	X				X			X	X	X
Substances from fermented foods	X	X			X			X		

A - anticancerogenous
B - antimicrobial
C - antioxidant
D - antithrombotic
E - immunomodulatory
F - anti-inflammatory
G - blood pressure regulation
H - cholesterol level lowering
I - blood glucose lowering
J - digestive improvements

Table 3. Examples of food ingredient functionality (adapted from Watzl & Leitzmann, 1999)

more efficient than bread or other cereal products (Jones & Jew, 2007). With future studies and increased understanding on the normal human GI system, it will be possible to develop further functional foods with enhanced nutritional value, bioavailability and specific health beneficial functionality (Salminen et al., 1998).

2.3 Gastrointestinal physiology of functional foods

A most crucial part in the utilisation of food and bioactive ingredients, after the required processing and formulation of food, is the behaviour of food during gastrointestinal passage, i.e. the way it is processed by the human body. Due to initial chewing and further movements of the food matter, the structure is exposed to physical fragmentation, and in addition to pH changes, changes in the moisture content and exposure to acid /enzymatic activity, and bioactive ingredients are more or less released from the structure and made available for absorption into the blood stream. The time of digestion is depending, among other factors such as the amount consumed, on the original food structure and its breakdown, as these factors determine how fast digestive enzymes can penetrate into the food matrix. Furthermore, the food composition plays an important role, as the digestion rate will also depend not only on the physical food structure, but also on the type and concentrations of food molecules (e.g. proteins, carbohydrates, minerals, dietary fibre, etc.), how these are interacting with each other, allowing the release of fragments or bioactive ingredients (Norton et al., 2006).

To demonstrate novel methods of approaching a better controlled availability of nutritional components, a study by Augustin et al. (2011) showed that the use of microencapsulation for improved bioavailability of marked bioactive compound (radiolabeled [14C]-trilinolenin or [14C]-tributyrin and [3H]-resveratrol) did not alter the time needed for the gastrointestinal passage, but significantly improved the absorption into the blood system. This proved the potential suitability of microencapsulation as a delivery vehicle for bioactive substances. Parallel to the approach of delivering certain bioactive phytochemicals to a specific point of release in the human GI tract, another similar approach lies in the delivery of certain microorganisms. As the human body and health status is not only influenced by the nutrient supply, but also by the GI microbial flora, based on the ability of these organisms to transform food components, influence the absorption of bioactive molecules, it may be of interest to release certain organisms into the GI tract after their stomach passage, e.g. in microencapsulated form. The need for further studies on how the normal human intestinal flora with all aspects, e.g. fermentation, immune system, functions, and the effect of wanted modifications via the use of prebiotics and probiotics, was also stated by Salminen et al. (1998).

2.4 Food processing

2.4.1 Development of centralized food production

The human diet undergoes traditionally changes, both in nutritional composition and preparation methods, and these circumstances require linking the understanding of health beneficial nutrition to the development of novel food products. Throughout history, several significant steps of variations can be noted, starting with anthropological investigations on the diets of our ancestors, and leading to the analysis of a modern diet driven by health concerns and fashion. As humans developed on a natural diet on the basis of a hunter and gatherer society, the sourcing and preparation of suitable food was the major activity on a day-to-day basis. Obtained goods had to be prepared and mostly consumed immediately, as prolonged storage was often not possible.

The appearance of settlements, larger societies, and a further specialisation of individuals on their activities for the community, also determined the need for food preservation and storage, and so an enablement of the group to plan future activities, e.g. procurement of winter supplies. These early food processing methods all aimed at the preservation of food and consisted of simple activities, e.g. sorting suitable items, cleaning, drying, salting, smoking, packaging and if possible, a controlled storage to minimize spoilage. These inventions were made solely by observation, by trial and error. Also the application of biotechnology in food production and preservation can be ascribed to the early stages of modern societies.

Since those times, food production methods had to be more and more specialised on different products and food groups. The rise of these specialised activities formed traditional professions, e.g. bakers, butchers, cheese makers, or brewers, in addition to supporting professions, e.g. coopers, the first engineers to produce food processing equipment and packaging technology.

As the first heat sterilization plant in France was developed by Appert in the early 1800s (Throne, 1986), the second half of the 19th century saw another major step and development of food processing, the birth of technologies and food factory set-ups as we know them today. The industrial revolution, as an example in Central Europe, with its relocation of large parts of the population from country side rural life style into city dwellings and full time activities apart from food sourcing, and also the need for a country to be able to supply the food needs of large armies, needed new methods and procedures, both in food preparation and packaging.

A major achievement of these developed production methods, with many processes adapted from other industrial fields, was certainly the increased food safety. However, the application of industrial processes with their sometimes harsh conditions in terms of handling, cleaning or temperature treatments, also led to a certain amount of depletion of nutrients from the so preserved food products. The industry was able to produce safe food, with good storage properties, however, the need of guaranteeing a healthy and nutritional balanced diet still lied within the consumer's responsibility and malnutrition could occur due to a consumer's lack of knowledge when limiting the diet to highly processed products and ingredients, e.g. refined sugar, white flour, convenient products such as canned vegetable and fruits. Nowadays, the average consumer is much better informed and with the wide options of food products and choices available to us, with growing concern about malnutrition and the knowledge of related health risks, the need and demand for a diet of nutritionally high value is greater than ever.

2.4.2 Overview on food processing

An overview about common food processing technologies is presented in Table 4. According to the desired products, several production steps are necessary in order to produce a food product from raw ingredients, change the physical and chemical appearance of the product, ensure food safety, consistent quality, shelf-life and supply. A typical processing step, to be found in many solid and liquid products is heat treatment, which can be applied either to prepare the product (i.e. cook the product for added bioavailability of nutrients, denaturise proteins, modify carbohydrates and starches), to develop desired flavours, aroma and colour components (e.g. Maillard reaction), modify the food structure (e.g. texture changes due to ingredient modifications or drying processes), or to preserve or sterilize the food by heat induced inactivation of microorganisms, toxins and enzymes (e.g. heat sterilisation of canned food products, blanching of vegetables, to inactivate enzymes). As can be expected, the heat treatment regimes, and also other principal food processing steps, often lead to a loss of bioactivites of native ingredients, which are essential for the human diet. Fruit and vegetable products, as these are a major source of important phytochemicals, have to be protected during processing, packaging and storage, to ensure their availability for the human diet.

Beside the method of adding bioactive ingredients specifically to food items and so creating or re-creating the desired nutritional value of the product, which is explained later in this chapter, the modern way of food processing aims at preserving native bioactive ingredients in the raw food as much as possible. In order to achieve this, novel food processing methods have been developed, are under deployment, or are in the investigative stage and close to an industrial application. Examples for these novel methods are listed in Table 5, and their aim of preserving and protecting food nutrients and native bioactive ingredients can easily be understood, especially in the area of non-thermal treatments. In this approach, the advantages of a heat treatment, e.g. microorganism inactivation or a textural modification, are achieved by non-thermal methods, e.g. ultra high pressure treatment or the use of enzymatic reactions. Nott et al. (2000) and Zhong et al. (2004) demonstrated that through ohmic heating and microwave technology, a suitable food product could be manufactured with the same level of safety as by production with conventional heat treatment processes, but with improved organoleptic properties.

The reduction of quality degradation due to food processing by high pressure processing and application of pulsed electric fields, both methods characterized as non-thermal treatments, was shown by Matser (2004). These approaches in novel food processing have certainly advantages in terms of human nutrition and health, as well as improving food quality. However, having also to consider the other reasons for food processing, such as guaranteeing sufficient inactivation of spoilage microorganisms, these methods have their limitations, including the ability to process large volumes, processing costs.

2.4.2.1 Functional food processing

By definition, functional foods are food products, both natural occurring or processed food, which contain bioactive compounds with a functionality beyond the essential daily nutritional requirements to improve the human health. Fruits and vegetables are well recognized functional foods, however, their beneficial ingredients also can be extracted, purified and used as dietary supplements and consumed in concentrated form, or after addition to a different food product exert an added dietary value. However, these methods of a simple addition may result in an unwanted and negative change of sensory and structure of food products. With the aid of novel processing technologies (such as microencapsulation), these effects could be minimized or avoided, e.g. through microstructural modifications (Palzer, 2009). These novel technologies differ from the traditional food processing methods and have certain advantages in their capacity to prevent the inactivation of bioactive ingredients. In order to provide a greater amount and variety of functional foods, beside the traditional natural products, food manufacturing companies are working continuously on the development of novel products. This can either be in the form of modified raw ingredients, e.g. vegetables with increased amount of phytochemicals, or in the form of adding desired bioactive ingredients to other food. The fortification of food is a well established production method and can be found in application in numerous products, for example breakfast cereals with added vitamin (e.g. folic acid), minerals or fruit juices fortified with ω3-fatty acids. Producers have to consider, if the product is able to simple contain the added ingredient within its natural matrix, or if further process modifications are needed (e.g. encapsulation). This approach could include delivery of the protected bioactive ingredients to their target site and release under certain trigger factors (enzymes, pH, salts, etc.) (Chen et al., 2006).

	Aim of the process	Principle of operation	Typical application examples	Equipment examples
Mechanical Processes				
Size classification	Dividing a mix of particles according to size	Sieving and size classification	Grain processing / milling applications	Sifting machines / air separator
Sorting	Dividing a mix of particles according to other characteristics than size, e.g. specific density	Sorting according to differences in density, magnetic susceptibility, electric conductivity	Grain processing / processing of herbs and other plant parts	Stone separator / magnetic separator
Filtration	Separation of a liquid / solids mixture	Separation of solid particles from suspension by a filtration media	Beverage industry / dairy industry / ingredient manufacture	Filtering machines, e.g. fixed bed filtration / membrane filtration unit
Centri-fugation	Separation of suspensions with smallest particle sizes	Centrifugation forces	Dairy industry / beverage industry / fruit and vegetable processing / oil manufacturing	Centrifuge / separator
De-foaming	Removal of unwanted stable foam during process operations	De-stabilisation of foam by mechanical fixtures, separating gas / liquid	Beverage industry / dairy industry	Mechanical fixtures within process machinery / tanks
De-dusting	Removal of fine solid particles from gaseous phase, e.g. avoidance of dust explosions	Centrifugation forces / filtration media	Milling-, baking-, powder-, ingredient industry	Aerocyclone / air seperator
Flotation	Separation of solid particles from liquids	Attachment of particles to gas bubbles, followed by foam separation	Beverage industry	Flotation reaction vessel
Mixing	Homogenous mix / particle distribution in solid/solid or solid/liquid mixes	Mechanical mixing	Applications throughout the industry	Various mixing vessels and machines
Dispersion	Homogenous mix of liquid/liquid system	Mechanical mixing / homogenisation	Oil/water mixtures, e.g. mayonnaise	Mixing vessels

	Aim of the process	Principle of operation	Typical application examples	Equipment examples
Disag-gregation	Production of smaller particle sizes, solids	Mechanical combination	Milling of food ingredients	Milling / crushing machines
Spraying	Production of smaller particle sizes, liquids	Spraying of liquids through nozzles by pressure	Dairy industry, ingredient industry	Spray dryers / coating applicators
Ag-glomeration	Production of larger particles from powder mixes	Affinity of particles	Ingredient industry, pellet and tablet production	Pelletisation drum / tablet press
Thermal Processes				
Heating	Support of other processes by change of rheological/chemical properties; pasteurisation; sterilisation; denaturation; flavour development	Food is exposed to heat energy, in different applications and by various methods (steaming, boiling, roasting, indirect heating, micro-wave, etc)	Throughout all food production processes	Cooking vessels, autoclaves, reaction vessels, continuous liquid sterilisation (UHT), drying machines
Cooling	Temperature control of product	Removal of heat energy, by active or passive cooling	Throughout all food production processes	Equipment similar to heating applications
Evaporation	Reduction of liquid phase / increase of solid content	Heating (under modified pressure) to evaporate solvent (water, etc.)	Beverage industry / powder and ingredient manufacture	Evaporation tower
Crystal-lisation	Separation of solids from liquids	Temperature changes induces crystallisation of solid in high concentrations	Sugar industry / ingredient industry	Crystallisation reactor
Osmosis	Separation of liquid solid mixes	Pressure difference across a separation membrane	Fruit and vegetable juices / protein and lactose production	Ultrafiltration unit / reverse osmosis

Table 4. Mechanical and thermal processes common to food processing

2.4.2.2 Functional food design and safety

When considering the aspects of functional food development, one need to understand the food structure and the related product characteristics, as the structure primarily determines the behaviour of food within the human GI tract. (Davis & Gordon, 1982). The challenge in functional food product development lies in maintaining the desired and traditional sensory attributes, in terms of flavour and texture, while maintaining the functionality of the added active ingredient. The targeted consumer group and their consumption habits have to be taken into account and a tailored product needs to fulfil the consumer's requirements, in order to achieve a positive market response. If this addition of the ingredient is achieved with a satisfying result on the product's appearance, the next crucial step lies in the assurance of the ingredients' functionality. Besides in vitro and in vivo (animal and human) studies on the bioavailability of the bioactive compounds, researchers and scientists can apply mathematical models (Ottino, 2005), including current understandings of ingredient activity, release mechanisms and human metabolism, to follow and predict the ingredients' functionality and hence justify the its addition to the product. Currently, research activities in product microstructure design for process modelling in mouth behaviour and other sensorial characteristics, such as, taste development or physical sensation (e.g. mouth feeling) (Malone et al., 2003) are building up support data for these functionality studies. A recently developing area in functional food development considers the interactions between foods and nutrient supplements and an individual's genome and its effect on nutritional needs, (nutrigenomics). The main idea of nutrigenomics, is that a dietary recommendation for one individual might be inappropriate to another. This emerging science might be able to provide individual tailored nutrition for population groups or individuals, and could be a major step in reducing diet related illnesses and resulting health care costs. Functional ingredients, either embedded as ingredients in other food products or as enriched nutritional supplements, are generally considered as safe, due to their origin from natural food sources, however, as shown earlier for the β-carotene supplements, even dietary supplements ought to be evaluated carefully.

2.5 Microencapsulation in functional foods

Similar to traditional encapsulation methods where a shell material protects a sensitive core, e.g. chocolate covered peanuts to reduce exposure to surrounding oxygen and occurring rancidity, microencapsulation shows the same technological advantages for ingredients on a much smaller scale. Also here, the same technological advantages which are achieved by microencapsulating certain ingredients, are that the capsule protects the core material against degradation, or reaction with other ingredients, and at the same time protects the food from any unwanted flavour of the ingredient. Furthermore, encapsulated ingredients, either in dry or liquid form (e.g. solid dry powders or structured emulsions), can easily be added to food items during most processes and achieve a homogeneous distribution. Another technological advantage is the possibility of a delayed or controlled release of the ingredient from the capsule, which can be triggered by various methods (e.g. time, pressure, temperature, pH, water activity, physical force / chewing, time, enzymes, etc) (de Vos et al., 2010). Food structures can be used directly as a delivery vehicle for a functional ingredient, if the bioactive substance can be included and is protected sufficiently. If further protection is required, the active molecule can first be encapsulated (in solid or liquid form) and then be added to the food system (Ubbink & Krüger, 2006). However, the use of these functional ingredients should not alter the product's structural properties, and so a specifically developed ingredient is needed for each individual product. (de Vos et al.,2010). In order to overcome the technological issues related to added bioactive ingredients in food products, microencapsulation is now a state-of-the-art technology for manufacturers. The successful encapsulation of active nutrients or non-nutrients in powder or liquid form, as it has been initially developed by the pharmaceutical and chemical industry to protect pharmaceutical active ingredients or other chemical compounds, has been demonstrated (Palzer, 2009).

2.5.1 Overview on microencapsulation technology

In the same way as food packaging methods are designed to protect a food item against any unwanted spoilage, and microencapsulation is essentially a packaging technology on a smaller scale, a microencapsulated food ingredient has to be adapted and developed to suit the product conditions, in terms of chemical composition, processing factors, application and storage methods. With the range of different encapsulating methods and materials available, it is possible to develop and manufacture a great range of functional ingredients. Table 6 gives an overview of shell materials applied in common microencapsulated ingredient products. By applying different shell materials and encapsulation methods, the microcapsules can be produced with specific attributes, e.g. particle size, shape, point and trigger of intended core release (Kirby, 1991). However, it is typical that the capsulation process for each intended application has to be designed specifically, and this is determined by the functionality of the ingredient and the surrounding matrix. The food product itself, the suitable coating material, the point and trigger of core release, the environmental conditions the capsule must be able to withstandand the suitable size of the capsule determine a technological viable application (Chen, 2004). Being widely applied in the pharmaceutical industry, the much smaller profit margins in the food industry are also playing an important role when considering the use of an encapsulated ingredient. The coating material itself has to demonstrate certain properties, which allows it to be applied in a coating process in terms of temperature resistance and rheological properties (Drush, 2011).

Furthermore the encapsulant, i.e. the coating material needs to be inert to the core material and at the same time be able to protect the core during processing or storage. One of the aspects of highest importance when choosing a coating material is its acceptance in a food product, in other words, the coating material should not lead to a negative effect on the food product and it has to be food grade throughout its production and application (Gibbs, 1999).

Technologies	Benefits	Applications	Limitations
High pressure treatment / ultra high pressure treatment	No formation of unwanted compounds (products from heat treatment); Best preservation of natural nutritional value, flavour, appearance, texture; possible production of safe food	Successful treatment (sterilisation) of products which are highly prone to spoilage and damage through processing (e.g. fresh dairy products, meat, seafood, fruit and vegetable)	High process costs (Investment and maintenance); Available equipment mostly batch processes

	(microbiological safety) with highest quality (nutritional and sensorial)		
Freeze drying	Good preservation of products characteristics and applicable to a large product range	Applied in products with fragile texture where natural ingredients (e.g. flavours) need to be protected. E.g. fruits as ingredients for high value applications	Formation of certain heat induced substances; Limited microbiological safety; High production costs
Pulsed electric field	Gentle processing, though cell disintegration (if desired); No formation of unwanted compounds (products from heat treatment);	Applications for wide product range, solid and liquid product processing	Missing heat treatment limits microbiological safety (spores); Missing enzyme inactivation; Cooled storage is required
Membrane filtration / ultrafiltration	No formation of unwanted compounds (products from heat treatment);	As additional treatment in milk pasteurisation (overall smaller temperature treatment possible)	Only for liquid products; High energy costs

Table 5. Examples for novel food processing methods

Ingredient category	Coating materials	Widely used methods	References
Carbohydrates	Starch, maltodextrins, chitosan, corn syrup solids, dextran, modified starch, cyclodextrins	Spray and freeze drying, extrusion, coacervation, inclusion complexation	Godshall,1988; Flink & Karel, 1970; Reineccius & Coulter, 1989; Reineccius, 1989 Reineccius, 1991
Cellulose	CMC (Carboxymethylcellulose, methyl cellulose, ethylcellulose, celluloseacetate-phthalate	Coacervation, spray drying, edible film coatings	Greener & Fennema, 1989a; 1989b;
Gums	Gum acacia, agar, sodium alginate, carrageenan	Spray drying, gel beads	Dziezak, 1991
Lipids	Wax, paraffin, beeswax, diacylglycerols, oils, fats	Emulsion, liposomes, film formation	Kampe & Fennema, 1984; Kim & Baianu, 1991
Protein	Gluten, casein, gelatin, albumin, peptides	Emulsion, spray drying	Ono, 1980

Table 6. Examples for applied coating materials used for functional food ingredients (adapted from Desai & Park, 2005)

In the production of these ingredients, certain attention has to be paid to the encapsulation itself and how the coating material is applied onto the core. Only an intact capsule can protect and hold the core sufficiently and often this shell is applied in multiple layers. In order to form this suitable shell, the coating material has to be applied by a specific process, designed for the coating material (Kim & Baianu, 1991). Typical processes for encapsulating functional micronutrients and bioactive ingredients are listed in Table 7. In general, the applied processes of encapsulation can include emulsification and extrusion, coacervation, liposome entrapment, spray drying, spray cooling, spray chilling, fluidized bed coating, etc. (Desai et al., 2005). The variety of food ingredients can range from ingredients which are important for the food processing and production, e.g. microorganisms, flavours, sweeteners, colorants, lipids, micro / macro nutrients, enzymes, but also more and more novel ingredients with positive effects on human health are produced for functional foods. Functionality of time delay can also be desired to achieve a slow release.

2.5.2 Physical structure of microencapsulated food ingredients

The appearance of an encapsulated food ingredient can be described as a core material (the functional ingredient, usually purified and/or stabilised by another carrier substance), which is surrounded by a mono or multi layer shell of a suitable material (Bakan, 1973; Shahidi & Han, 1993). The bulk ingredient can be in liquid, paste or solid / dry powder form, according to the nature of the core ingredient, the shell material, the applied production method and intended usage, and range in size from submicrometer to several millimetres. The protection is given by certain membranes and a multiple protective layer. In this case, the shell protects the inner core and environment to some extent against mechanical damage and also against environmental conditions. These microcapsules may have a multitude of different shapes, depending on the materials and methods used to prepare them (Dziezak, 1998).

An example for the technical process of encapsulation in the dry form is the encapsulation of bioactive compounds into powder form by spray granulation. Herein, the bioactive compounds, e.g. carotenes as colourant, are in liquid solution and sprayed onto inert core particles, e.g. sugar crystals with a specific size classification, where the solution then dries and forms layers around the solid particle. An additional protection layer can be applied. For the encapsulation of lipophilic vitamins (A, D, E, K), spray chilling is widely used. The bioactive compounds, being dissolved in oil and emulsified with a gelatine solution, is sprayed onto a powder bed, fluidized by cooled air, and solidified around the particles. The obtained particles are then further dried in a fluidized bed dryer/cooler. Other ingredients, e.g. hydrophilic substances, demand different processing conditions where heat energy is applied in order to form a dry coating of the sprayed liquid onto the particles by removal of the solvent, e.g. water. In the production of these various particles, also agglomeration has to be controlled, as it has to be avoided or it is intended to take place. For instance, the addition of fine starch powder into the fluidized bed after encapsulation does function against agglomeration. Beside these methods for producing dry powder particles, also liquid / semisolid capsules have their applications and one common example for their production is coacervation. In order to achieve this, the bioactive ingredient is mixed in liquid phase and blended with a second liquid, containing a shell-forming material such as alginates and other ingredients which are prone to gel formation. The spontaneously formed capsules or micro spheres are for further processing by spray chilling or spray drying for easy handling (Desai & Park, 2005).

2.5.3 Microencapsulation of polyphenols and carotenoids

With the development of microencapsulation and its adaption to the food industry, polyphenols and carotenoids as bioactive ingredients are readily available for a wide range of products and with various purposes. The main trend for growing and at this time relevant application of bioactive compound encapsulation is to deliver health-beneficial ingredients to functional foods. An important role for consumer health, disease prevention and even treatment is being played by nutraceuticals of a natural source of especially phytochemicals (Howells, et al., 2007).

In the determination of a suitable encapsulation method for carotenoids, the specific food system determines the initial basis on which to decide. The specific food production process has to be taken into account, as well as the food characteristics (dry / liquid product), sensory aspects, production costs, bioavailability of the carotenoids after storage, and consumption. Furthermore, the market requirements and local legislation have to be taken into account. Several encapsulation methods have demonstrated their suitability to encapsulate carotenoids (including β-carotene, lycopene, astaxanthin, zeaxanthin and lutein, which are common ingredients used for food coloration and added nutritional value due to their antioxidant behaviour), e.g. various emulsification processes, high pressure homogenisation, liposome entrapment, micro beads production, spray drying and freeze drying (Ribeiro et al., 2010). Studies have shown that the encapsulation of polyphenols does protect their functionality and stability, and furthermore can induce a health benefit by tailoring the encapsulation / release mechanism for an increased bioavailability (de Vos et al., 2010). For example, Lycopen has been encapuslated using emulsion technology to enhance to solubility and bioavailability. As with other functional food ingredients, which are currently in the focus of attention due to their health beneficial properties, most encapsulation methods available have been adapted successfully to polyphenols and an extensive summary of these methods is given by Fang & Bhandari (2010). These produced ingredients result in a wide variation of possible morphologies, structures and characteristics and so yield a great range of possible applications.

Polyphenols, carotenoids, and also other food components with the potential to constitute further upcoming nutraceuticals, in relation to the growing knowledge on their functional properties and health benefits, will play an important role in the development of novel food products with an added health benefit. Further improvements in their manufacturing technologies, stabilization and controlled bioavailability will aid to a growing number of health food products in the future. Research and development activities will most likely focus on adapting and exploiting further delivery methods, encapsulated products with ingredient combinations tailored to nutritional needs, and a cost efficient production for a mass market.

2.5.4 Controlled release mechanism and delivery of bioactive ingredients

The advantage of supplying a food, fortified with a bioactive ingredient, has been described earlier and the advantage of a controlled release has been stated. As the extensive

Ingredient Category	Substance examples / Food ingredient	Preferred encapsulation method	Food applications / Benefits
Acidulants	Lactic acid, glucono-δ-lactone, vitamin C, acetic acid, potassium sorbate, sorbic acid, calcium propionate, sodium chloride	Fluidized-bed coating, extrusion	Applied to assist in the development of colour and flavour in meat emulsions, dry sausage products, uncooked processed meats and other meat containing products. Acids and baking soda is used in wet and dry mixes to control release of carbon dioxide during processing in the bakery industry.
Flavouring agents	Citrus oil, mint oils, onion oil, garlic oil	Inclusion complexation, extrusion, spray drying	To transfer liquid flavourings into easy to handle and stable free flowing powders.
Sweeteners	Sugars, artificial sweeteners (e.g. aspartame)	Co-crystallization, fluidized-bed coating	To reduce hygroscopicity, resulting in an improved flowability. Prolonged sweetness perception.
Colorants	Annatta, β-carotene, turmeric	Extrusion, emulsion	For easier handling and improved solubility. Protection from oxidation and improved application in dry mixing.
Lipids	Fish oil, linolenic acid, rice brain oil, egg white powder, sardine oil, palmitic acid	Spray drying, freeze drying, vacuum drying	To prevent ingredient from oxidation during storage and processing.
Vitamins and minerals	Vitamin A, D, E, K (fat soluble) Vitamin C, B1, B2, B6, B12, niacin, folic acid (water soluble)	Coacervation, inclusion complexation, spray drying, liposome	Reduction of off-flavours. To control release time. To enhance stability. To protect from interaction with other ingredients.
Enzymes and microorganisms	Lipase, invertase, *Brevibacterium linens, Penicillium roqueforti*	Coacervation, spray methods, liposome	To improve stability and control time and point of release.

Table 7. Examples for microencapsulated food ingredients (adapted from Desai & Park, 2005).

knowledge base generated by the pharmaceutical industry during the development of drug delivery systems is of great advantage to the development of food grade delivery systems, these systems can be specifically designed in order not to negatively affect the food product properties. At the same time, they will be able to deliver the desired compound to the specific point within the human body, e.g., mouth, stomach, small intestine or colon (McClements et al., 2008; McClements et al., 2009a; Ubbink et al., 2008). As bioactive ingredients, encapsulation materials and the food matrix can alter considerably during processing, storage, consumption and digestion, e.g. changes caused by pH, ionic strength, surface activities, enzymatic activities (lipases, proteases, amylases), force and flow profiles (pressure, disruption, agitation) associated with chewing, stomach and intestine passage (Armand et al., 1997, 1999; Van Aken, 2007), the ingredient may benefit from associate changes such as degradation due to acid pH, cleavage by enzymes such as deglycosylase, complexation by other dietary ingredients, and other related physiochemical changes. Furthermore, one of these digestive impacts can be applied as the trigger for the release of the encapsulated bioactive ingredient. For example, pancreatic proteases may act as an agent dissolving a capsule designed to resist acidic pH and pepsin in the stomach.

To describe these food and drug delivery methods, several mathematical models have been developed (Pothakamury & Barbosa-Canovas, 1995; Siepmann, J. & Siepmann, F., 2008) and with detailed knowledge of parameters such as particle size, active ingredient concentration within the particle and in the surrounding matrix, and diffusion coefficients, these models can be applied to aid in the understanding of the process kinetics. The amount of empirical data available for drug delivery is vast, however, similar databases for food ingredients are still missing and relevant data is just emerging (Sereno et al., 2009). Controlled release of phytochemicals from encapsulated stages has been reported (Augustin et al., 2001; Augustin, M. A. & Sanguansri, L., 2008; Chen et al., 2006; Narayanan et al., 2009; Weiss et al., 2008; Dziezak, 1998).

Designing food structure to control stability, digestion, release and absorption of lipophilic food components and the available release mechanisms allow a wide range of applications in the development of functional food ingredients (McClements et al., 2008). the context is with the previous and next sentence together, that these are all points for future consideration in product development. The developed knowledge, together with deeper understanding of the relations between food properties and bioactive ingredient adsorption, is aiding in the design of food materials and encapsulation methods which, after protecting the ingredient, give controlled release at specific points in the gastrointestinal tract (Augusin et al., 2011; Hejazi & Amiji, 2003; McClements, et al., 2009b).

3. Conclusion

The world continues to face the increasing burden of dietary and life-style related diseases and of the increasing aging population. This results in increased interest and consumer demand for fortified healthy food products. Phytochemicals as natural sources of health-promoting ingredients have been extensively studied by scientists. Great opportunities for developing foods fortified with these active ingredients are rapidly arising for food manufacturers, including ADD. Although, these future food products, with fortified bioactive ingredients, can improve and maintain a nutritional balance, the use of phytochemicals, nutraceuticals and functional foods requires a deep knowledge and understanding of the complex physicochemical processes that occur within food and on effective strategies to design foods that can increase the bioavailability of valuable bioactive ingredients. Scientists, food and pharmaceutical industries are not only required to address quality and stability of functional foods, but also to improve the consumer education on the efficiency and safety of dietary supplements and functional foods, which claim to be health promoting. These are still existing issues that functional food designers and manufacturers need to face.

From the perspective of food technology, and as well as far as functional food products are concerned, the future research lies in novel food processing methods, food design and the understanding of the relations between bioactive ingredients release (specifically aimed at microencapsulation development for bioactive compounds to meet specific needs of food applications), personalized nutrition, processing technology with improved efficiency, sustainable production, while environmental friendly packaging also is of concern. The key findings of these activities provide an opportunity to engineer functional foods with increased health-promoting benefits.

4. References

Armand, M., Pasquier, B., Andre, M., Borel, P., Senft, M. & Peyrot, J. (1999). Digestion and Absorption of 2 Fat Emulsions with Different Droplet Sizes in the Human Digestive Tract. *American Journal of Clinical Nutrition*, Vol. 70, No. 6, pp. 1096-1106

Armand, M., Pasquier, B., Borel, P., Andre, M., Senft, M. & Peyrot, J. (1997). Emulsion and Absorption of Lipids: the Importance of Physicochemical Properties. OCL e Oleagineux Corps Gras. *Lipides*, Vol. 4, No. 3, pp. 178-185

Augustin, M. A. & Sanguansri, L. (2008). Encapsulation of Bioactives. In: *Food Materials Science – Principles and Practice*, J. M. Aguilera & P. J. Lillford (Eds.), pp. 577–601. Springer. NewYork.

Augustin, M. A., Sanguansri, L., Margetts, C. & Yong, B. (2001). Microencapsualtion of Food Ingredients. *Food Australia*, Vol. 53. pp. 220-223

Augustin, M.A., Abeywardena, M.Y., Patten, G., Head, R., Lockett, T., De Luca, A. & Sanguansri, L. (2011). Effects of Microencapsulation on the Gastrointestinal Transit and Tissue Distribution of a Bioactive Mixture of Fish Oil, Tributyrin and Resveratrol. *Journal of Functional Foods*, Vol. 3, Issue 1, (January 2011), pp. 25-37

Bakan, J.A. (1973). Microencapsulation of Food and Related Products. *Food Technology*, Vol. 27, No. 11, pp. 34–38

Bell, L.N. (2001). Stability testing of nutraceuticals and functional foods. In : *Handbook of nutraceuticals and functional foods*, R.E.C. Wildman (Ed.), pp. 501-516, CRC Press, New York, ISBN 978-0849364099

Block, G., Patterson, B. & Subar, A. (1992). Fruit, Vegetables, and Cancer Prevention: a Review of the Epidemiological Evidence. *Nutrition and Cancer*, Vol. 18, Issue 1, pp. 1-29

Bohn, T. (2008). Bioavailability of non-provitamin a carotenoids. *Current nutrition & Food Science*, Vol. 4, No. 4, pp. 240-258

Buchwald-Werner, S., Gärtner, C., Horlacher, P. & Schwarz, G. (2008). Fortification with Substances Other than Vitamins and Minerals (Polyphenols, Carotenoids, Fatty Acids and Phytosterols), In: *Food Fortification and Supplementation - Technological,*

Safety and Regulatory Aspects, P.B. Ottaway, (Ed.), CRC Press; 1 Edition, (March 20, 2008), ISBN 978-0849364099

Chen, L., Remondetto, G.E. & Subirade, M. (2006). Food Protein-Based Materials as Nutraceutical Delivery Systems. *Trends in Food Science & Technology*, Vol. 17, pp. 272-283

Chen, X.D. (2004). Heat-Mass Transfer and Structure Formation During Drying of Single Food Droplets. *Drying technology*, Vol. 22, No. 1 & 2, pp. 179-190
Comprehensive Reviews in Food Science and Food Safety, Vol. 7, pp. 29 - 49

Davis, E.A. & Gordon, J. (1982). Food Microstructure, an Integrative Approach. *Food Microstructure*, Vol. 1, pp. 11-12

de Vos, P., Faas, M. M., Spasojevic, M. & Sikkema, J. (2010). Encapsulation for Preservation of Functionality and Targeted Delivery of Bioactive Food Components. *International Dairy Journal*, Vol. 20, pp. 292-302

Desai, K.G.H. & Park, H.J. (2005). Recent Development in Microencapsulation of Food Ingredients. *Drying technology*, Vol. 23, pp. 1361-1394

Drush, S. (2011). Interfacial Engineering for Optimisation of the Functionality of Microcapsules. *Bioencapsulation Innovations*, (March-April 2011), pp. 29-32

Dziezak, J.D. (1991). Focus on Gums. *Food Technology*, Vol. 45, No. 3, pp. 116–118

Dziezak, J.D. (1998). Microencapsulation and Encapsulated Food Ingredients. *Food Technology*, Vol. 42, pp. 136–151

Fang, Z. & Bhandari, B. (2010). Encapsulation of Polyphenols - a Review. *Trends in Food Science & Technology*, Vol. 21, pp. 510-523

Faulks, R.M. & Southon, S. (2008). Assessing the Bioavailability of Nutraceuticels, In: *Food Fortification and Supplementation*, P.B. Ottaway, (Ed.), Woodhead Publishing, ISBN 978-1-84569-144-8

Frei, M. & Becker, K. (2004). Agro-Biodiversity in Subsistence-Oriented Farming Systems in a Philippine Upland Region: Nutritional Considerations. *Biodiversity and Conservation*, Vol. 13, pp. 1591-1610

Gibbs, B. F., Kermasha S., Alli I. & Mulligan C. N. (1999). Encapsulation in the Food Industry: a Review. International *Journal of Food Sciences and Nutrition*, Vol.50, No. 3, pp. 213-224

Goodman, G.E., Thornquist, M.D., Balmes, J., Cullen, M.R., Meyskens, Jr., F.L., Omenn, G.S., Valanis, B. & Williams, Jr., J.H. (2004). The Beta-Carotene and Retinol Efficacy Trial: Incidence of Lung Cancer and Cardiovascular Disease Mortality During 6-Year Follow-Up After Stopping Beta-Carotene and Retinol Supplements. *Journal of the National Cancer Institute*, Vol. 96(23), pp. 1743-1750

Gresele, P., Cerletti, C., Guglielmini, G., Pignatelli, P., deGaetano, G. & Violi, F. (2011). Effects of Resveratrol and Other Wine Polyphenols on Vascular Function: an Update. *The Journal of Nutritional Biochemistry*, Vol. 22, Issue 3, (March 2011), pp. 201-211

Hejazi, R. & Amiji, M. (2003). Chitosan-Based Gastrointestinal Delivery Systems. *Journal of Controlled Release*, Vol. 89, pp. 151-165

Herring, T. & Albrecht, J. (2005). Functional Foods. *Extension*, Published by the Board of Regents of the University of Nebraska, 2005, Available from http://www.ianrpubs.unl.edu/live/ec473/build/ec473.pdf

Hertog, M. G. L., Feskens, E. J. M., Kromhout, D., Hollman, P. C. H. & Katan, M. B. (1993). Dietary Antioxidant Flavonoids and Risk of Coronary Heart Disease: the Zutphen Elderly Study. *The Lancet*, Vol. 342, Issue 8878, (23 October 1993), pp. 1007-1011

Howells, L.M., Moiseeva, E.P., Neal, C.P., Foreman, B.E., Andreadi, C.K., Sun, Y., Hudson, E.A. & Manson, M.M. (2007). Predicting the Physiological Relevance of In Vitro Cancer Preventive Activities of Phytochemicals. *Acta Pharmacologica Sinica*, Vol. 28, No. 9, pp. 1274-1304

Jones, P.J. & Jew, S. (2007). Functional Food Development: Concept to Reality. *Trends in Food Science & Technology*, Vol. 18, Issue 7, (July 2007), pp. 387-390

Kim, H.H.Y. & Baianu, I.C. (1991). Novel Liposome Microencapsulation Techniques for Food Applications. *Trends in Food Science and Technology*, Vol. 2, pp. 55-60

Kirby, C.J. (1991). Microencapsulation and Controlled Delivery of Food Ingredients. *Food Science and Technology Today*, Vol. 5, No. 2, pp. 74-80

Lattanzio, V., Kroon, P.A., Linsalata, V. & Cardinali, A. (2009). Globe Artichoke: A Functional Food and Source of Nutraceutical Ingredients. *Journal of Functional Foods*, Vol. 1, Issue 2, (April 2009), pp. 131-144

Lea, M.A. (2010). Fruit Phenolics in Colon Cancer Prevention and Treatment, In : *Bioactive Foods and Extracts : Cancer Treatment and Prevention,* R.R. Watson (Ed.) & V.R. Preedy (Ed.), 415-425, CRC Press, ISBN 9781439816196

Leifert, W.R. & Abeywardena, M.Y. (2008a). Cardioprotective Actions of Grape Polyphenols. *Nutrition Research*, Vol. 28, Issue 11, (November 2008), pp. 729-737

Leifert, W.R. & Abeywardena, M.Y. (2008b). Grape Seed and Red Wine Polyphenol Extracts Inhibit Cellular Cholesterol Uptake, Cell Proliferation, and 5-Lipoxygenase Activity. *Nutrition Research*, Vol. 28, Issue 12, (December 2008), pp. 842-850

Lesmes, U. & McClements, D.J. (2009). Structure - Function Relationships to Guide Rational Design and Fabrication of Particulate Food Delivery Systems. *Trends in Food Science & Technology*, Vol. 20, pp. 448-457

Liu, R.H. (2003). Health Benefits of Fruit and Vegetables are from Additive and Synergistic Combinations of Phytochemicals. *American Journal of Clinical Nutrition*, Vol. 78, No. 3, (September 2003), pp. 517-520

Liu, R.H. (2004). Potential Synergy of Phytochemicals in Cancer Prevention: Mechanism of Action. *Journal of Nutrition*, Vol. 134, pp. 3479-3485

Malone, M.E., Appelqvist, I.A.M. & Norton, I.T. (2003). Oral Behaviour of Food Hydrocolloids and Emulsions. Part 1: Lubrication and Deposition Considerations & Part 2: Taste and Aroma Release. *Food Hydrocolloids*, Vol. 17, pp. 763-784

Matser, A.M., Bregje Krebbers B., van den Berg, R.W. & Bartels, P.V. (2004). Advantages of High Pressure Sterilisation on Quality of Food Products. *Trends in Food Science & Technology*, Vol. 15, pp. 79-85

McClements, D.J., Decker, E. A. & Park, Y. (2009a). Controlling Lipid Bioavailability Through Physicochemical and Structural Approaches. *Critical Reviews in Food Science and Nutrition*, Vol. 49, No. 1, pp. 48-67

McClements, D.J., Decker, E.A., Park, Y. & Weiss, J. (2008). Designing Food Structure to Control Stability, Digestion, Release and Absorption of Lipophilic Food Components. *Food Biophysics*, Vol. 3, No. 2, pp. 219-228, ISSN 15571858

McClements, D.J., Decker, E.A., Park, Y. & Weiss, J. (2009b). Structure Design Principles for Delivery of Bioactive Components in Nutraceuticals and Functional Foods. *Critical Reviews in Food Science and Nutrition*, Vol. 49, No. 6, pp. 577-606

Narayanan, N., Nargi, D., Randolph, C. & Narayanan, B.A. (2009). Liposome Encapsulation of Curcumin and Reseratrol in Combination Reduces Prostate Cancer Incidence in PTEN Knockout Mice. *International Journal of Cancer*, Vol. 125, pp. 1-8

Norton, I., Moore, S. & Fryer, P. (2006). Understanding Food Structuring and Breakdown: Engineering Approaches to Obesity. *AIChE Journal*, Vol. 52(5), pp. 1632–1640

Nott, K.P., Halla, L.D., Bows, J.R., Hale, M. & Patrick, M.L. (2000). MRI Phase Mapping of Temperature Distributions Induced in Food by Microwave Heating. *Magnetic Resonance Imaging*, Vol. 18, pp. 69-79

Ottaway, P.B. (2008). Food Fortification and Supplementation - Technological, Safety and Regulatory Aspects, CRC Press; 1 Edition, (March 20, 2008), ISBN 978-0849364099

Ottino, J.M. (2005). New Tools, New Outlooks, New Opportunities. *AIChE Journal*, Vol. 51, No. 7, pp. 1839-1845

Paine, J.A., Shipton, C.A., Chaggar, S., Howells, R.M., Kennedy, M.J., Vernon, G., Wright, S.Y., Hinchliffe, E., Adams, J.L. & Silverstone, A.L. (2005). Improving the Nutritional Value of Golden Rice Through Increased Pro-Vitamin A Content. *Nature Biotechnology*, Vol. 23, pp. 482–487

Palzer, S. (2009). Food Structures for Nutrition, Health and Wellness. *Trends in Food Science & Technology*, Vol. 20, Issue 5, (May 2009), pp. 194-200

Parker, R.S. (1997). Bioavailability of carotenoids. *European Journal of Clinical Nutrition*, Vol. 51, pp. 86 - 90

Pothakamury, U. R. & Barbosa-Canovas, G. V. (1995). Fundamental Aspects of Controlled Release in Foods. *Trends in Food Science & Technology*, Vol. 6, No. 12, pp. 397-406

Ribeiro, H.S., Schuchmann, H.P., Engel, R., Walz, E. & Briviba, K. (2010). Encapsulation of Carotenoids. In: *Encapsulation Technologies for Active Food Ingredients and Food Processing*, N.J. Zuidam & V.A. Nedović (Eds.), Springer Science Business Media, LLC 2010). pp. 211-252, Springer, New York, ISBN 978-1-4419-1007-3

Rimando, A.M. & Suh, N. (2008). Natural Products and Dietary Prevention of Cancer. *Molecular Nutrition & Food Research*, Vol.52, Sup 1, pp. S5

Sajilata, M.G., Singhal, R.S. & Kamat, M.Y. (2008). The carotenoid pigment zeaxanthin – a review.

Salminen, S., Bouley, C., Boutron-Ruault, M.-C., Cummings, J. H., Franck, A., Gibson, G. R., Isolauri, E., Moreau, M.-C., Roberfroid, M. & Rowland, I. (1998). Functional Food Science and Gastrointestinal Physiology and Function. *British Journal of Nutrition*, Vol. 80, Suppl. 1, pp. 147–171

Schwarz, S., Obermüller-Jevic, U.C., Hellmis, E., Koch, W., Jacobi, G. & Biesalski, H.K. (2008). Lycopene inhibits disease progression in patients with benign prostate hyperplasia. *Journal of Nutrition*, Vol. 138, pp. 49 – 53

Sereno, N. M., Hill, S. E., Taylor, A. J., Mitchell, J. R. & Davies, S. J. (2009). Aroma Permeability of Hydroxypropyl Maize Starch Films. *Journal of Agricultural and Food Chemistry*, Vol. 57, No. 3, pp. 985-990

Shahidi, F. & Han, X.Q. (1993). Encapsulation of Food Ingredients. *Critical Reviews in Food Technology*, Vol. 33, No. 6, pp. 501–504

Siepmann, J. & Siepmann, F. (2008). Mathematical Modeling of Drug Delivery. *International Journal of Pharmaceutics*, Vol. 364, No. 2, pp. 328-343

Slattery, M.L., Benson, J., Curtin, K., Ma, K.N., Schaeffer, D. & Potter, J.D. (2000). Carotenoids and Colon Cancer. *American Journal of Clinical Nutrition*, Vol. 71(2), pp. 575-582

Tang, G., Qin, J., Dolnikowski, G.G., Russell, R.M. & Grusak, M.A. (2009). Golden Rice is an Effective Source of Vitamin A. *American Journal of Clinical Nutrition*, Vol. 89, pp. 1776-1783

Terao, J. (2009). Dietary flavonoids as antioxidants. In : *Food factors for health promotion. Forum of nutrition. Basel*, Karger, Vol. 61, pp. 87-94

Thorne, S. (1986). The History of Food Preservation. Barnes & Noble Books, Totowa, N.J., ISBN 0389206393

Ubbink, J. & Krüger, J. (2006). Physical Approaches for the Delivery of Active Ingredients in Foods. *Trends in Food Science & Technology*, Vol. 17, pp. 244-254

Ubbink, J., Burbidge, A. & Mezzenga, R. (2008). Food Structure and Functionality: a Soft Matter Perspective. *Soft Matter*, Vol. 4, No. 8, pp. 1569-1581

Van Aken, G. A. (2007). Relating Food Microstructure to Sensory Quality. In: *Understanding and Controlling the Microstructure of Complex Foods*, D. J. McClements (Ed.), pp. 449-482, CRC Press, Boca Raton, FL., ISBN 9781845691516

Verhoeven, D.T.H., Goldbohm, R.A., vaPoppel, G., Verhagen, H. & vanderBrandt, P.A. (1996). Epidemiological Studies on Brassica Vegetables and Cancer Risk. *Cancer Epidemiology Biomarker & Prevention*, Vol. 5, pp. 733-748

Watson R.R. & Preedy V.R. (2010). Bioactive Foods and Extracts: Cancer Treatment and Prevention, CRC Press, ISBN 978-1-4398-1619-6, pp. 423

Watzl, B. & Leitzmann, C. (1999). Bioaktive Substanzen in Lebensmitteln. Hippokrates Verlag, Stuttgart, ISBN 978-383045308

Weiss, J., Decker, E.A., McClements, D.J., Kristbergsson, K., Helgason, T. & Awad, T. (2008). Solid Lipid Nanoparticles as Delivery Systems for Bioactive Food Components. *Food Biophysics*, Vol. 3, pp. 146–154

Wildman, R.E.C. (2007). Handbook of Nutraceuticals and Functional Foods, Second Edition, CRC Press, 2007, ISBN: 978-0-8493-6409-9

Yamaguchi, M. & Uchiyama, S. (2004). β-cryptoxanthin stimulates bone formation and inhibits bone resorption in tissue culture in vitro. *Molecular Cell Biochemistry*, Vol. 258, pp. 137 - 144

Ye, X., Al-Babili, S., Kloti, A., Zhang, J., Lucca, P., Beyer, P. & Potrykus, I. (2000). Engineering the Provitamin A (Beta-Carotene) Biosynthetic Pathway Into (Carotenoid-Free) Rice Endosperm. *Science*, Vol. 287, pp. 303–305

Zhong, Q., Sandeep, K.P. & Swartzel, K.R. (2004). Continuous Flow Radio Frequency Heating of Particulate Foods. Innovative Food Science & Emerging Technology, Vol.4, No.5, pp. 475-483